Customizing SAP S/4HANA® with SAP® Cloud Platform

 PRESS

SAP PRESS is a joint initiative of SAP and Rheinwerk Publishing. The know-how offered by SAP specialists combined with the expertise of Rheinwerk Publishing offers the reader expert books in the field. SAP PRESS features first-hand information and expert advice, and provides useful skills for professional decision-making.

SAP PRESS offers a variety of books on technical and business-related topics for the SAP user. For further information, please visit our website: *www.sap-press.com*.

Bilay, Gutsche, Krimmel, Stiehl
SAP Cloud Platform Integration: The Comprehensive Guide (3rd Edition)
2020, 906 pages, hardcover and e-book
www.sap-press.com/5077

Acharya, Bajaj, Dhar, Ghosh, Lahiri
SAP Cloud Platform: Cloud-Native Development
2019, 489 pages, hardcover and e-book
www.sap-press.com/4713

Manohar, PavanKumar, Ravindranathan
Internet of Things with SAP: Implementation and Development
2020, 497 pages, hardcover and e-book
www.sap-press.com/5013

Herzig, Heitkötter, Wozniak, Agarwal, Wust
Extending SAP S/4HANA: Side-by-Side Extensions with the SAP S/4HANA Cloud SDK
2018, 618 pages, hardcover and e-book
www.sap-press.com/4655

Ivan Femia
SAP Cloud Platform Certification Guide: Development Associate Exam
2019, 477 pages, paperback and e-book
www.sap-press.com/4825

Paresh Mishra, Vipin Varappurath

Customizing SAP S/4HANA® with SAP® Cloud Platform

Designing a Future-Ready Enterprise Architecture

Editor Meagan White
Acquisitions Editor Hareem Shafi
Copyeditor Melinda Rankin
Cover Design Graham Geary
Photo Credit Shutterstock.com: 161983700/© Andrii Horulko
Layout Design Vera Brauner
Production Graham Geary
Typesetting III-satz, Husby (Germany)
Printed and bound in the United States of America, on paper from sustainable sources

ISBN 978-1-4932-2008-3
© 2021 by Rheinwerk Publishing, Inc., Boston (MA)
1st edition 2021

Library of Congress Cataloging-in-Publication Data
Names: Mishra, Paresh (Writer on business computing), author. |
  Varappurath, Vipin, author.
Title: Customizing SAP S/4HANA with SAP cloud platform : designing a
  future-ready enterprise architecture / Paresh Mishra, Vipin Varappurath.
Description: 1st edition. | Bonn ; Boston : Rheinwerk Publishing, 2020. |
  Includes index.
Identifiers: LCCN 2020036127 (print) | LCCN 2020036128 (ebook) | ISBN
  9781493220083 (hardcover) | ISBN 9781493220090 (ebook)
Subjects: LCSH: Cloud computing. | Business enterprises--Computer networks.
Classification: LCC QA76.585 .M575 2020 (print) | LCC QA76.585 (ebook) |
  DDC 004.67/82--dc23
LC record available at https://lccn.loc.gov/2020036127
LC ebook record available at https://lccn.loc.gov/2020036128

Contents at a Glance

Dear Reader,

Most of us, I think, have had the pleasure of going to IKEA at least once in our lives. I don't know about you, but my first reaction upon walking through the doors is to feel a little overwhelmed by all of the choices. Looking for a light fixture? Here are 123 options (not to mention the different bulbs)! Need a bedside table? 8 options, but with 3-5 color variants each! Regardless of the shopping list I walked in with, I always somehow end up with half a dozen extra items to spruce up my home.

For the uninitiated, SAP Cloud Platform services have something of the same feel. There are *so many* of them, they do so many useful things, and I suspect it is not unusual for a company to go to SAP with a shopping list of three services and end up with an extra one or two, because wow—SAP Cloud Platform Launchpad is going to look great in your living room (err... help your company centralize your apps).

Like the giant map on the wall of IKEA, this book should help serve as a guide to the SAP Cloud Platform services that will help you customize your SAP S/4HANA landscape. Your tour guides, Paresh and Vipin, are experts in navigating the terrain, and will lead you safely to your destination, whatever it may be.

What did you think about *Customizing SAP S/4HANA with SAP Cloud Platform: Designing a Future-Ready Enterprise Architecture*? Your comments and suggestions are the most useful tools to help us make our books the best they can be. Please feel free to contact me and share any praise or criticism you may have.

Thank you for purchasing a book from SAP PRESS!

Meagan White
Editor, SAP PRESS

meaganw@rheinwerk-publishing.com
www.sap-press.com
Rheinwerk Publishing · Boston, MA

Contents

PART II Future Aware

3 Architecture and Design Considerations for a Future-Ready Landscape 65

4 Master Data Distribution 85

5   Process and Data Integration   111

6   Business Process Automation and Optimization   153

7   User Experience and Mobile Consumption   177

8   Security   203

PART III   Future Creation

9   Developing SAP S/4HANA Extensions   235

10 Continuous Integration and Delivery — 275

11 Continuous Monitoring — 307

12 Developing Cloud-Native Applications — 323

PART IV   Final Steps

13 Trailblazing Enterprise Innovations — 345

Contents

Preface

Every organization and industry is going through a transformation process in which existing business processes and models are challenged. The speed of transformation has been so rapid that the need for an agile and flexible information system and landscape is greater than ever. SAP S/4HANA is the latest generation of enterprise resource planning (ERP) system from SAP, which has revolutionized how business processes and data are managed. However, there is always a need to customize SAP S/4HANA to meet your business demands, processes, and models, and the systems and landscapes must be flexible and adaptable for any required future changes. To make your landscape future ready, you need to start the process to transform your enterprise systems and landscape early and have access to the right tools.

Organizations are increasingly moving to the cloud, which requires a major rework of enterprise architecture, security and application design, integration patterns, automation, and intelligent technologies. There are multiple factors driving the need for revamped technical landscapes; one of the major factors is that user experience expectations are now driven by modern web designs and social media—ease-of-use is now paramount to user adoption. Users are demanding this shift and organizations that make the move attract and retain employees well-suited to this new business environment where rapid change is the norm. Throughout all of this, we see a shift in how systems and data are accessed and utilized. One aspect of being future-ready means being agile enough to adapt to all of these developments.

SAP Cloud Platform is a platform-as-a-service (PaaS) offering from SAP. This book will explore how you can use SAP Cloud Platform as a foundation and constant companion for your transformation journey to SAP S/4HANA, providing tools and services to build a flexible and agile future-ready landscape. We will explore the solution through four stages of your transformation journey to SAP S/4HANA: foundation, future aware, future focused, and trailblazer. We'll introduce different services and principles from SAP Cloud Platform in these stages to build your future-ready intelligent enterprise.

Target Audience

The target audience for this book includes various categories of professionals. We hope this book will equally benefit SAP consultants, SAP project managers, different business process owners and architects, technical architects, and solution architects. Even beginners in the SAP world should find a lot of useful information to jumpstart their SAP careers, either as users or as consultants. Beginner readers will gain a fundamental understanding of how to approach an enterprise transformation journey and learn principles to build a flexible, future-ready landscape.

We'll cover different aspects and steps in SAP S/4HANA customization. Therefore, this book should be an invaluable help for senior and junior technical and solution architects and consultants alike. For consultants that are just starting out with SAP S/4HANA transformation or are looking to optimize existing systems and landscapes to meet future business demands, this book will reveal the secrets of utilizing SAP Cloud Platform along with your existing SAP S/4HANA system and will guide you through a step-by-step process to implement transformation processes to build a future-ready intelligent enterprise. Senior consultants who already have a few transformation projects under their belt should also find this book beneficial for learning the approach to using a platform as a service along with SAP S/4HANA to build flexible and agile architecture. These more experienced consultants will also learn about great advancements in SAP Cloud Platform and cloud-native technology and how these advancements benefit and are crucial to your organization. This resource should be a useful reference for business architects and for technical and solutions consultants, especially those coming from an SAP ERP background.

We hope this book will also be invaluable not only during implementation, but also during production to support continuous improvement processes.

Another major target audience for this book consists of various business process owners and users of the system. Many SAP books on the market focus either on the technical side of implementing SAP process areas or on the functional side of using these solutions. This book provides a unique blend of both approaches: we will not only teach you how to approach the transformation journey of your organization to SAP S/4HANA using SAP Cloud Platform, but also provide an approach to analyzing and optimizing your existing business processes and identifying business improvement and optimization processes.

This book also targets middle and high-level management interested in optimizing the business processes and landscape in an organization, using SAP S/4HANA to become a future-ready intelligent enterprise. Management should be constantly informed about the current state of the business processes and landscape of the company, especially to make the organization adaptable to meet future business demands. Managers will have an invaluable advantage and be able to stay ahead of the challenges in today's highly competitive business environment. This book will teach you how to utilize these capabilities in the most effective way.

Finally, this book's target audience includes project managers in SAP S/4HANA implementations. Implementing SAP S/4HANA is a complex undertaking. Project managers must be strong from a technical point of view and must understand the functionalities being implemented and plan different detailed steps during the implementation. As such, project managers will hopefully find this book useful as they will learn to handle complexity in the implementation and find an able coach in this book to guide them through the implementation project. We'll also discuss the typical questions that must be asked, design principles to be involved, and tasks that must be performed to become a future-ready intelligent enterprise using SAP S/4HANA. Thus, project managers will be able to properly assess the required effort for implementing various process and understand the tools that are required as an input to implement a project plan.

Objective of This Book

Customizing SAP S/4HANA with SAP Cloud Platform is the ultimate guide to understanding how you can customize SAP S/4HANA, following best practices to meet your current and future business demands. The objective of this book is to provide a comprehensive guide both to approaching your business and landscape transformation/optimization and to using various capabilities in SAP Cloud Platform along with SAP S/4HANA to complement and advance your business processes. This book should benefit SAP consultants, architects, and business process owners. The objective of this book is to provide integrated content as we walk you through various design and architectural principles to make your landscape future ready. We'll explain how you can identify your current as-is technical landscape and business process and outline a process to transform this to the future landscape and business processes. Along the journey, we will identify potential bottlenecks in the landscape and the process and design principles to be followed to overcome them. The book ends by walking you through process to identify your industry-specific innovations, prioritizing and strategizing them for maximum impact for your business.

If this is your first book on customizing SAP S/4HANA, we hope it also helps you jump-start your SAP Cloud Platform journey with SAP S/4HANA. We aim to provide concise, yet robust, knowledge on the various capabilities in SAP Cloud Platform and how it complements and supports customizing SAP S/4HANA or sets the foundation to start your landscape and business process optimization journey.

This book is a combination of a technical guide describing how to optimize, set up, and develop cloud-native applications, extensions, and integrations for SAP S/4HANA, and a presentation of guidelines for how to improve your business process efficiencies and engagement with your end users/vendors/suppliers through different capabilities in SAP Cloud Platform.

This book also aims to share our enormous experience in enterprise transformation garnered from many customer projects and consulting. You'll benefit from best practices derived from projects in various industries as they pertain to the transformation journey. You'll learn about the best tools to use to meet various technical and business requirements for your organization,

This blend of technical guide and business process textbook will hopefully serve as your ultimate resource for your exciting journey into the fascinating world of future-ready intelligent enterprise in SAP S/4HANA!

> **Note**
>
> Any transformation journey has people as one of its foundational pillars—but this element is not covered in the scope of this book.

Structure of This Book

In this book, we'll teach you how to transform and customize your technical landscape and business processes to be future ready with SAP S/4HANA. The book is organized into the following chapters:

- **Chapter 1: Future-Ready Enterprise Architecture**
  In this chapter, you'll learn what a future-ready enterprise is. You'll understand existing approaches and frameworks for managing enterprise architecture and business processes, then learn why a new concept is necessary. This chapter will explore characteristics for future-ready enterprises and explain why SAP Cloud Platform is an essential tool for SAP S/4HANA architecture design.

- **Chapter 2: Understanding Your Current Business Processes and Landscape**
  In this chapter, you'll learn about your current business processes. You'll learn about various tools and services to evaluate existing business processes and identify optimization potential to increase your organizational effectiveness and efficiency.

- **Chapter 3: Architecture and Design Considerations for a Future-Ready Landscape**
  In this chapter, we'll explain in detail about the design and architectural frameworks you'll use to create a future-ready enterprise. Readers will learn about practical considerations that must be taken into account while optimizing the SAP system landscape. You'll be guided to come up with relevant design considerations in your organization to start your future-ready transformation journey.

- **Chapter 4: Master Data Distribution**
  Availability of consistent master data across an organization is a critical ingredient in your transformation journey. This chapter will teach you how to use master data services in SAP Cloud Platform to design an SAP S/4HANA master data distribution strategy for data integration, enrichment, validation, and retention.

- **Chapter 5: Process and Data Integration**
  In this chapter, we'll explain various facets of system integration in an enterprise and apply correct integration design principles across your system landscape. You'll learn about various SAP Cloud Platform tools and best practices that can help you set up your complex integration landscape in a flexible and agile way.

- **Chapter 6: Business Process Automation and Optimization**
  This chapter will teach you how to optimize SAP S/4HANA business processes in your organization using SAP Cloud Platform Workflow Management and robotic process automation tools.

- **Chapter 7: User Experience and Mobile Consumption**
  This chapter will teach you how to design a consistent and engaging user experience for your SAP S/4HANA end users, enabling them to access your system anywhere and everywhere.

- **Chapter 8: Security**
  In this chapter, you'll learn about applying various security strategies in your organization for managing SAP Cloud Platform in alignment with SAP S/4HANA. You'll learn about various tools and services on SAP Cloud Platform to help manage your enterprise applications, extensions, and integrations for SAP S/4HANA.

- **Chapter 9: Developing SAP S/4HANA Extensions**
  In this chapter, you'll learn how to extend your SAP S/4HANA business process to differentiate your business. You'll learn about various tools and technologies that you can use to easily extend your business processes and best practices for using them.

- **Chapter 10: Continuous Integration and Delivery**
  In this chapter, you'll learn how to set up your landscape to deliver continuous innovations for your business needs without disruptions in an agile and flexible approach.

- **Chapter 11: Continuous Monitoring**
  In this chapter, you'll learn how to set up continuous monitoring for applications, extensions, and integrations running in SAP Cloud Platform to support your SAP S/4HANA business process and set up a remediation process to act immediately in case of alerts.

- **Chapter 12: Developing Cloud-Native Applications**
  In this chapter, you'll learn how to develop cloud-native applications for SAP S/4HANA on SAP Cloud Platform. You'll learn about various best practices and resiliency patterns to include in your applications to ensure that they're built elastically to meet your future needs and requirements.

- **Chapter 13: Trailblazing Enterprise Innovations**
  In this chapter, you'll learn how future-ready enterprises can use innovation to address whitespaces and gaps in the business, using industry-standard guidelines for specific industries.

- **Chapter 14: Measuring Success**
  In this chapter, you'll learn about various approaches to measure success and offer value during a future-ready enterprise design project. You'll learn about approaches to implement a transformation strategy.

Terminology

We'll present common SAP Cloud Platform terminology in this section. These terms will be used throughout the book, and it's important for you to have an understanding of them before you begin.

Global Account

When you sign up for SAP Cloud Platform, a global account is assigned/created to you by default. A *global account* is a logical container used to provision all resources licensed under SAP Cloud Platform (see Figure 1). For different commercial models, separate global accounts are created and delivered as of the time of writing this book. As an SAP customer, you can have multiple global accounts based on license structure.

> **Note**
>
> Global accounts provided in a consumption agreement would differ from global accounts provided under a subscription-based agreement.

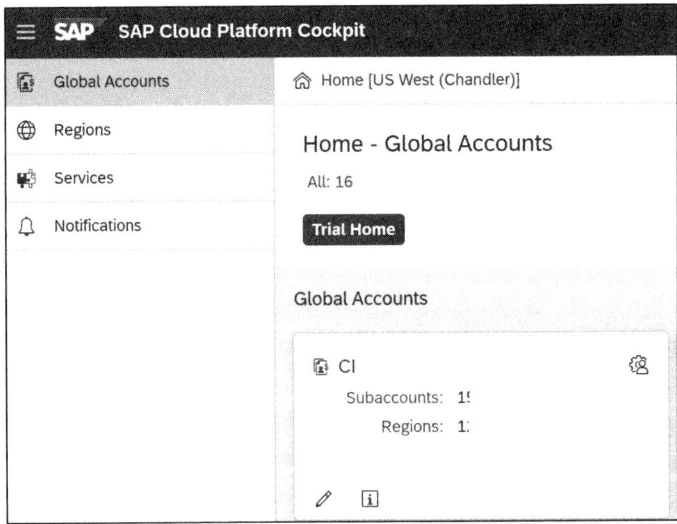

Figure 1 SAP Cloud Platform Global Account

The provisioned and available services are available under **Entitlements** in the global account, as shown in Figure 2.

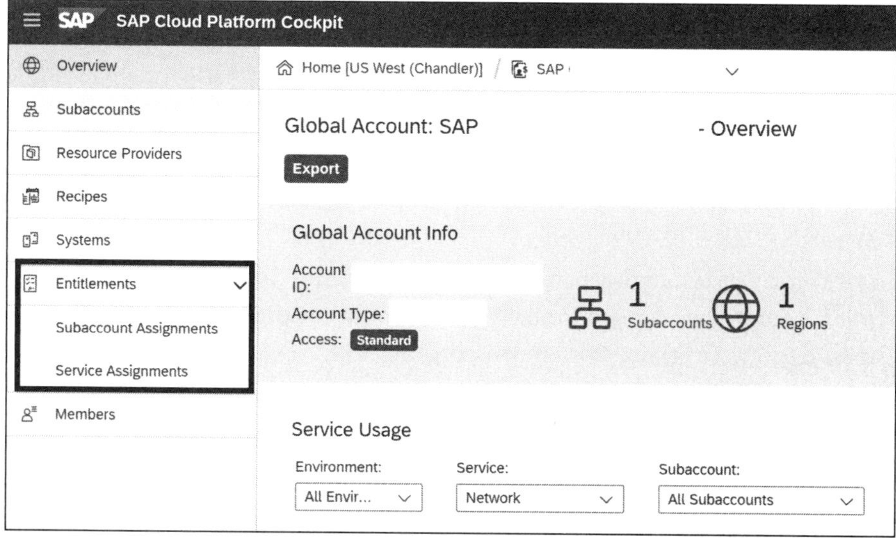

Figure 2 Entitlements for SAP Cloud Platform

Subaccounts

Subaccounts are further structures that you create inside a global account to segregate your projects. You can create multiple subaccounts in a global account, independent of each other, as shown in Figure 3. Each subaccount is associated with a *region*, which is the physical location where your application and data reside. You assign the licensed quotas from a global account to each subaccount based on how and where you want to use them.

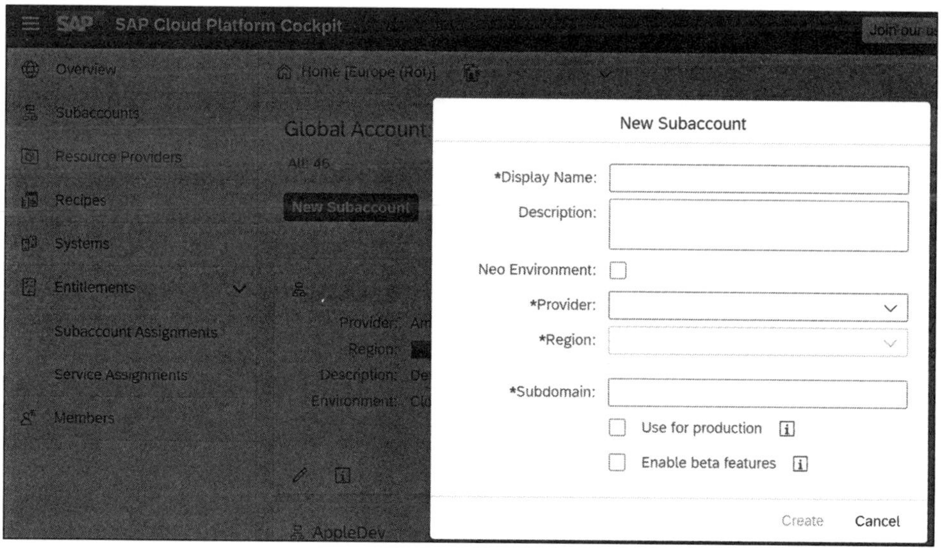

Figure 3 Subaccount in SAP Cloud Platform

Creation of a subaccount is an important aspect to consider in your project lifecycle planning and setting up the environment.

Directories (Beta)

At the time of writing this book, directories are in beta release and expected to be generally available toward the end of 2020.

Directories provide additional capabilities to group, organize, and manage your subaccounts. For example, you can group all your subaccounts belonging to a particular line of business (LOB), project, and so on in your organization.

With the introduction of directories, you can monitor usage and costs and provide permission to manage entitlements at this level, as shown in Figure 4. You also can manage authorizations for directories to set who can administer and view them.

You also have the option to set any custom properties and tags at the directory level, which helps you better manage and filter the cost and usage at this level.

You can edit existing subaccounts to move them under a specific directory.

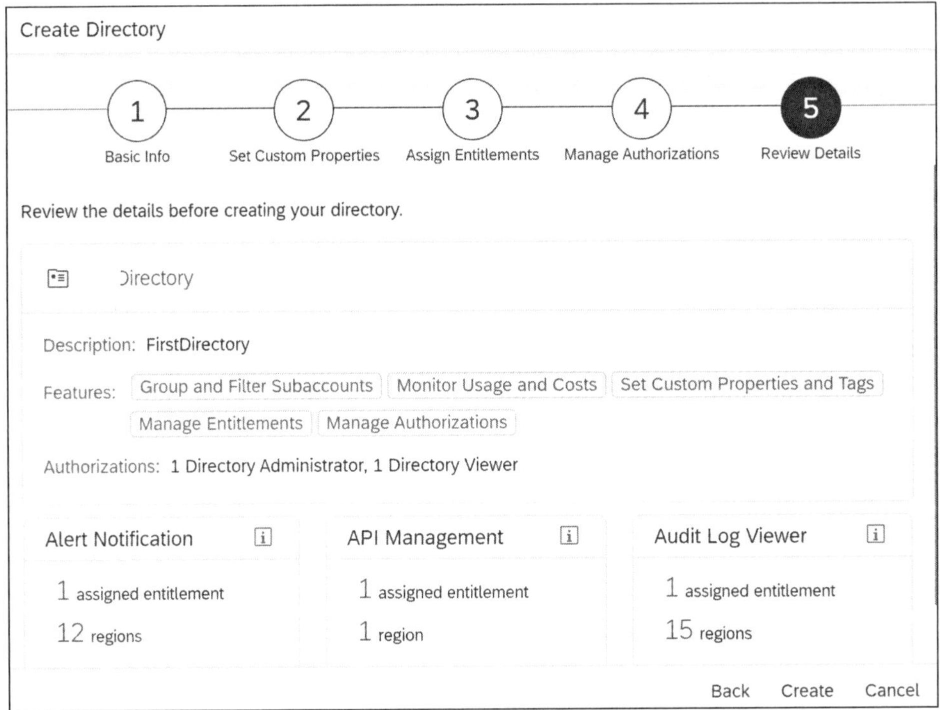

Figure 4 Create Directory

Cloud Foundry Organization

For subaccounts created to use the Cloud Foundry runtime, you enable the Cloud Foundry environment, which creates an organization as shown in Figure 5.

Once you have enabled Cloud Foundry and created an organization, you will see the respective API endpoint for the region, as shown in Figure 6.

There is a 1:1 relationship between a subaccount and an organization.

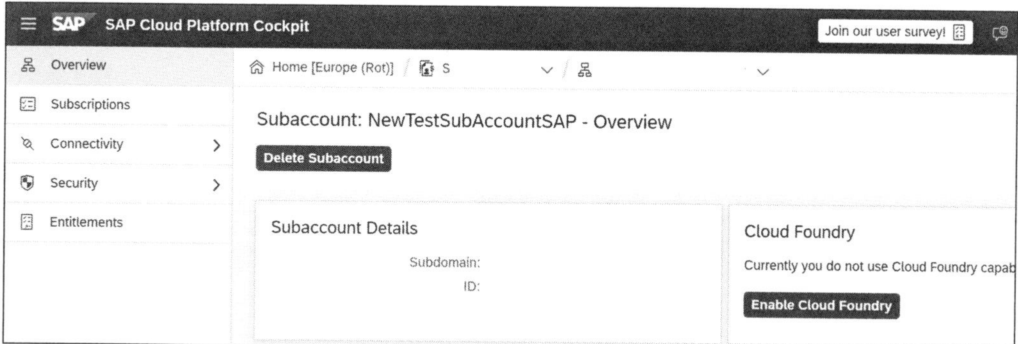

Figure 5 Enabling Cloud Foundry

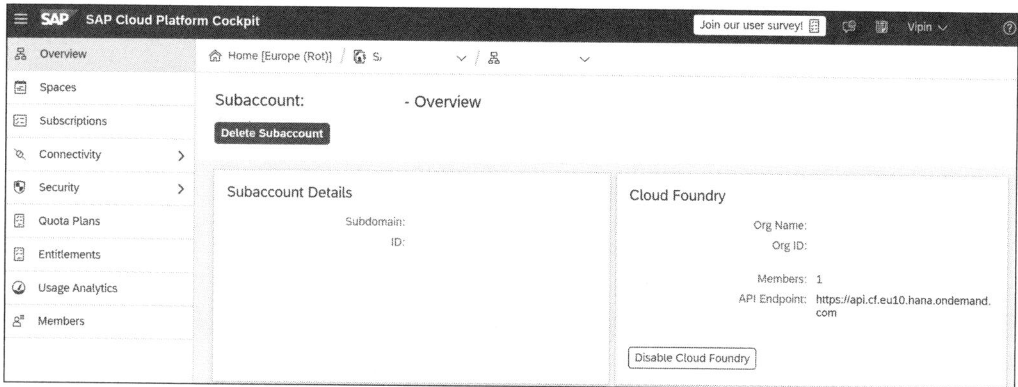

Figure 6 API Endpoint for Cloud Foundry-Enabled Subaccount

Spaces

Spaces are further structures available within the SAP Cloud Platform Cloud Foundry subaccount/organization. You can create multiple spaces within a subaccount and assign quotas to spaces using quota plans, as shown in Figure 7.

Creation of spaces is another important aspect to consider while developing a plan to design your landscape.

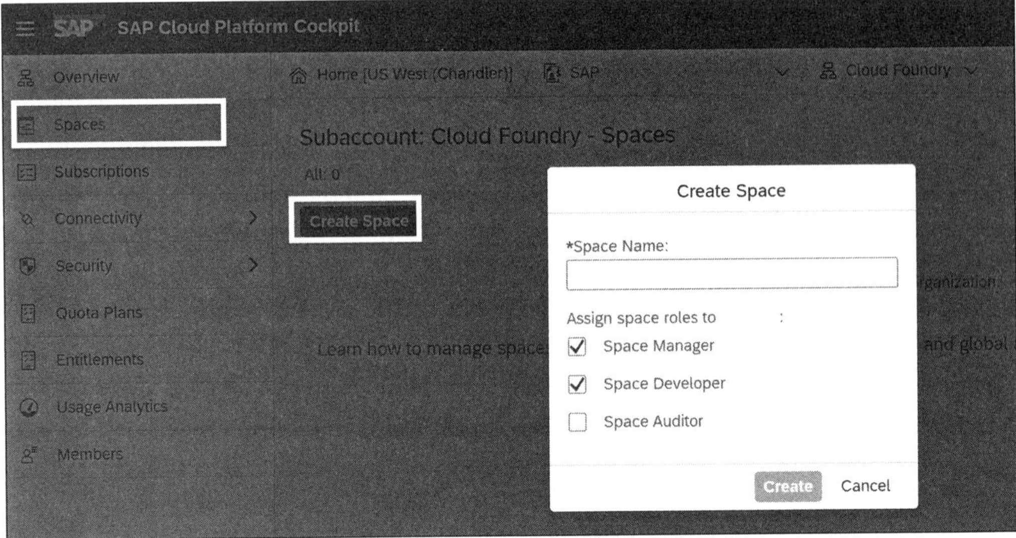

Figure 7 Spaces and Quota Plans in Spaces

Relationships among Global Accounts, Subaccounts, Organizations, and Spaces

The relationships among global accounts, subaccounts, organizations, and spaces are depicted in Figure 8.

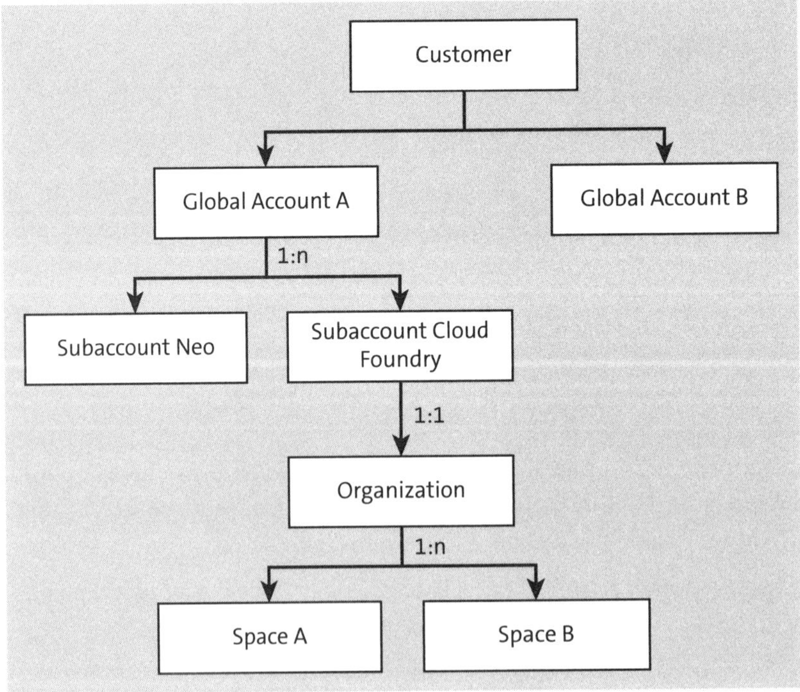

Figure 8 Relationships among Global Accounts, Subaccounts, Organizations, and Spaces

With directories, you can have a directory as a parent for a subaccount. A directory can have multiple subaccounts under it. This is shown in Figure 9.

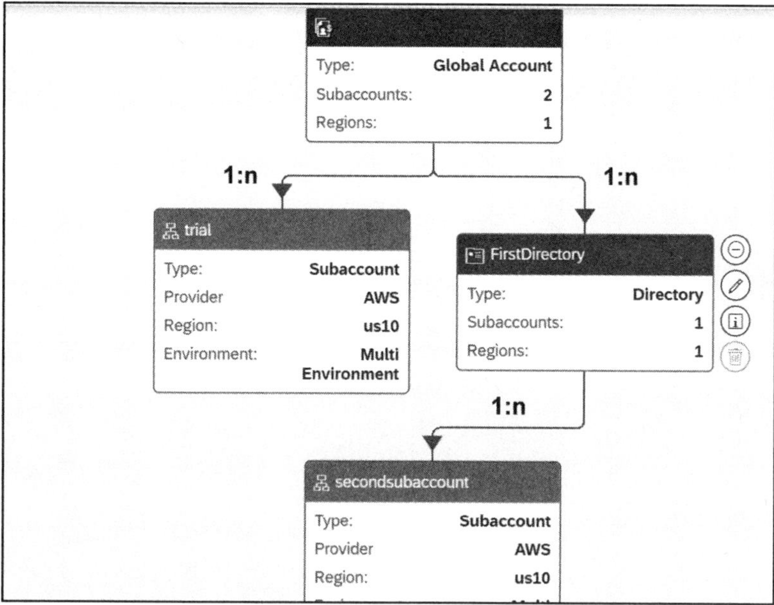

Figure 9 Relationships with Directories

Acknowledgments

We would like to thank our wonderful colleagues, team members, friends, and families for their amazing support while completing this book.

Special mentions go to the following colleagues:

- Michael Bain, global VP of customer engagement, for his support through the writing process
- Ashok Rathor and Tridib Chaudhuri, senior customer engagement executives, for their contributions
- Jason Bell, Lakshmi Narasimhan Neelakantan, Arun Rajamani, Allan Van Lelyveld, and Daniela Lohr for their inputs and suggestions
- Carsten Kress, VP of the customer engagement EMEA team, for his support
- Friends and colleagues from the SAP product development team

Conclusion

As noted, the objective of this book is to help teach you how to build a future-ready enterprise by transforming your technical landscape and optimizing your business

processes. You'll learn in detail about various tools and services that are essential partners in your transformation journey. You'll get a good understanding of these various
tools and processes and identify the best practices for using them by identifying use
cases and projects in your organization.

Without further ado, let's start our future-ready enterprise transformation journey!

Foundation

Chapter 1
Future-Ready Enterprise Architecture

Technology-driven digital transformation for enterprises has been the big story lately, and it allows organizations to open doors to many markets. The next stage of enterprise evolution is not just to be resilient in difficult times, but to demonstrate agility, speed, time to value, and achieving the ultimate nirvana state of being future ready.

Our world is changing continuously, and enterprises are facing new business and technology models. The legacy systems and ERP systems of the old world simply cannot keep pace. To drive the transformation journey with SAP S/4HANA, technology plays the role of key enabler, and SAP Cloud Platform is the common fabric and the tool that plays a part in all the stages of this journey.

In the past, enterprise transformation meant different things to different organizations:

- It could mean reimagining business processes, thus using technology to reinvent and create flexible and responsive systems to enable current and future business models.

- It could also mean to streamline and simplify operations, thus using technology to increase efficiency through process and task automation.

- It could improve the experience of your key stakeholders, thus using technology to create new ways for buyers and suppliers to reach each other and collaborate in an interconnected digital network. This could facilitate consistent, accurate, and timely communication via contact centers, portals, and mobile applications.

- It could enhance access to accurate information and make integrated, enhanced data available for informed decisions by internal and external stakeholders in real time, thus using technology to provide data-driven insights.

- Finally, it could help a company achieve regulatory compliance every time and always, thus using technology to adapt to every changing regulatory requirement while being cost-effective.

The purpose of this chapter is to explain how to use technology to build a future-ready enterprise, what stages and realms exist when considering the technology architecture-to-be, and how SAP Cloud Platform has a role to play in building such architecture.

Specifically, in the following sections, we'll start by helping to define what a future-ready enterprise is. From there, we'll discuss SAP Cloud Platform's role in creating a future-ready enterprise. Finally, we'll end the chapter by providing you with the stages, or realms, of becoming a future-ready enterprise. We have organized the book around these stages, so pay careful attention!

Note

You shouldn't look at SAP Cloud Platform as *only* an integration and extension toolset during the actual SAP S/4HANA project implementation. SAP Cloud Platform also has a role to play when building your business case for SAP S/4HANA transformation—not only mapping existing business processes, but also reimagining those business processes (for example, evaluating how to structure your project or considering the technology options that are available to you today).

1.1   What Is a Future-Ready Enterprise?

The idea behind a future-ready enterprise is the need to create an enterprise in which the technology platform provides the ability to withstand internal and external disruptions and further enables the agility to adapt and respond to changes in market dynamics. Thus, the enterprise continues to innovate and improve the performance of the company with the help of the technology platform.

Note

There are multiple transformation models to support innovation coming from strategic consulting firms, such as McKinsey's three horizons of growth:

1. Horizon 1 describes the need to keep the lights on for the business, with slight adjustments to the existing process.
2. Horizon 2 describes the need to extend your current business model.
3. Horizon 3 describes the need to introduce new elements to your existing business model.

For more information, see *https://www.executestrategy.net/blog/mckinseys-three-horizons-of-growth*.

There are some models that describe that road to become future-proof, and they move through the following steps, eventually leading to a future-ready enterprise:

1. Optimization
2. Digitization
3. Business process improvement
4. Future-ready enterprise

The transformation model SAP proposes is a four-stage journey to becoming future ready. These stages (discussed in more detail in Section 1.3) are as follows:

- *Foundation* is the first stage of the journey and can be mapped to SAP S/4HANA's *discover* and *prepare* project phases.
- *Future aware* is the second stage and can be mapped to the *explore* phase for an SAP S/4HANA implementation project.
- *Future focused* is the third stage and can be mapped to the *realize, deploy,* and *run* phases for SAP S/4HANA projects.
- Finally, *trailblazer* is a stage for continuous innovation.

> **Note**
>
> SAP Cloud Platform is the fabric that spans all the stages/realms when an enterprise decides to drive its transformation journey as we've described. It has a role to play in all four stages, not just the future focused stage or the realize phase in an SAP S/4HANA transformation project.

1.2   Why Is SAP Cloud Platform Essential for a Future-Ready Enterprise?

Companies have a strategic need for the right technologies that can help them become intelligent enterprises in the experience economy.

To help enterprise thrive in the experience economy, SAP provides the Business Technology Platform which contains the capabilities they need to adapt their business processes and turn data into action.

Figure 1.1 shows how SAP Cloud Platform provides solutions across four key technology markets—data management, analytics, application development and integration services, and intelligent technologies—thus addressing all organizations' needs in the data-to-action value chain.

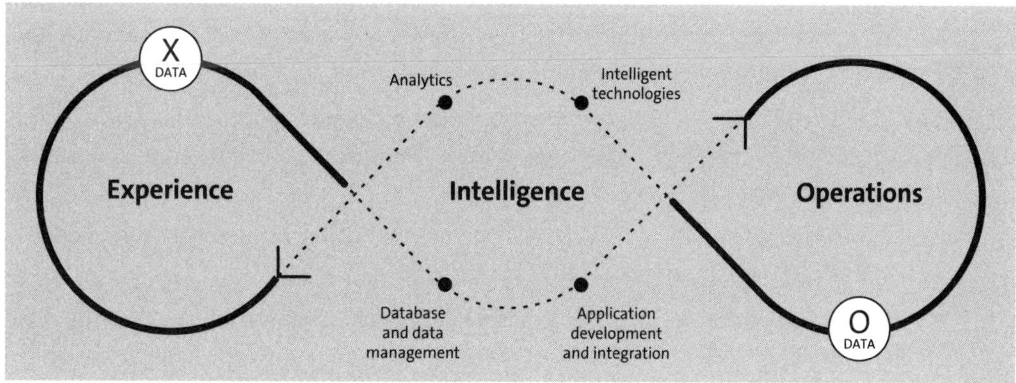

Figure 1.1 Intelligent Enterprise

SAP Cloud Platform is the key pillar of the Business Technology Platform and is positioned centrally as a core product to cater to the main extension, integration, and innovation scenarios for cloud and on-premise core systems.

SAP Cloud Platform is often referred to as the central nervous system of a company's evolving landscape with its multiple cloud and on-premise offerings from SAP. This is because SAP Cloud Platform's integration and extension capabilities bind various solutions via the offered services.

Let's drill down into the main value proposition and role of SAP Cloud Platform and how it can help organizations to bring together applications, data, and business processes. SAP Cloud Platform draws maximum value from existing application investments, and this is what companies need in order to survive and thrive.

1.2.1   Role of SAP Cloud Platform

SAP Cloud Platform provides a broad set of technical and business services that allow organizations to integrate data and processes from SAP and third-party systems; empower new, data-driven innovation scenarios; and offer flexible deployment strategies.

The ability to access, connect, and analyze both business-critical operational data and more intangible, experience-based data provides companies with actionable data. Additional cloud offerings like SAP Analytics Cloud provide end users with access to this data. SAP Cloud Platform also allows you to create integrated and digital experiences for users across all relevant touchpoints (such as web, mobile, portal, and conversational).

The ability to extend and integrate existing on-premise and cloud applications, without disrupting core business processes, is essential to for agile innovation. One of the goals of SAP Cloud Platform is to allow you to make process changes more quickly, in a scalable and cost-effective way, and act with greater agility to take advantage of new business opportunities.

Companies on their journeys toward becoming intelligent enterprises need to extend and integrate their applications and digitize their business processes to meet their unique needs. They must differentiate themselves, boost employee productivity, and empower their developers to build innovative apps in an agile manner.

SAP Cloud Platform is an integration and extension business platform that offers multiple capabilities in a single place, resulting in less effort and fewer implementation challenges for organizations.

SAP Cloud Platform enables organizations to obtain business outcomes faster and elevate the enterprise experience, as follows:

- Simplify and accelerate integration across your critical applications, improving your value chain.

- Improve efficiency and productivity, cut costs, and greatly increase your organization's ability to deal with change.

- **Accelerate development of your application extensions.** Enhance and adapt SAP solutions for your unique needs with methodologies, enterprise-grade tools and services, security features, and ready-to-use business content.

- **Achieve tailored value with a robust ecosystem.** Reduce development investments by leveraging SAP's open partner ecosystem that delivers prebuilt solutions to solve line-of-business- and industry-specific needs.

SAP Cloud Platform is offered in a multicloud environment via its Cloud Foundry environment and Neo environment. Cloud Foundry is a strategic platform for all service offerings, managing the delivery of business services across various data centers around the globe. SAP uses Cloud Foundry technology to deliver SAP Cloud Platform. Cloud Foundry is open source, and SAP is one of the key platinum partners of the Cloud Foundry Foundation.

SAP Cloud Platform is available in most regions across the globe (see Figure 1.2). You can choose between data centers hosted by SAP and those run by various infrastructure providers, which are SAP's infrastructure-as-a-service (IaaS) partners:

- Microsoft Azure

- Amazon Web Services

- Alibaba Cloud

- Google Cloud Platform

> **Note**
>
> Details of the various data centers available across regions and service availability based on data centers can be found at *http://s-prs.co/v515700*.

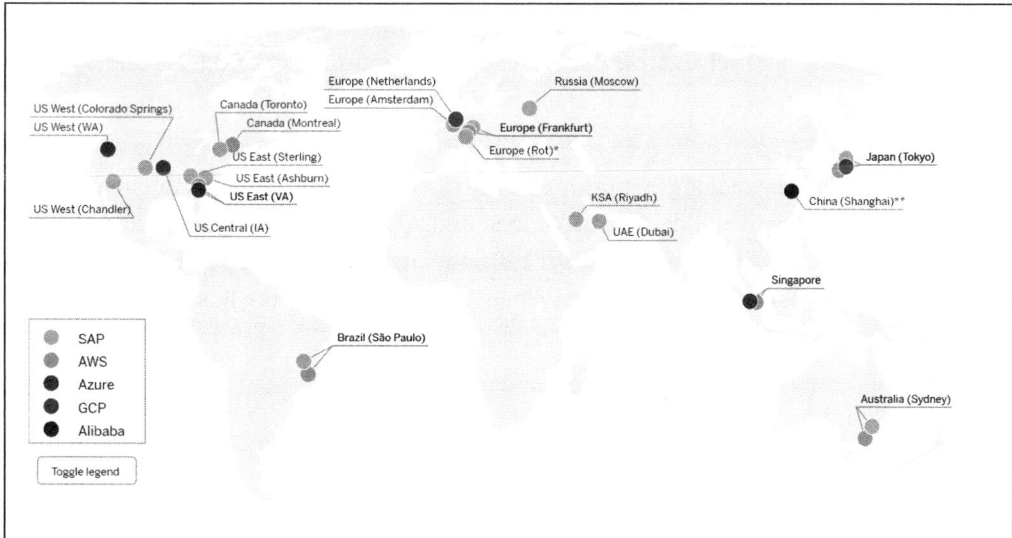

Figure 1.2 SAP Cloud Platform Regions (September 2020)

> **Note**
>
> Service description guides for various services offered by SAP can be found at *https://www.sap.com/about/agreements/policies/cloud-platform.html.*

1.2.2 SAP Cloud Platform and the Intelligent Enterprise

As organizations are moving from the digital era to the intelligent era, these enterprises are exploring and utilizing the tools that can help them build a robust base for tomorrow. SAP Cloud Platform is a foundation for an agile and intelligent enterprise, enabling the rapid transformation of data into insights—ensuring seamless process automation, innovation, and ideal experiences.

SAP Cloud Platform has a comprehensive catalog of services, spanning from integrations, extensions, and building new apps to data intelligence, robotic process automation, and conversational AI. These services offer an agile enterprise all the capabilities it needs to succeed in this fast-paced market. Emerging technologies, such as machine learning and artificial intelligence (AI), the Internet of Things (IoT), and chatbots, are getting ingrained into our daily lives, with customized experiences in all aspects of business and personal life.

Customers, employees, and business partners are used to these technologies, and therefore market demands evolve rapidly, customer expectations shift equally as fast, entry barriers for new competitors are lowered, and new business models appear—in some cases, impacting entire industries.

Business leaders are faced with permanent and accelerated changes in business and technology. In this competitive environment, with blurring industry borders, it's more important than ever for companies to focus on their core business on one hand, but differentiate themselves from their competitors and provide products and services that deliver immediate value to their customers with a top-notch experience on the other.

The Business Technology Platform, with key capabilities like analytics, intelligent technologies, database and data management, and application development and integration, provides most of the enterprise capabilities needed from the platform perspective to help an enterprise succeed in the cloud. The key pillars of the Business Technology Platform are shown in Figure 1.3.

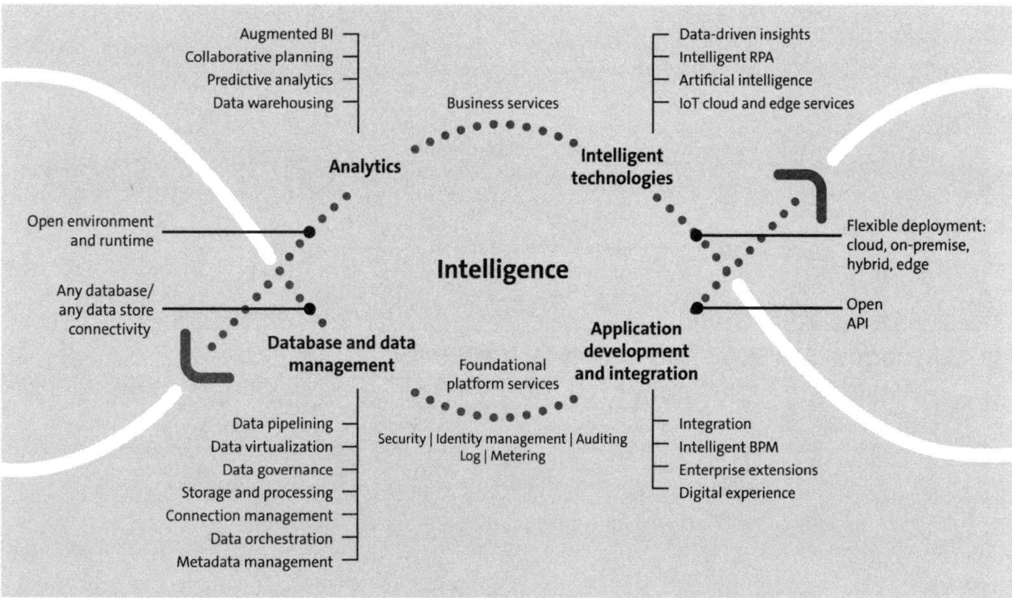

Figure 1.3 SAP Cloud Platform as Business Technology Platform Key Pillar

1.2.3   Elevate Your Enterprise Integration and Extension Experience

SAP Cloud Platform lets you achieve business process excellence, deliver engaging digital experiences, and simplify data-driven innovation via multicloud architecture. SAP Cloud Platform provides two suites that support you in your business transformation, as shown in Figure 1.4:

1. SAP Cloud Platform Integration Suite accelerates enterprise-grade integration of heterogeneous landscapes, using out-of-the-box content to seamlessly integrate end-to-end processes, data, people, and devices.

2. SAP Cloud Platform Extension Suite provides an easy and fast way to enhance and adapt SAP solutions, to meet a company's unique needs, and to gain a competitive advantage.

These suites provide a comprehensive set of the following elements:

- Best-run methodologies
- Enterprise-grade, secure tools and services
- Ready-to-use business content

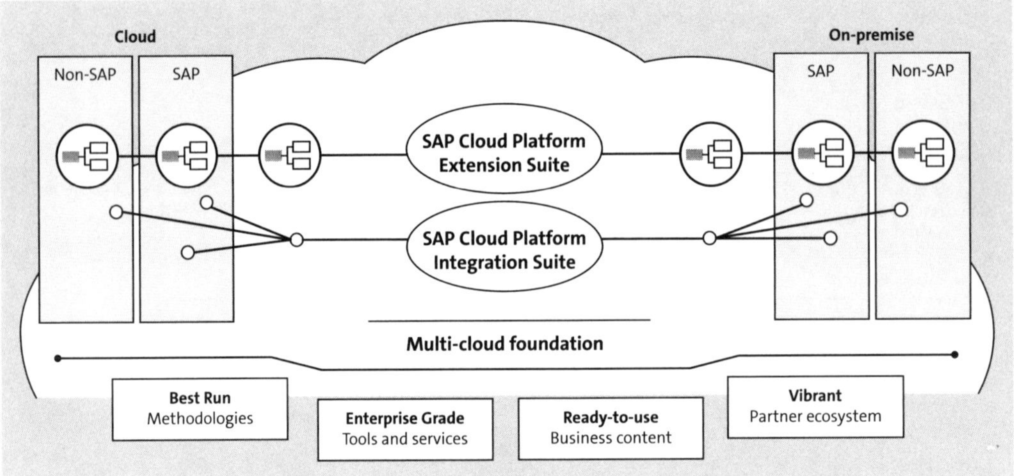

Figure 1.4 SAP Cloud Platform Integration Suite and Extension Suite

This combination allows you to deliver value for your business with speed, agility, and efficiency. While these suites focus on solving dedicated business needs, the underlying foundation offers a secure, multicloud, future-proof underpinning that embraces all major hyperscaler infrastructures and cloud-based, native technologies. Let's now look at these suites in more detail:

- **SAP Cloud Platform Integration Suite**
  SAP Cloud Platform Integration Suite capabilities (see Figure 1.5) help to simplify *enterprise integration* needs.

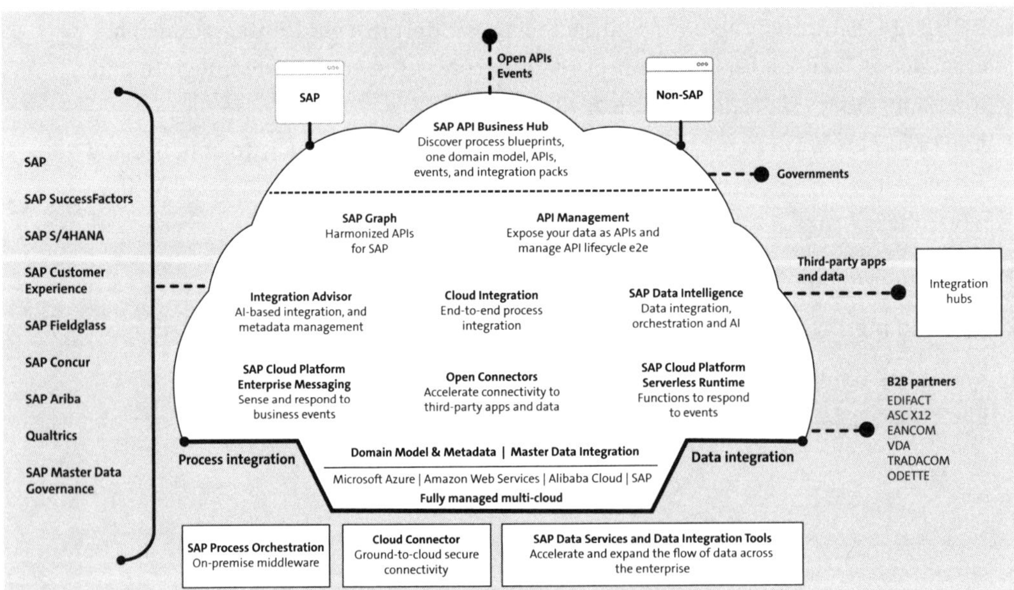

Figure 1.5 SAP Cloud Platform Integration Suite

The suite provides prebuilt integration content and out-of-the-box business events. It uses an API-first approach to assemble, publish, and monetize its APIs. For more information, see Chapter 5.

- **SAP Cloud Platform Extension Suite**
  SAP Cloud Platform Extension Suite capabilities (see Figure 1.6) help create *enterprise application extensions* to accelerate business value and help businesses become intelligent enterprises. For more information, see Chapter 9.

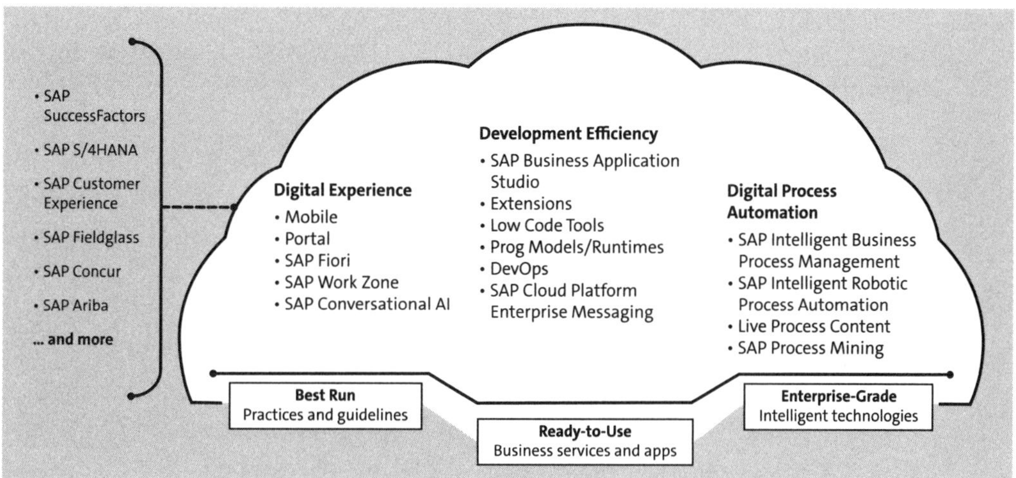

Figure 1.6 SAP Cloud Platform Extension Suite

With the rich portfolio of ecosystem applications that support and complement the existing SAP application portfolio provided by these two suites, you should be able to spend less of your budget on developing and maintaining homegrown applications and more on innovation and growth. Quicker innovation drives more success, which frees up more budget to help fund further innovation.

1.2.4   Pricing and Packaging

SAP Cloud Platform makes it easy to find, try, buy, and consume cloud services via following the options:

- **Trial system**
  Enterprises can test-drive SAP Cloud Platform and can build and deploy applications. Various tutorials are available for building and deploying apps that suit a business's needs.

- **Consumption-based commercial model**
  SAP Cloud Platform Enterprise Agreement (SAP Cloud Platform EA) allows for the consumption of cloud credits based on actual usage. This model provides the highest

flexibility and access to many services. With this service, the enterprise can monitor its service usage and spend on a real-time basis.

- **Subscription-based model**
  This option allows enterprises to choose a fixed set of services for a fixed rate, regardless of actual consumption.

An estimator tool is available to check pricing based on services required for the consumption-based model. The tool can be accessed at the following link: *https://www.sap.com/products/cloud-platform/pricing/estimator-tool.html.*

Example pricing based on a single service, SAP Cloud Platform Application Runtime, is shown in Figure 1.7.

Figure 1.7 Pricing Estimator Tool Example

1.2.5   Service Catalog

Organizations on the path of future-ready enterprises must explore various services to check how they can be utilized for integration, extension, and application development.

More than 60 services are currently offered as part of SAP Cloud Platform. A few of these many services are shown in Figure 1.8. They can be explored at the following link: *https://discovery-center.cloud.sap/#/serviceCatalog.*

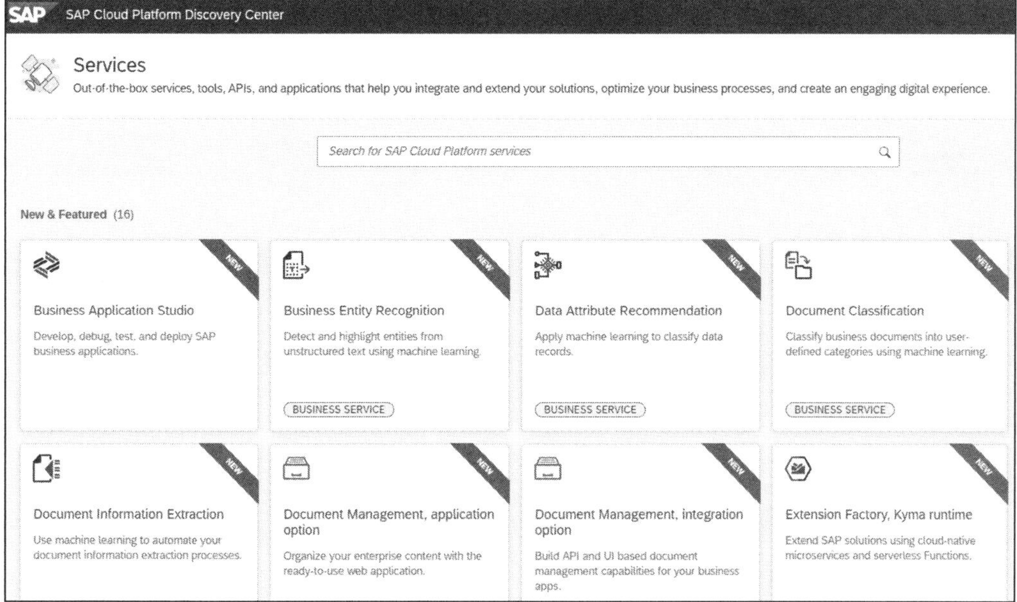

Figure 1.8 SAP Cloud Platform Discovery Center

1.3   Realms in the Future-Ready Enterprise

As technologies, business practices, and market demands change over time, the concept of future readiness will also evolve. To maintain future readiness, organizations cannot be complacent. They will need to think of future-readiness as an ongoing journey and stay nimble, adaptable, and innovative to maintain their competitive edge.

Companies are turning to SAP to drive innovation and to transform the business in today's data-driven, digital, and service-oriented economy. SAP has adopted the bimodal IT model in its support of organizations, with SAP S/4HANA as the system of record and SAP Cloud Platform as the system of innovation. You can connect the entire enterprise with SAP S/4HANA, then tailor, integrate, and expand the core with SAP Cloud Platform.

This bimodal IT approach is shown in Figure 1.9. SAP S/4HANA core-mission-critical applications are decoupled from line-of-business edge applications, which are required to address a unique requirement for a business or to address a whitespace. Such applications are developed on SAP Cloud Platform, allowing companies to innovate both the core and the edge applications, each at their own pace. This decoupling allows line-of-business applications to be deployed instantly based on the constantly evolving requirements of the business unit, while upgrades to the core can be done quickly and effortlessly because it isn't bogged down with customizations and custom objects required to run the edge applications.

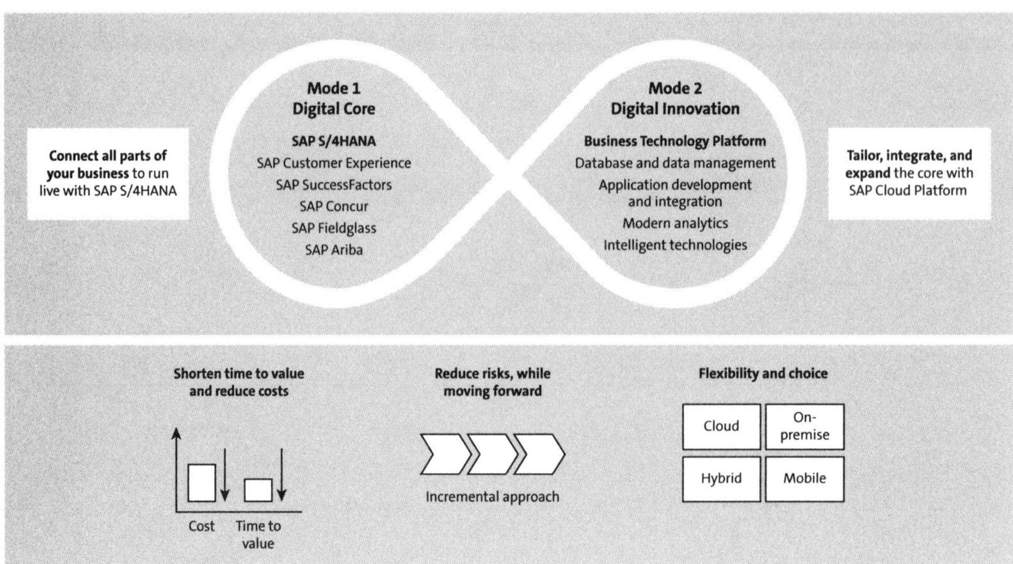

Figure 1.9 Digital Innovation Layer

> ### Example
>
> For example, if a chief marketing officer (CMO) wants to deploy a customer loyalty mobile application to track customer behavior and better tailor marketing content and campaigns, that app could be quickly built on SAP Cloud Platform, leveraging machine learning capabilities as well, without the need to alter or touch any code in SAP S/4HANA. Because SAP Cloud Platform natively supports SAP HANA, the CMO can be assured that the new loyalty app will integrate with his or her backend SAP S/4HANA core. In addition, when the time comes for a new SAP S/4HANA update or enhancement, the IT department can be confident that the new loyalty application will not be impacted because it's decoupled from the core, allowing IT to complete the SAP S/4HANA update quickly and with less effort.

There are multiple models for transformation available. These have been presented by McKinsey, Gartner, and Boston Consulting, and large enterprises have followed them.

Regardless of your chosen transformation option, the way you set up and manage your project will determine your ability to turn SAP's innovations driven by SAP Cloud Platform into your company's advantage. It will also decide if, a few years from now, your new system will become another legacy landscape. Or will it provide the business agility and speed to continue to outplay the competition through the full automation of business processes; technology architecture, combining application landscapes, data, integration, security, and user and mobile experiences on new data types; and new levels of insights?

In the following sections, we'll present a transformation model in the context of a technology platform, as shown in Figure 1.10. The stages of this model are as follows:

❶ Foundation

❷ Future aware

❸ Future focused

❹ Trailblazer

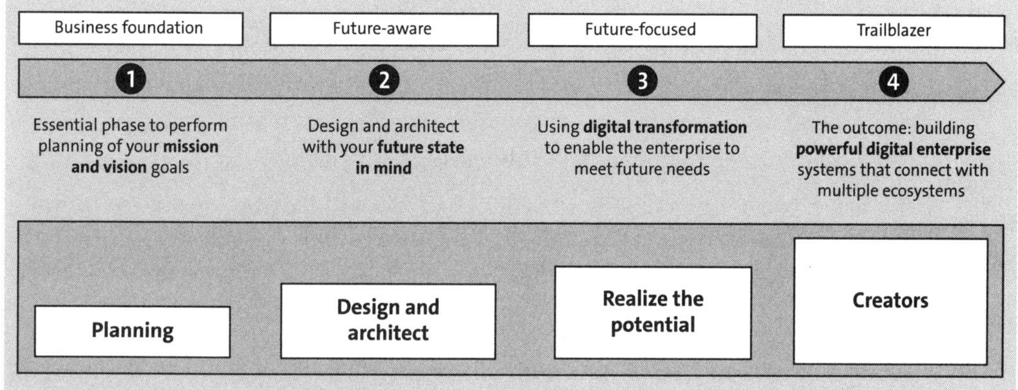

Figure 1.10 Stages in Future-Ready Enterprises

1.3.1   Business Foundation: Understanding Your Current Process

The business foundation phase is the first and most essential phase in the transformation journey, performing the planning for a transformation with a defined mission, vision, and clear-cut goals and looking specifically across business processes. One option is to take an evolutionary approach to transformation.

In the evolutionary approach, an enterprise can choose to take an incremental approach to transformation, in which an enterprise doesn't have to rebuild from scratch; instead, the organization can use the Business Technology Platform to improve on the efficiency of existing business processes and adapt them to changing market requirements. This may work very well for large enterprise.

This is a stage in which IT and the business have an opportunity to align, as they often have diverging views. It sees benefits in terms of costs, efficiency, and control, whereas business stakeholders see benefits in terms of simplicity, self-service, and new and improved business scenarios in applications. As the transformation increases, however, cost and complexity also often increase, at least in the short term.

In simpler terms, this can be called the *planning stage* and can be compared to the discovery and prepare phases of project implementation. This stage is all about understanding your as-is processes.

The business foundation layer helps to map the current state and visualize the target landscape as follows:

- It describes the organizational process.
- It defines your baseline and target process architectures.
- It defines process gaps to help you figure out your requirements.

1.3.2   Future Aware: Architecting and Designing

The second stage of the transformation journey the enterprise will take is mostly about designing and architecting your landscape. During this stage, an organization defines the data, application, and landscape architecture. A technical as-is landscape analysis exercise is performed to understand your existing IT landscape.

In this stage, you define a high-level overview of your IT landscape and of the communication interface between the IT landscape and the business processes supported by it. It should typically try to address the following questions:

- Which IT solutions have you deployed?
- Which business processes are supported by the IT solutions, either completely or partially?
- Which business processes are carried out manually?
- What are the communication and integration mechanisms between various IT systems?

1.3.3   Future Focused: Cloud Application Development

In this stage, the focus is on realizing the solution architected in the previous stage as part of your future-ready transformation journey. Thus, this centers on deploying a single transformation platform for business success, from which you can create intelligent, mobile-ready apps to better serve customers, modernize business processes, and compete.

In this stage, you simplify the development of your application in order to connect with, customize for, and empower end users (internal employees, customers, development users). Thus, using a cloud platform with access to cloud-native technologies allows you to do the following:

- Provide consistent and engaging experiences. Improve people's engagement and productivity through personalized digital workplaces, prebuilt business templates, and conversational bots.
- Automate business processes with flexibility. Increase process efficiency and performance with digital workflows, robotic process automation bots, and predefined, live process-content packages.

- Develop and manage with agility. Accelerate application development and seamlessly manage enterprise extensions with ready-to-use services and built-in tooling.

1.3.4 Trailblazer: Manage Innovations

This stage provides tools and frameworks for enterprises in their future-ready transformation journeys to use innovations to address whitespaces and gaps in their business. This allows them to speed up innovation and develop new solutions faster, leveraging connected processes and data.

Enterprises that are agile enough may choose to take a revolutionary approach to transformation and use technology to drive continuous innovation and be on the leading edge. This stage provides a framework for an enterprise to investigate the latest technology trends and the impact on the industry in which that enterprise operates.

This is a continuous cycle, and the intention is to help a company realize greater value from its investment and position its IT team as an agile innovator within the organization, thereby realizing greater value from end-to-end business processes.

1.4 Summary

For enterprises to be agile, cross-functional, and ultimately future ready, we recommend that they move through the four stages of the journey presented here, becoming highly successful in the years to follow.

The foundation stage defines the business process optimization layer as the first step in this journey, with the second, future aware stage driving the replatforming topic with the right designs and architecture and the future state in mind. The third stage, future focused, is the actual execution of the transformation journey. Finally, a continuous innovation culture stage, trailblazer, ensures continuous evolution, to remain always in a future-ready state.

SAP Cloud Platform has capabilities to support enterprises in all stages of their transformation journey, especially when on their journey to SAP S/4HANA to implement the digital core. SAP Cloud Platform's scope and capabilities should not be just restricted to extension and integration platforms, but seen holistically and used by enterprise architects and solution architects for business process optimization and for optimizing the IT landscape. This is a continuous necessity to improve time to value and as an innovation layer for incorporating new capabilities while keeping the enterprise agile.

Chapter 2

Understanding Your Current Business Processes and Landscape

Transformation does not address change from one location to another, but a journey to a future-ready enterprise. In this chapter, we'll discuss how to kickstart this journey with the improvement of existing business processes.

Existing enterprises are not looking for massive, two-year transformation projects. Currently, these types of projects are compressed to six to twelve months, and they're not just focused on productivity gains but also time to value.

It's becoming more important to have an agreed-upon outcome and define what success looks like, which can be a moving target depending upon who you ask within the organization. Digital transformation might mean multiple projects happening at the same time, side by side for different stakeholders. It's important not to spawn silo initiatives, but measure success across projects and measure them as one program moving toward becoming a future-ready enterprise.

CIOs are looking for looking for quick productivity gains and specific outcomes—perhaps through optimizing the process and the impact of the systems. Some examples of these outcomes are reducing churn or recapturing past customers.

For enterprises to survive in these difficult times, with a global pandemic in play, leaders must change their business process to fit the new normal. Businesses that make the choice to not simply revert back to their old ways but instead embrace new ways of doing business and optimize their business processes will survive and come out as winners. SAP's Business Technology Platform has the capabilities and the tools to drive this specific business outcome, helping enterprises become future ready.

Reimagining business process; adding new technologies, systems, or tools; rethinking recruiting, hiring, retention, and training strategies; and even rethinking business models can all be part of creating your future-ready enterprise.

As part of preparing enterprises to be future ready, in Chapter 1, we discussed the four stages of the transformation model. In this chapter, we'll discuss the foundation stage in more detail.

The foundation of a transformation consists of two main elements: functional processes and the technical landscape. Enterprise and business architects need to map both their as-is landscape in terms of current business processes and the current system landscape.

"

In this chapter, we'll focus on current business processes before moving on to some of the tools and frameworks SAP provides, evaluating both. We'll end the chapter with a quick look at some of the outcomes you can expect from this process.

2.1 Current Business Processes

Many organizations still use legacy on-premise systems. Many of them have expanded their IT landscapes over the last few decades by buying point solutions to solve specific business needs. This has resulted in fragmented stacks, poorly integrated software, disconnected duplicate master data, and an inconsistent user experience (UX), all of which frustrate users and slow down innovation.

These organizations will move to the cloud at different speeds, and they rightly expect both a continuous level of innovation and security of their investments. Thus, the move to SAP S/4HANA is not just a technical replacement; it's about reinventing how a business operates, delivering net-new customer and employee experiences, and establishing new business models. SAP Cloud Platform allows businesses the time they need to reengineer business processes and systems, helping them tap into the full business opportunities of SAP S/4HANA.

Enterprises that want to drive transformation with SAP S/4HANA can leverage SAP's Business Technology Platform and SAP Cloud Platform to drive the SAP S/4HANA journey, starting from the discovery phase or business case for transformation.

Enterprises do not just need a culture change that emphasizes innovation; even changing, enhancing, or standardizing business processes can be a driver for transformation.

It's important to understand whether your current business processes support your long-term strategy. A strategic redesign of business processes may be required. The intention is to define a common language and structure through business capabilities (for impact, maturity assessment, etc.).

Businesses have always been under pressure to innovate their product offers, improve productivity, reduce time to market, and increase profitability, and one way to achieve that is to drive continuous process improvements.

Process that aren't optimized result in the following:

- Increased cycle time, higher operating costs, and decreased sales due to bottlenecks and deviations in processes.
- Margin loss, as it's estimated that enterprises lose 10% to 20% of their margin due to a failure to find and eliminate inefficiencies.
- Noncompliant processes or fraud, which can have monetary and legal consequences.
- Inefficiencies and deviations need correction before a system migration project and can cause delays.

The first step of becoming a future-ready enterprise is to optimize existing business processes. To understand current business processes, business process architecture modeling needs to be done to visualize best-practice processes for the target landscape and orchestrate their integration with existing processes. The Business Technology Platform provides capabilities such as SAP Process Mining by Celonis and Spotlight by SAP, which help determine whether redesign of existing business process is required.

This drives the need for organizations to invest in and transform their business. The SAP solution map for the process excellence methodology is highlighted in Figure 2.1.

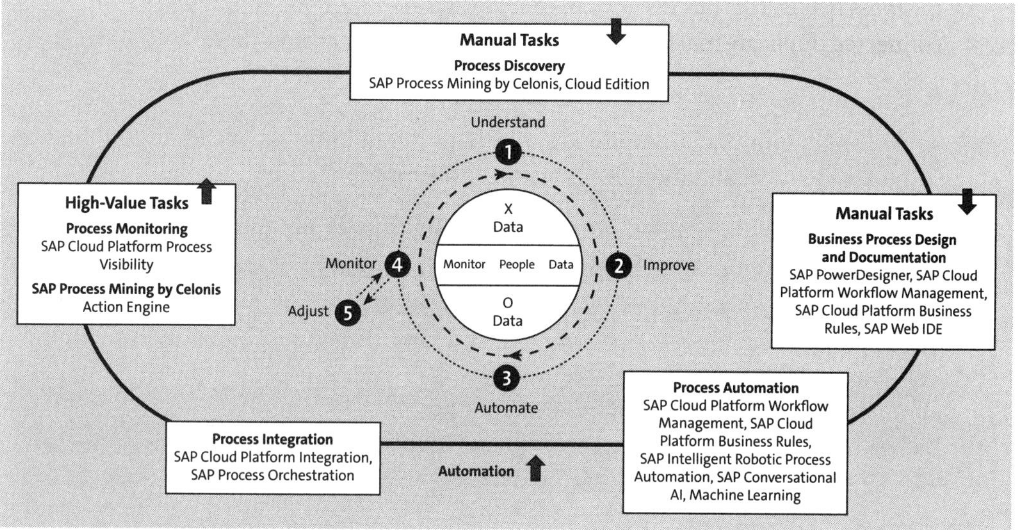

Figure 2.1 SAP Process Excellence

Process mining is the first step, which helps organizations to realize greater success and value with process improvement projects. Large organizations and enterprises need process excellence tools such as SAP Process Mining, which provides the following advantages:

- *Insight to design*, enabling complete data-based transparency into process execution so you can detect deviations or bottlenecks and drive efficiency improvements

- *Insight to action*, providing an activity tracker for in-flight business processes, thus enabling real-time end-to-end visibility into ongoing processes so you can detect and respond to emerging situations and drive agile business operations

2.2   Future-Aware Business Processes

Design thinking is an iterative process used by teams to define and tackle problems that are unknown or not completely developed. Teams go through defined process phases—empathize, define, ideate, prototype, and test—in which they determine users' needs and demands, challenge current assumptions and status quos, redefine problems, and aim to quickly create a prototype to gather end user feedback. The methodology has proven successful for organizations to understand unmet customer demands and develop new insights into business processes, allowing them to develop innovative and creative business outcomes for their industries.

The outcomes are as follows for companies:

- Identify and eliminate redundancies
- Identify whitespaces and fill them with SAP Cloud Platform extensions
- Identify automation opportunities and optimize workforce utilization to increase efficiencies while reducing operational costs
- Build resilient solutions that are future proofed
- Build solutions that are ready to scale, that cater to business dynamics, and that are flexible to meet adoption needs via the fastest methods

As a part of future-ready enterprise solutions, execution of a design thinking program is done in a series of workshops together with partners and SAP experts to address problem statements, business challenges, adoption of future industry trends, and so on. The identified items are interpreted to generate ideas and experiment with solutions to deploy, as illustrated in Figure 2.2.

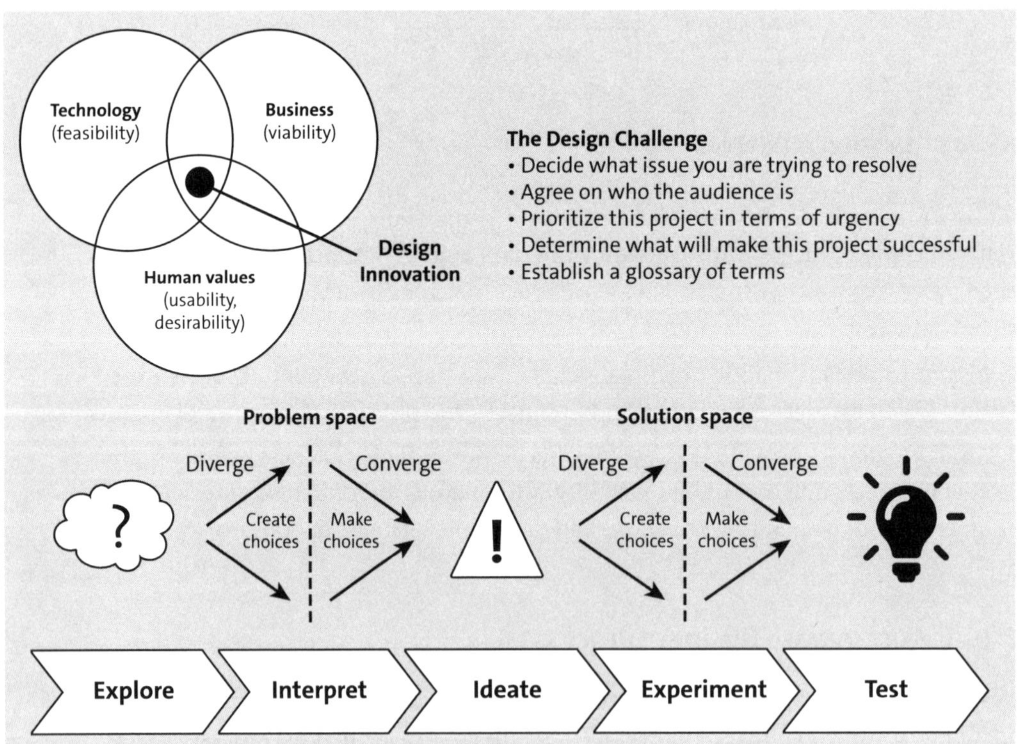

Figure 2.2 Design Thinking Workspace

Another approach to unravel and uncover challenges in the existing business processes and identify user pain points is through a business process discovery workshop. The workshop starts by asking the following questions of decision makers, business process architects, and solution architects in the organization:

- Where do you want your business to be?
- Are you happy with the current processes in your organization?
- Are the current processes in your organization serving the best interests of your employees and your end customers?
- Are your business processes in the organization aligned with the overarching strategy of your organization?
- Do your business processes in the organization fit industry best practices and standards?
- Do your business processes in the organization help differentiate you from your competitors?

If the answer to any of these questions is no or you're unsure, you have an opportunity in your organization to reimagine your business processes and further optimize them to meet your future strategy and demands.

Business process discovery workshops aim to capture business processes in an organization and thus facilitate process improvement projects by identifying and targeting the right projects to make the most impact and drive your business. The approach uses data and analytical insight, using tools and services such as those discussed in Chapter 2 to capture as-is business processes as an input. It uses these insights to further drive process improvements that are the best fit for your industry transformations.

Business process discovery workshops are ideally done in three phases:

- Phase 1 targets capturing data for your business and industry, existing business processes, and personnel information. Phase 1 also captures information on short- to medium- to long-term goals for your business to transform your organization.
- Phase 2 builds on Phase 1 and targets facilitation of the workshop by bringing teams together to help define new business processes that will be used in the future.
- Phase 3 includes presentation of the findings from Phase 1 and business processes created in Phase 2.

The outcomes of both approaches discussed so far are analyzing your current business processes and identifying, strategizing, and prioritizing the business processes that can have the biggest impact on your business to measure success.

2.3   Tools and Frameworks

SAP provides a number of tools that help you explore your current business process and landscape and see if redesigning your existing business process is required (or not). In this section, we'll walk through four of them:

1. SAP Process Mining
2. Spotlight by SAP
3. SAP API Business Hub
4. Ruum by SAP

Let's now jump into evaluating how SAP's Business Technology Platform provides tools to drive business process improvements and insights for enterprises.

2.3.1   SAP Process Mining by Celonis

SAP Process Mining uses process mining technology to help you visualize and understand the operative processes running within your SAP solutions and third-party applications.

Process mining is a field used to extract knowledge from the event logs stored and recorded by an information system like SAP ERP and SAP S/4HANA. The information stored in these event logs is used to analyze the underlying processes and their variants to visualize how a business is running currently and then identify potential areas of improvement.

SAP Process Mining extracts all the data about business process execution from systems and builds up intuitive graphical representations. This helps you to get a clear picture of when tasks are performed, their efficiency, and other aspects like the manual effort used to complete them. With this information, businesses can quickly understand how processes work and work on remediation of inefficient and cost-intensive processes.

By identifying process weaknesses, process owners can focus on potential areas of improvement by utilizing various SAP Cloud Platform tools for process automation and visibility, leading to better outcomes and optimized processes.

Note

Celonis is an SAP partner, and this solution extension is available as part of the SAP App Center at *https://www.sapappcenter.com/en/product/display-0000012977_live_v1*.

In the following sections, we'll look at some of the key tools available as part of SAP Process Mining. We'll also describe the next steps after using SAP Process Mining.

Process Explorer

The *process explorer* is the core element of SAP Process Mining, which makes it possible to visualize and analyze processes. In Figure 2.3, you can see an example of the process explorer and its different features, which will be explained throughout the following sections.

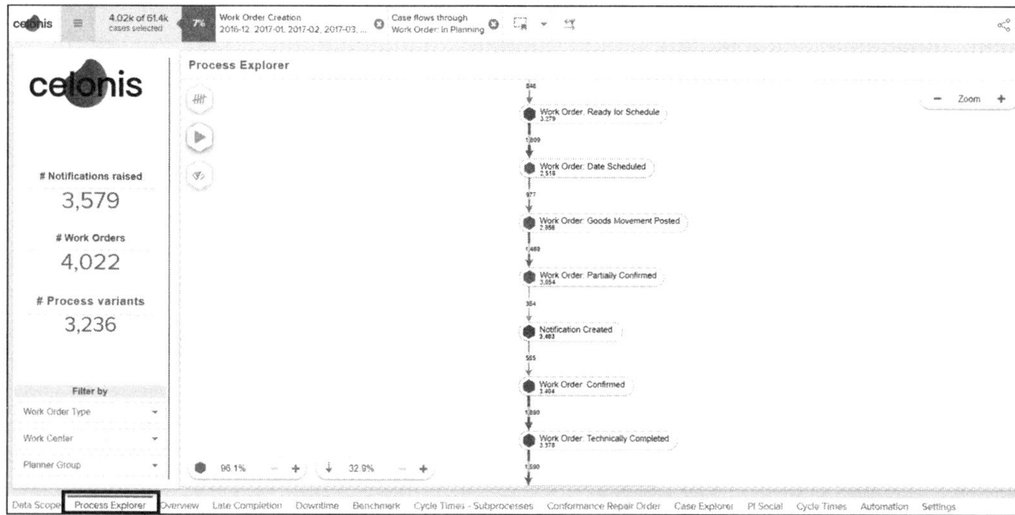

Figure 2.3 Process Explorer

Case Explorer

The *case explorer* is a tool for examining single cases and their associated activities. *Cases* are specific scenarios within business process that you want to highlight and work on. The case details show all activities that are passed by this case, in the right order. Figure 2.4 shows an example of a case and its details.

CASE ID	NUMBER OF ACTIVITI...	DURATION	ORDER	ERNAM	CLIENT	ORDER TYPE	ERDAT	AEDAT
2E2D46CF142D1B...	12	10M	0313175AD000DA...	Lang Winkler 1437	100	02	2017-01-14 00:00:00	2018-04-07 00:00:00
60E3430F1CD746E...	17	3M	07D8713E9B7707C...	Ward Weckerly 1401	100	02	2017-03-26 00:00:00	2017-06-25 00:00:00
21A765858D6AC3F...	15	2M	07DB210C6ACA09...	Karri Ptacek 1042	100	02	2017-07-05 00:00:00	2017-07-11 00:00:00
788AA335161C87B...	13	10M	07DB730F0D8EAE...	Karri Ptacek 1042	100	02	2016-12-26 00:00:00	2017-10-31 00:00:00
4A2B883518492F3...	14	2M	0C1786768FE1373...	Kristen Eckhoff 1001	100	02	2017-10-22 00:00:00	2017-12-26 00:00:00
5CC90BDB646808...	15	3M	0C4810A5EEBF5B...	Lang Winkler 1437	100	02	2017-07-31 00:00:00	2017-09-04 00:00:00
34CB7BA3F9C4BA...	20	1M	10913719FC82156...	Lang Winkler 1437	100	02	2017-08-08 00:00:00	2017-08-14 00:00:00
B7C9587CC7C7A4...	18	19d	109E60ECC220354...	Dominica Kirkley 986	100	02	2017-10-14 00:00:00	2017-10-22 00:00:00
EBEAA0FBC67851...	20	9M	10A8BA318592728...	Melissia Kreidler 598	100	01	2017-06-04 00:00:00	2018-02-21 00:00:00
12700413B19F462...	19	4M	10AE435450C619F...	Anna Hulings 640	100	01	2017-07-16 00:00:00	2017-10-13 00:00:00
278862DA957AD2...	19	1y	1988D9ECAED741...	Summer Fusaro 503	100	01	2017-03-01 00:00:00	2018-02-05 00:00:00
23CAAA780041EC...	16	6d	199E577593BDE40...	Tyree Decastro 1453	100	02	2017-08-22 00:00:00	2017-08-28 00:00:00
346A70F956B8EC...	16	1y	1DD5AEBD05ADA...	Toby Narvaez 1301	100	02	2017-04-19 00:00:00	2017-12-13 00:00:00
47BDFDF480CE11...	18	1M	224F3F5B43BC293...	Lawerence Pintor 1...	100	02	2017-04-27 00:00:00	2017-05-23 00:00:00
349BBDEF4BA702...	20	1M	269B4C4911706AF...	Tyree Decastro 1453	100	02	2017-10-24 00:00:00	2017-11-28 00:00:00
1952F188C9BAC49...	18	8M	26A785A615D86EA...	Emilio Sones 697	100	01	2017-09-11 00:00:00	2018-05-27 00:00:00
5C07427691146F8...	17	1y	2F02195A9049E0D...	Emmie Schuldt 1488	100	02	2017-06-07 00:00:00	2017-10-23 00:00:00
8396313897CFB13...	19	4M	33B06437BF96C6A...	Rachel Onorato 504	100	01	2017-03-02 00:00:00	2017-06-04 00:00:00
D25F0BF4352C94...	13	1y	37AE0487F4B2D71...	Emmie Schuldt 1488	100	02	2016-12-20 00:00:00	2017-10-23 00:00:00
4D39E43B1B34AA...	20	5M	37CB16A5A3578D...	Lawerence Pintor 1...	100	02	2017-09-10 00:00:00	2018-02-06 00:00:00
6EFCCEE5E52A89...	23	1y	44A0619E32A4C7B...	Anna Hulings 1509	100	02	2017-04-19 00:00:00	2017-10-31 00:00:00
F175EC7C88ECDF...	17	6d	44ADEC180E49B5...	Ward Weckerly 1401	100	02	2017-06-28 00:00:00	2017-07-04 00:00:00
9DF37933641BC27...	6	9M	48EC9320D9A3DB...	Ward Weckerly 641	100	01	2017-07-17 00:00:00	2018-04-14 00:00:00

Figure 2.4 Case Explorer and Details Page

Case details can also be accessed with a flow of events that have happened to understand the specific instance and related activities in detail, as shown in Figure 2.5.

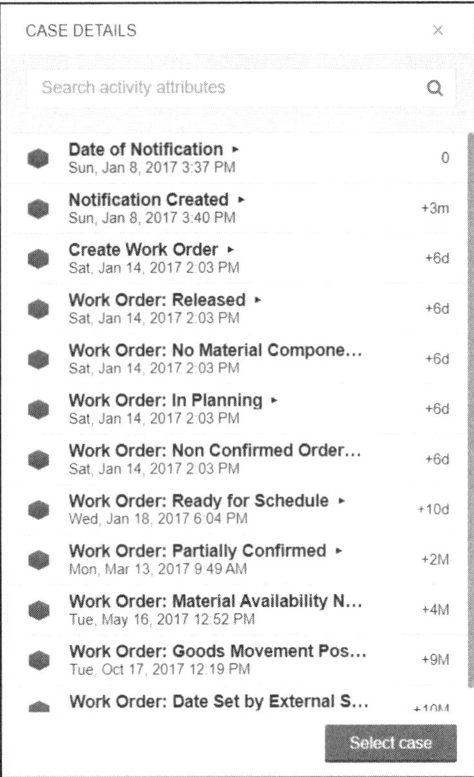

Figure 2.5 Case Details

Analysis

In SAP Process Mining, process analysis is structured in projects, each of which can contain a near-endless number of analysis documents. Analysis documents always include at least one analysis sheet—but there is no maximum number of allowed analysis sheets. The following options are available:

- Blank sheet
- Benchmarking
- Case explorer
- Process explorer
- Process conformance

Data models and proactive insights are some other tools worth mentioning to get insights from processes and recommendations from SAP Process Mining.

For any analysis, the **Data Scope** tab provides details of various transactions, time frames, types, groups, and so on, to filter the scope of analysis and analyze the process and cases in details.

For example, in Figure 2.6, analysis of plant maintenance-related objects can be done with the filter criteria such as type, work center, planner group, and so on.

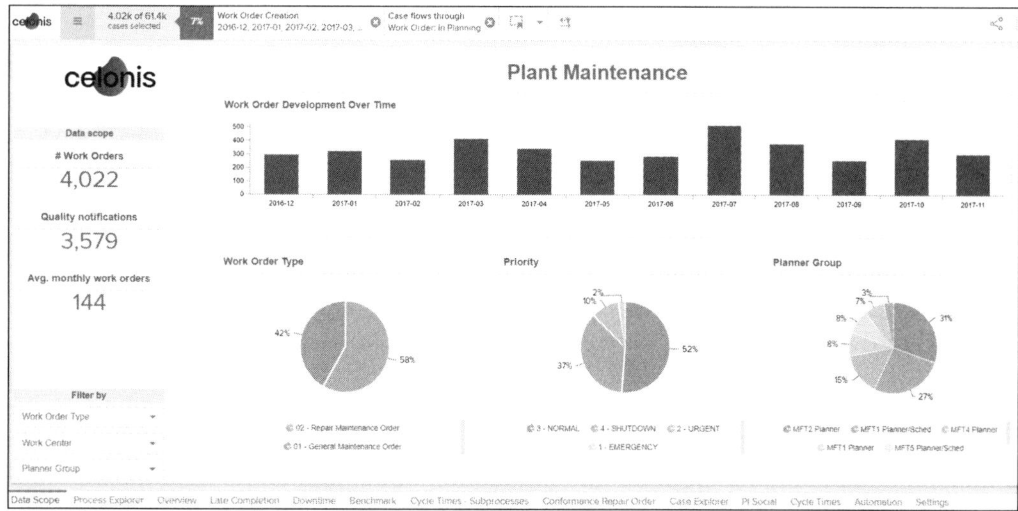

Figure 2.6 Analysis Scope

Next Steps

Process discovery is key aspect of future-ready enterprise preparedness and transparency to ensure that large organizations understand how they operate at a detailed level, what variations exist in system, and which areas are key laggards and to analyze them in detail. This paves a path for process optimization and a plan for improvement based on the analysis outcome.

As shown in Figure 2.7, you can use SAP Process Mining to visualize the actual processes and optimize the business processes by getting business data from multiple SAP and non-SAP system processes, and even user interaction data stored in spreadsheets and web apps, to provide optimized outcomes.

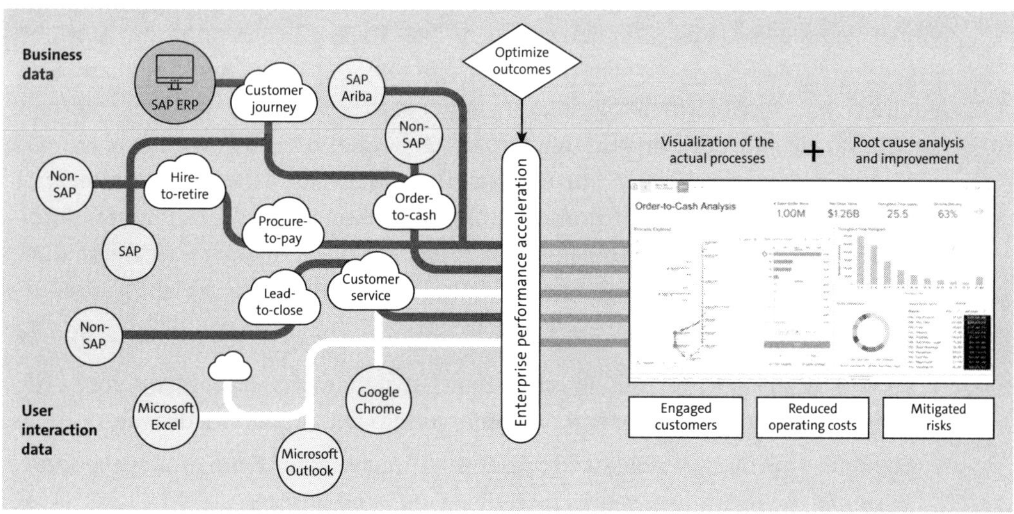

Figure 2.7 Visualize Business Processes

2.3.2   Spotlight by SAP

Spotlight by SAP provides transparency for data-driven actions. It helps to integrate and annotate operation data from SAP and non-SAP systems to understand usage, key performance indicators (KPIs), process performance indicators (PPIs), and user behavior, and then maps this data to process taxonomy. Other features include the following:

- Compare operations performance across units, geographies, and systems
- Leverage SAP's data to benchmark yourself against your peers
- Understand your process performance and view recommendations, helping you prioritize what to fix first
- Get actionable insights for your transformation projects
- Get help getting started with preconfigured transformation scenarios
- Define your own scenarios where necessary

Spotlight by SAP is in beta (*https://www.getspotlight.io/*) for the enterprise version. It currently provides a one-time report and provides different levels of view than SAP Process Mining.

Spotlight by SAP helps you define your own transformation journey to a future-ready enterprise by providing insights into current process performance and recommendations for how to fix them, as well as relevant automation opportunities for your enterprise. Spotlight by SAP also allows you to do the following:

- Identify technologies as enablers for LOB processes
- Prioritize and map intelligent capabilities in and beyond SAP S/4HANA
- Ascertain how to start your path to the intelligent enterprise

2.3.3   SAP API Business Hub

SAP provides end-to-end (E2E) process blueprints. These E2E blueprints are now published and available on SAP API Business Hub (*api.sap.com*). With them, SAP shares a reference architecture of which solutions are required to become an intelligent enterprise and the target state of enterprise architecture. Organizations can now explore how E2E processes can be broken down into applications, modules, business services, technologies, and integrations. All integration content is linked, such as APIs for integration. Based on this information, companies can plan their roadmaps to adopt the solution and to transition from their current solution landscapes to their target landscapes.

Four business process (see Figure 2.8) already exist on SAP API Business Hub, as follows:

1. *Lead to cash* provides an end-to-end scenario that manages all aspects of your experience, from the initial interaction to order fulfillment and service delivery.
2. *Source to pay* covers strategic and operational procurement, from finding the source of supply to procuring as well as receiving goods?and services.

3. *Hire to retire* is a subprocess of the *recruit to retire* scenario, covering the lifecycle of an internal employee.

4. *Travel to reimburse* simplifies the travel process, from trip planning to reimbursement, while ensuring compliance for employers and employees in various roles.

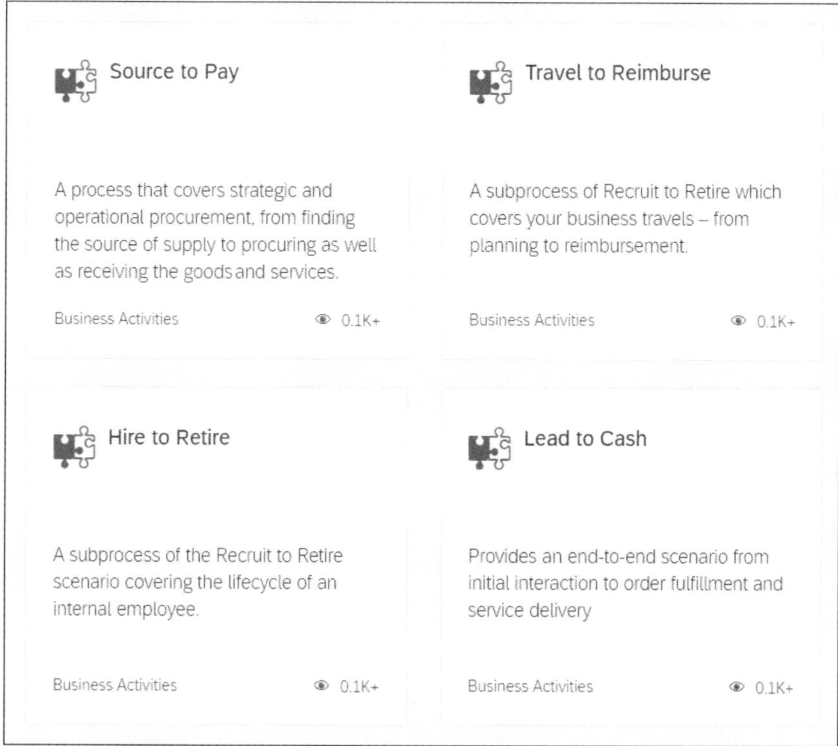

Figure 2.8 SAP E2E Business Processes

Note

For more information on SAP API Business Hub, see *https://api.sap.com/themes/BusinessProcesses*.

Looking at the details of each process, an enterprise can look to optimize its existing processes to suit its future transformation journey with SAP S/4HANA and further evaluate its extension and integration requirements.

For example, the lead to cash business process provides an end-to-end scenario from initial interaction to order fulfillment and service delivery. It provides a process template based on best practices, as well as common scenarios, and it can be adapted or extended based on your specific requirements. The process may vary depending on the industry type, the type of customers, and the sales channel used, whether e-commerce or direct sales. Figure 2.9 shows an example of the lead to cash process.

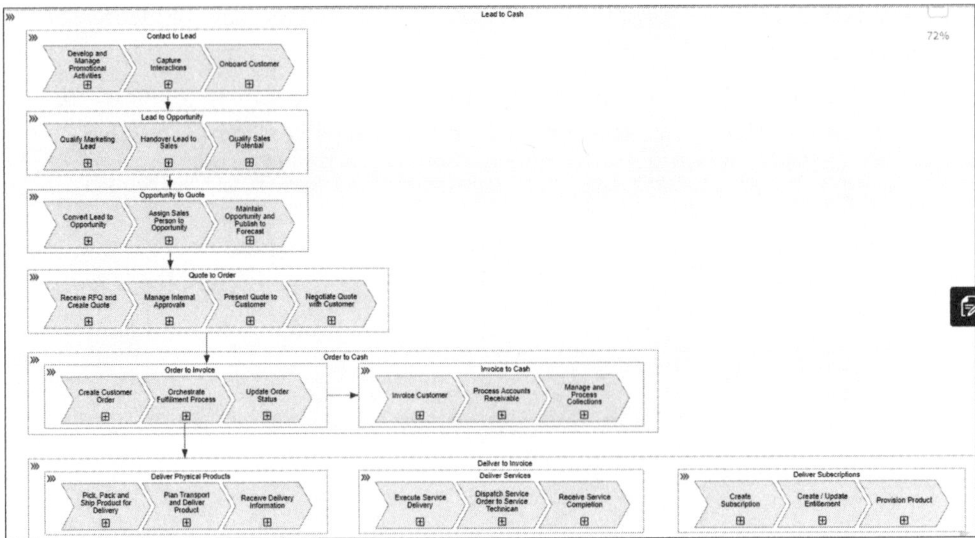

Figure 2.9 Lead to Cash Business Process

2.3.4   Ruum by SAP

Ruum by SAP is a work-management platform that allows organizations to build, automate, and execute processes on top of their existing architecture without the need to invest in coding. It provides the following functionalities:

- Allows anyone to model, automate, and run new processes without the need to change existing architecture

- Extensively integrated with SAP digital core, yet agile and adaptive at all times

- Includes enhanced collaboration, task tracking, project management, and file management features

- Is combined with SAP Intelligent Robotic Process Automation for automation capabilities

Ruum by SAP can help in bring people and data together and automate day-to-day tasks and approval processes to help business scale, as follows:

- Collaborate: Bring the right people and data together

- Manage: Set deadlines and monitor progress

- Automation: Empower business users with templates and project management AI

- Report: Get an executive view of your business to take decision at scale

Developers can build intelligent processes across connected applications, add tasks and approvals, collect inputs, and manage files on a shared canvas. Workflows can be enhanced with self-service automation, replacing manual, repetitive tasks with integrated bots; processes are live within minutes. Ruum by SAP provides a secure environment while giving a business the ability to solve 90% of its process-related problems itself.

Ruum by SAP provides full visibility across all open items. Extended features also allow team members to communicate, manage projects, and track open tasks. Administrators can monitor and review licenses, active users, and live usage analytics with their own dedicated Ruum by SAP dashboards.

Ruum by SAP provides templates, as shown in Figure 2.10, to quickly get you started in creating a lightweight process.

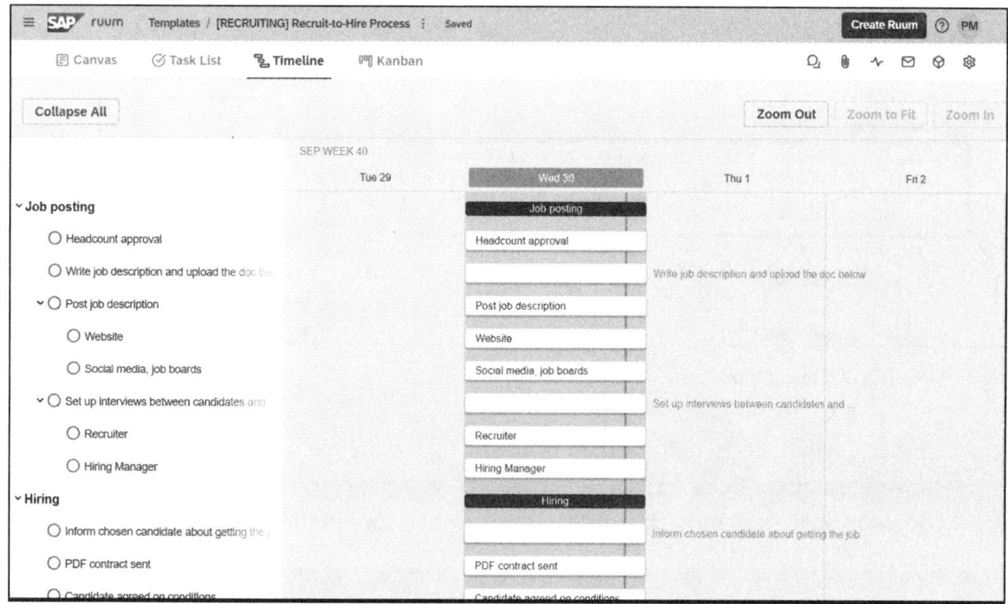

Figure 2.10 Process Building Using Ruum by SAP

For example, in Figure 2.11, we've chosen an existing template to manage renewal opportunities.

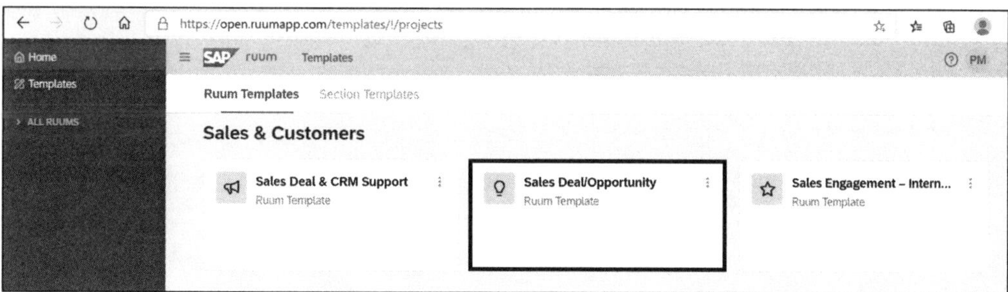

Figure 2.11 Ruum by SAP Templates

Once within the canvas, an end user can customize the sections, invite other users, assign task lists by priority and due date, add subtasks, add attachments and variables, and more, as shown in Figure 2.12. This helps you quickly create a customizable workflow for the new business process that has been created.

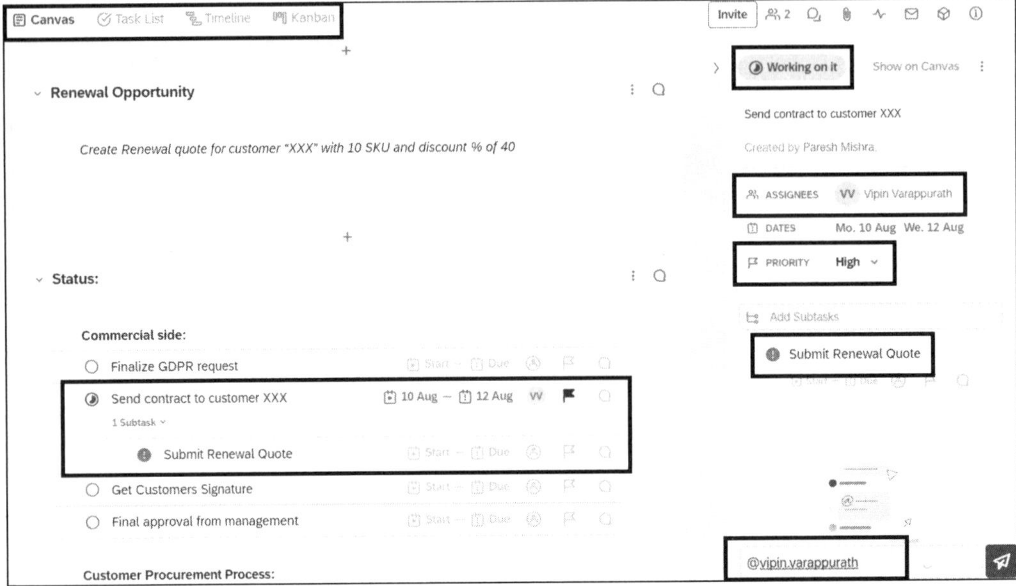

Figure 2.12 Sales Opportunity Example

Similarly, you can use additional templates for employee onboarding or create new custom templates from scratch. In Figure 2.13, you can copy the existing template into your workspace and publish and then customize it for your own needs.

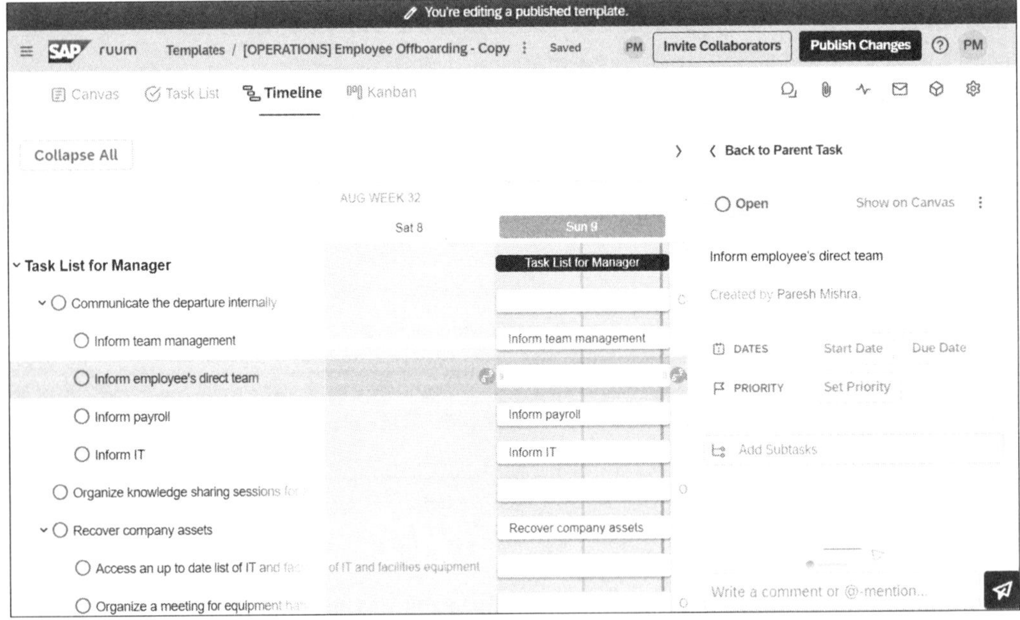

Figure 2.13 HR Template

Considering the current climate, we need IT to enable businesses to automate processes themselves. The market needs a better, faster, and cheaper solution for process automation, and this is where Ruum by SAP provides its value. Most enterprises want continuous improvement, and their business units are in constant need of new and better processes. Ruum by SAP can help significantly speed up and improve enterprise process delivery.

As shown in Figure 2.14, Ruum by SAP complements SAP's Business Technology Platform portfolio, compliments SAP Business Workflow, and responds to an unattended market need: the need to allow developers to build quick business processes without IT involvement.

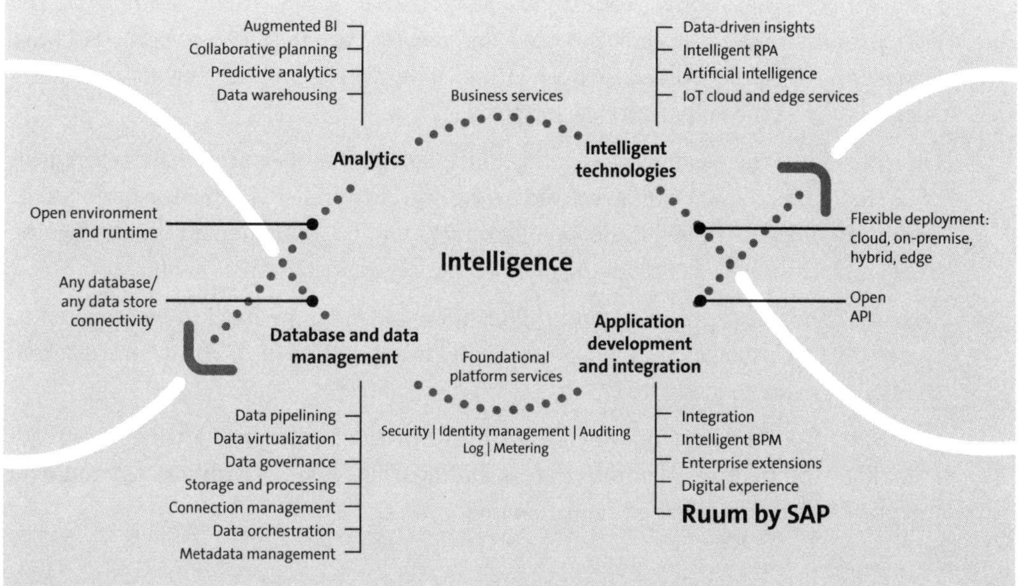

Figure 2.14 Business Technology Platform

Ruum by SAP provides business users with a process self-service and taps into the potential of automating all those processes that are today far from being digitized, documented, and automated.

This further creates a new level of transparency, and an enterprise still has full visibility into the number of open requests for a process that was never digitized. This will allow enterprises to have full ownership of their process landscapes and be able to gear up for being future ready.

2.4 Outcomes

In these times of uncertainty, enterprises can be motivated and driven by certain types of behavior: cost containment, improved margins, and how to drive transformation within the existing set of infrastructure and investment. Thus, today's CIO's job is not to just keep the lights on, but also to determine how to transform for a digital-only world as quickly as possible without running a full, massive transformation project.

One of the ways in which enterprises are continuously changing is by building new capabilities for existing business models—not spawning new projects and large-scale transformation projects, but looking into existing business processes and continuing to grow within that construct. They are looking for time to value: achieving an outcome quickly has a greater impact and can be done in a more cost-conscious way.

While some businesses have adapted quickly to things like remote work, digital commerce, and technology-enabled services, others have buckled under the pressure. For the companies that survive, leaders must adapt their strategies to the new normal. The businesses that make the choice to not simply revert back to their old ways—old ways which, to their credit, worked for a long time—but instead embrace new ways of doing business will be those that survive and thrive.

Enterprises will increasingly seek to optimize business models to remain relevant to their customers. Rapid changes caused by the current pandemic provide a good test of optimized business models, and key learnings must be implemented on an ongoing basis using the Business Technology Platform tools defined in this chapter.

Thus, both business processes and the technology landscape need to be revisited to understand the changes they will go through as part of the SAP S/4HANA transformation journey shown in Figure 2.15.

Start with process mapping, which leads into a broader enterprise architecture road-mapping exercise, thus aligning process and task automation with process and data integration and orchestration requirements.

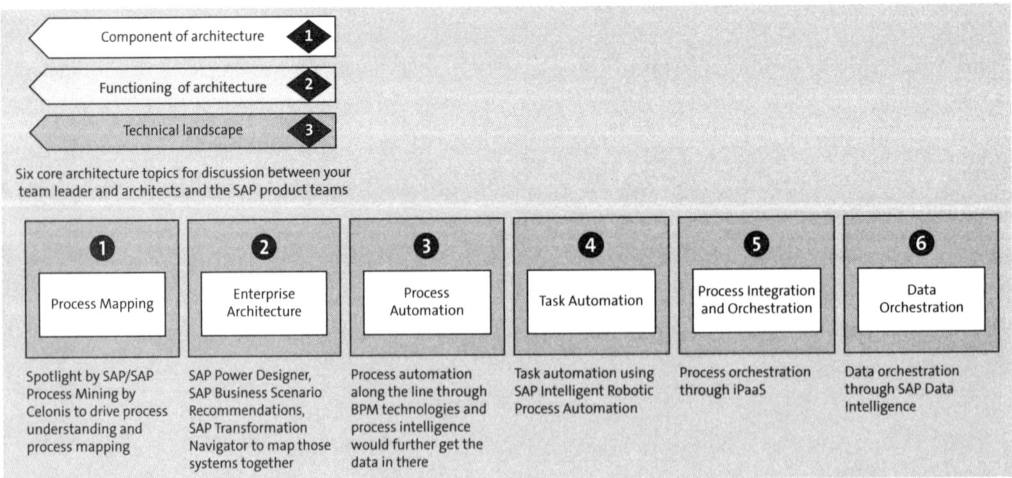

Figure 2.15 Future-Ready Enterprise

2.5 Summary

With the help of SAP's Business Technology Platform tools such as SAP Process Mining by Celonis, Spotlight by SAP, and Ruum by SAP, you can understand the impact of systems, understand and optimize processes, and gear up for the journey to the digital core with SAP S/4HANA.

SAP's Business Technology Platform has a role to play in this foundation stage of the future-ready transformation. It provides the following advantages:

- Reduces development costs and IT dependency, enabling business users to model, automate, and run processes without the need for development resources

- Eliminates process inefficiencies, allowing for ideal workflows to be designed, tested, and executed easily, thus removing manual interventions and the need to work across multiple applications

- Maximizes automation beyond the SAP core, tapping into the world of undigitized processes that occur outside the existing systems and offering an incomparable potential to automate manual and nondocumented workflows

- Ensures agility and accelerates innovation, providing a platform to pilot new processes that can be easily adjusted and remodeled where necessary

Process modeling is a challenge in *every* business, no matter the size, the industry, or where it is in terms of migrating to SAP S/4HANA. This is where SAP's Business Technology Platform helps during the foundation stage, leading to optimized business processes.

PART II

Future Aware

Chapter 3

Architecture and Design Considerations for a Future-Ready Landscape

In previous chapters, you've seen how to analyze your as-is business process and adapt them for your future-ready transformation. In this chapter, we'll dive into how to analyze your as-is technical landscape and architect and design it to be future ready, to meet your business requirements in an agile and flexible way. We'll look at different services from SAP Cloud Platform and various design outcomes as part of a transformation exercise.

The previous chapter discussed how you can get visibility and transparency into your current system landscape and business processes. In this chapter, we'll look at various architectural and design considerations that need to be applied on top of your as-is landscape while transforming it to its future-ready state.

We'll dive into various architecture principles that need to be taken into consideration while making the transformation journey in your organization with SAP S/4HANA and SAP Cloud Platform. The objective of this stage is to have a good understanding of your as-is technical landscape in order to drive the vision of the landscape-to-be, which is aligned with your business processes.

3.1 Future-Aware Landscape

Enterprises are faced with multiple challenges, such as the following:

- Fluctuating business strategies and significant merger, acquisition, and divestment activity
- Piecemeal approach to business integration and process automation
- Supported by multiple disparate ERP environments and supporting platforms, running on outdated technology platforms with limited business unit commonality and standardization
- Environments that have evolved over time, through a combination of business change, technology change, organic growth, and in recent times, acquisitions

Due to the size, scale and complexity of the business, your company may be looking for a future-ready state in which its landscape can transform to drive operational performance and sustainable growth and support core functional lines of business and processes. Some enterprises have reached a breaking point at which the gap between their business strategies and operational landscapes is now inhibiting their ability to realize future business goals.

Thus, they need to define a future-state landscape that provides the following:

- A unified operational landscape in line with marketplace competition
- A landscape that supports standardized, simplified, and streamlined common business processes running on top of it
- A system capability to support process execution
- A single source of truth across the landscape

The future-aware landscape should provide an efficient and effective operational platform that enables enterprises to realize their business goals.

It should be business-led, which means adopting industry and leading process from the previous state and changing current or new processes and activities for a new system/ landscape.

> **Note**
>
> SAP's Business Technology Platform provides the tools and framework to support enterprise architects; IT directors; and infrastructure, support, applications, and technical architects to come together and define the new state of the system landscape for their enterprises on their journey to be future ready.

The target state should provide landscape flexibility and cost saving across all the pillars. One tool that can help define the target state is SAP PowerDesigner.

SAP PowerDesigner can help you visualize, understand, and manage the impact of changes to your enterprise system. This end-to-end tooling software supports model-driven architecture (MDA) design with industry-standard modeling techniques, a powerful metadata repository, and unique link and sync technology. Its key functionality is as follows:

- Empower data, information, and enterprise architects with a powerful modeling solution
- Capture all business requirements, describe the interactions among them, and model artifacts with the related metadata
- Design for the future with as-is and future modeling, streamline implementation, and integrate business and technical viewpoints across the enterprise
- Communicate graphically to better realize the impact of changes

- Visualize the links among all architectural layers to understand the impact of changes
- Improve collaboration between business and IT with intuitive visualization and reporting tools
- Leverage data as a strategic company asset to drive greater value and efficiency

In SAP PowerDesigner, you can model your enterprise architecture, align enterprise architecture with business modeling, and carry out an architectural impact analysis in your landscape for potential architectural changes for the future.

3.2   Cloud Architecture and the TOGAF Standard

As stated earlier, one of the outcomes of the future-aware stage is to help an organization define a future-ready landscape. High-level outcomes of the future-aware state can be summarized as follows:

- Reduce complexity in the landscape via simple and scalable technology
- Deliver high-quality service with a stable and secure set of technology services
- Build a high-performing technology capability by optimizing business as usual and providing maximum flexibility for project delivery

To drive the overall transformation journey, every enterprise should do an assessment of its as-is landscape (current state architecture) and define its state-to-be (future-aware architecture). The future state of the architecture should define the following:

- Architecture vision and principles
- Conceptual architecture, logical architecture, and physical architecture
- Roadmap and strategy

An enterprise should assess what it currently has in the following areas:

- Application architecture
- Information and data management architecture
- Master data distribution architecture
- Integration architecture
- Security architecture
- Development architecture

You should also evaluate possible products/technologies required to deliver the future-aware architecture.

Many companies use the Open Group Architecture Framework (TOGAF) to deliver the enterprise architecture. TOGAF's architecture development model (ADM) describes a method for managing the lifecycle of an enterprise architecture and consists of ADM cycles. It's sometimes called the *wheel*.

> **Note**
>
> For more information on TOGAF, see the following links:
>
> - *https://www.opengroup.org/togaf*
> - *https://pubs.opengroup.org/architecture/togaf9-doc/arch/index.html*

Considering that TOGAF is a generic framework, it's often required to extend it to suit an individual enterprise's needs. SAP has its own SAP Enterprise Architecture Framework (SAP EA Framework), which is a methodology and toolset. It's based on TOGAF and is specifically designed to support packaged solutions.

TOGAF does not offer any recommendations or advice on which architecture (data or application) should be defined in Phase C of the ADM. SAP focuses on application, data, and technology architecture, which is what we focus on in Section 3.4, and its mapping to TOGAF's technical reference model (TRM).

> **Note**
>
> For additional information on the TRM, visit *https://pubs.opengroup.org/architecture/togaf8-doc/arch/chap19.html#tag_20_03*.
>
> For more information on Phase C of the ADM and application and data architecture considerations for information system architecture, visit *https://pubs.opengroup.org/architecture/togaf91-doc/arch/chap09.html*.

In addition, Section 3.4 also refers to the cloud computing reference architecture model defined by the National Institute of Standards and Technology (NIST). NIST defines five elements—the cloud consumer, cloud provider, cloud broker, cloud auditor, and cloud carrier—and addresses their roles in the realm of cloud computing.

> **Note**
>
> For more information on NIST, visit *https://www.nist.gov/publications/nist-cloud-computing-reference-architecture*.

3.3   Platform Advisory Methodology

We've discussed in previous sections how you can define future-state business processes and landscapes and introduced the TOGAF standard for cloud architecture. In this section, we'll take a deeper look at how you can transform your organization to a future-ready intelligent enterprise using SAP Cloud Platform in various future-state stages. We'll look at various capabilities on SAP Cloud Platform and provide a methodology with capabilities mapping that can be leveraged to transform your organization.

Note

We'll talk about all these features and functionalities in more detail, along with project use cases, in subsequent chapters.

In the following sections, we'll first introduce the platform capability model and then show you how it applies to each stage of the future-ready framework.

3.3.1   Platform Capability Model Overview

Figure 3.1 shows the platform capability model across the future-ready enterprise stages. This model can be used by your organization while transforming the business and landscape. This figure is available for download at *www.sap-press.com/5157*.

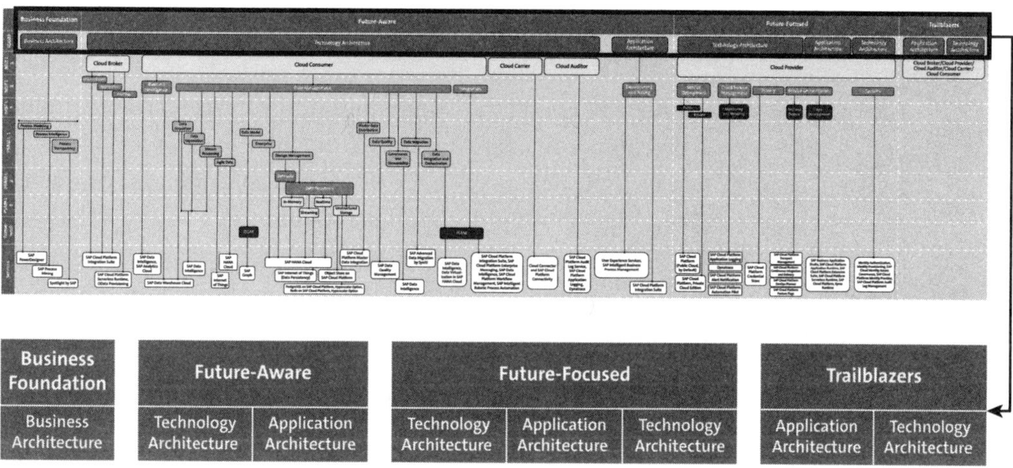

Figure 3.1 Platform Capability Model for Platform Advisory Methodology

In each stage of becoming future ready, we map your transformation journey via two architecture patterns, as described in TOGAF.

Application Architecture

Application architecture defines the behavior of applications and systems used in your organization to complete an end-to-end business process. We talk about how data is consumed and produced by different applications and how various applications need to work together to bring value to business.

This is defined based on functional and business requirements that need to be considered while building your future-ready landscape and meet business demands for now and the future.

Technology Architecture

Technology architecture defines the software and components required to support applications interacting with each other. Technology architecture provides a more concrete way for each application component to be realized, thus associating application components from application architecture with detailed technology components that need to be used to make it realize value.

Hence, technology architecture focuses on two broad things: realizing functional requirements and needs as defined by application architecture requirements, and meeting nonfunctional requirements associated with running your landscape and systems.

Within the platform advisory methodology, under the umbrella of application and technology architecture, various capabilities are mapped according to definitions and requirements set by NIST's cloud computing reference architecture, which considers the following elements:

- **Cloud consumer**

  The consumer is the stakeholder that the cloud computing service is created to support. It represents a person or organization that maintains a business relationship with and uses the service from a cloud provider. The cloud consumer may be billed for services provisioned. The following are examples of services available for a platform consumer:

  - Database

  - Application deployment

  - Integration

  - Development and testing

  - Business intelligence

- **Cloud provider**

  A cloud provider is an entity responsible for making services available to cloud consumers. A cloud provider builds the requested software/platform/infrastructure services, manages the technical infrastructure required for providing the services, provisions the services per agreed upon service levels, and protects the security and privacy of the services.

  For a cloud PaaS, the cloud provider manages the cloud infrastructure for the platform and provisions tools and execution resources for the platform consumers to develop, test, deploy, and administer applications. Consumers have control over the applications but cannot access the infrastructure underlying the platform, including the network, servers, operating systems, or storage.

Activities described for the cloud provider are following:

- Service deployment
- Service orchestration
- Cloud service management
- Security
- Privacy

- **Cloud broker**

  A cloud broker is the entity that manages the use, performance, and delivery of cloud services and negotiates relationship between the cloud consumer and provider.

  This is very important functionality: integrating systems and applications in a landscape can be very complex as organizations' landscapes are often heterogenous, and direct connectivity between the consumer and provider may not be manageable. The following service categories are provided by the cloud broker:

  - Service intermediation
  - Service aggregation
  - Service arbitrage

  Thus, a cloud broker provides capabilities to integrate different cloud offerings, using loose couplings and realizing business value via an ability to connect applications across different cloud vendors.

- **Cloud carrier**

  A cloud carrier is intermediary that provides connectivity and transport of cloud services between cloud consumers and cloud providers.

- **Cloud auditor**

  A cloud auditor is an entity that can conduct independent assessment of cloud services, information system operations, and the performance and security of a cloud implementation.

3.3.2 Foundation

The foundation stage is about understanding your current business processes and landscape and the systems interactions between them to meet business requirements, as shown in Figure 3.2.

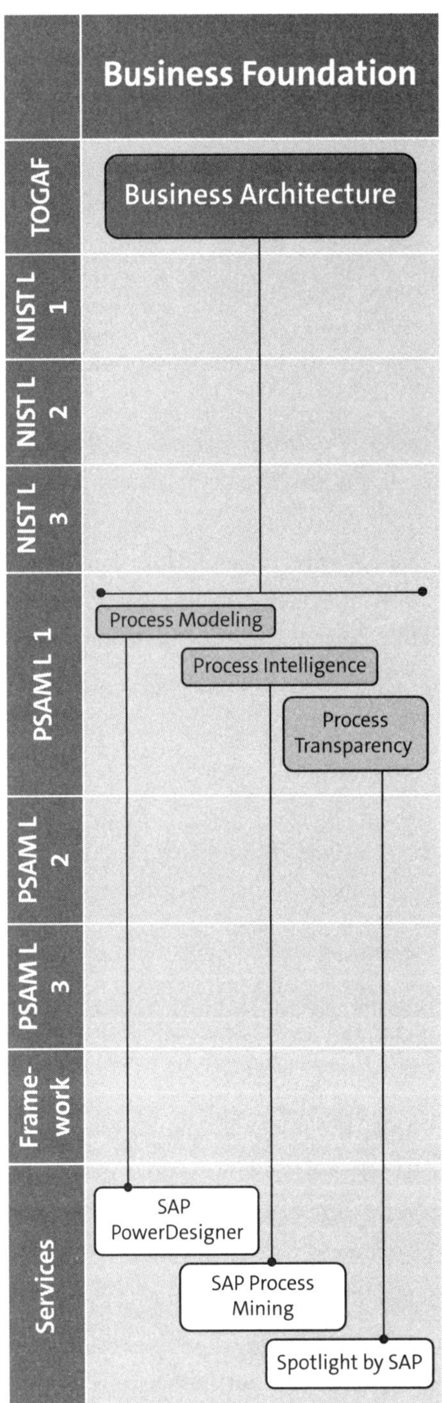

Figure 3.2 Platform Advisory Methodology: Foundation

Thus, in this stage, we focus on the following capabilities and use the advisory methodology to build the foundation:

- *Process intelligence* is about understanding about your current as-is business processes, determining deviations, and identifying areas to improve to build efficiency in process. Note that this needs to be planned as a continuous activity in your organization.

- *Process transparency* is about getting transparency into your business processes on a continuous basis and providing your business users with the ability and power to adapt your business processes to increase efficiency and improve employee engagement and customer experience. To reiterate, this is a continuous activity within the organization, *not* a one-time activity.

- *Process modeling* is about mapping and building an as-is enterprise architecture view of your systems and applications in your landscape. This is an important process in your foundation layer to accurately predict, design, and transform your organization by understanding various system interactions.

We have covered the capabilities and methodology in Chapter 2, and the future realms build on top of those.

3.3.3   Future Aware

In this section, we'll investigate various architectural considerations and capabilities you can leverage in SAP Cloud Platform to design your future application and technology reference architecture as defined in the platform advisory methodology.

Figure 3.3 shows the detailed outlining of the future-aware stage for cloud reference architecture development. In this stage, as you plan your transformation journey, the following advisory methodologies need to be followed:

- **Application architecture**

  At this stage, you must consider how business users would define interactions between systems and applications in your landscape. The success of your landscape transformation is defined based on how well the systems and applications are able to support end-to-end business processes and on how such application architecture can be made available and easily consumable for the business.

  Thus, at this stage, it's very important for your organization to define considering future-state business processes—how the business would interact with systems and applications.

  This capability is provided by the following SAP Cloud Platform services:

  - SAP Intelligent Robotic Process Automation (see Chapter 6)

  - SAP Cloud Platform user experience capabilities (see Chapter 7)

  - SAP Cloud Platform Integration Suite, API Management, as a secondary-level enabler for providing and exposing data to business (see Chapter 5)

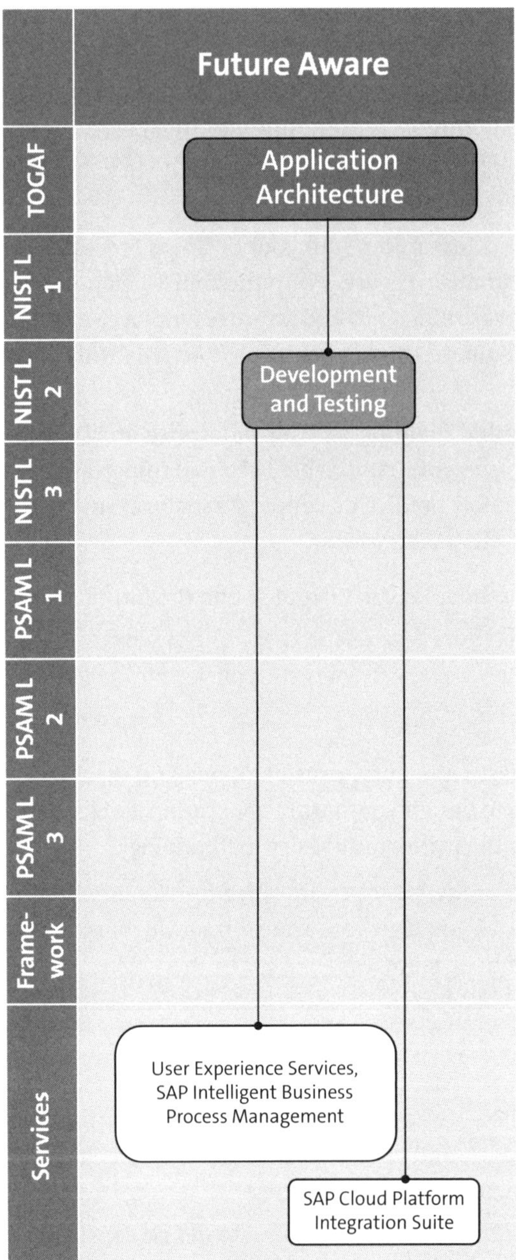

Figure 3.3 Platform Advisory Methodology: Future Aware (Application Architecture)

- **Technology architecture**
  Technology architecture design considerations include how systems and applications can interact with each other to meet functional and nonfunctional requirements. Figure 3.4 shows the technology architecture capability model within SAP Cloud Platform.

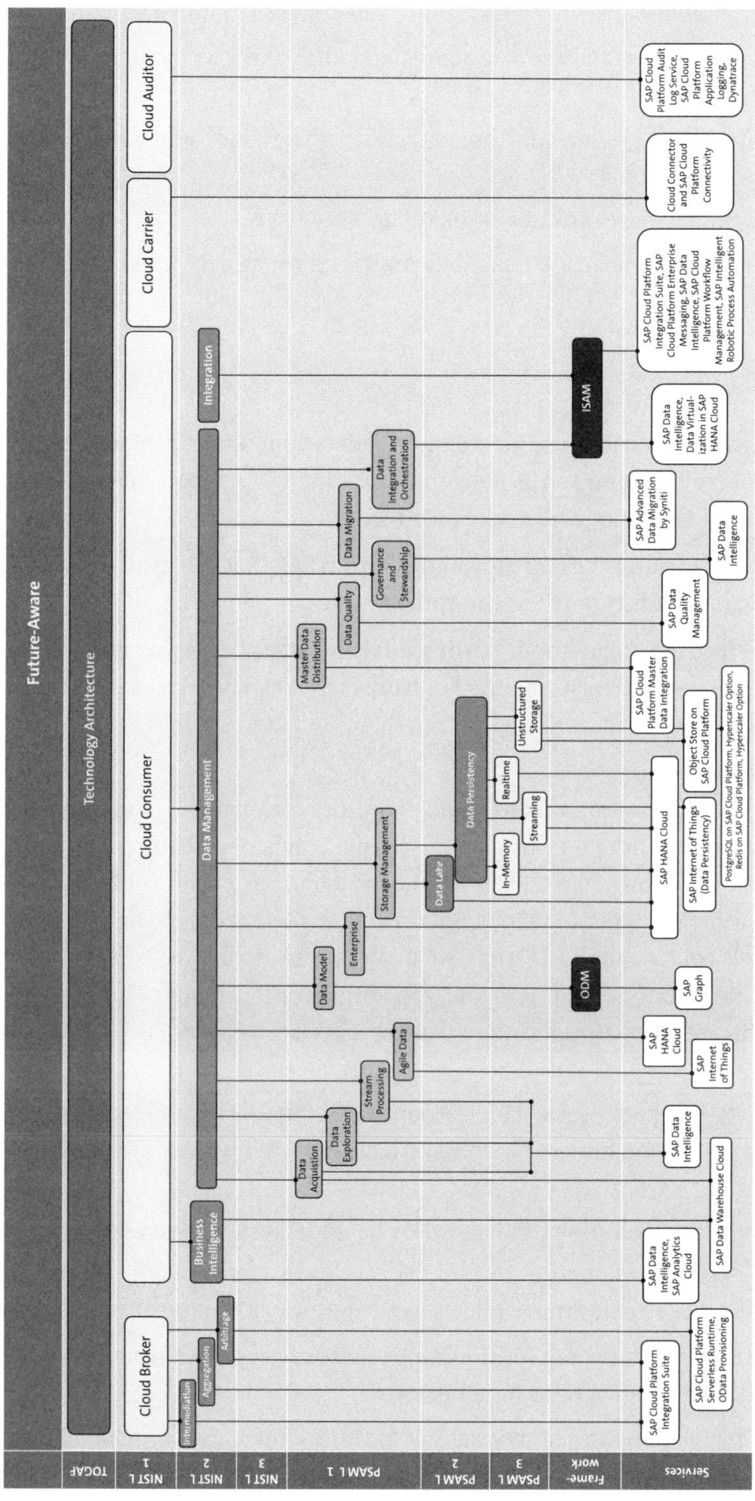

Figure 3.4 Platform Advisory Methodology: Future Aware (Technology Architecture)

The following capabilities are considered during technology architecture definition:

- Consistent availability of master data across landscape and data quality management (see Chapter 4)

- Data integration, integration and orchestration, governance and stewardship (Chapter 6 and Section 3.5)

- End-to-end system landscape integration (Chapter 6)

- Data modeling, storage management, data mart, enterprise data warehouse, and data lake (Section 3.4)

3.3.4   Future Focused

In this section, we discuss various architectural considerations and capabilities in the advisory methodology that you can leverage in SAP Cloud Platform to create your future application and technology reference architecture.

Figure 3.5 shows detailed outlining and capabilities within SAP Cloud Platform that you can leverage to create your future-ready enterprise.

Under this future-focused stage, the following advisory methodologies need to be heeded to design future-ready systems and landscapes that can be adapted to changing business requirements:

- **Application architecture**
  As you define your application architecture, you must consider the flexibility and agility your business needs to quickly adapt to changing business requirements and remain competitive in the market. Additional considerations may include skillsets available in your organization to help maintain and manage these changes in your organization on a continuous basis. Thus, we'll look at the capabilities of SAP Cloud Platform Extension Suite to help you and your organization remain flexible and agile and to define additional application requirements (see Chapter 9).

- **Technology architecture**
  Under technology architecture, we'll look at capabilities in SAP Cloud Platform that can help your organization remain flexible and agile to meet business and nonbusiness requirements.

  The following are considerations that need to be addressed while creating your future-ready enterprise:

  - Test and deploy your applications quickly in an agile way (Chapter 10)

  - Monitor and meter your system continuously to identify application performance issues and take necessary actions (Chapter 11)

  - Secure your applications and functionalities for future readiness (Chapter 8)

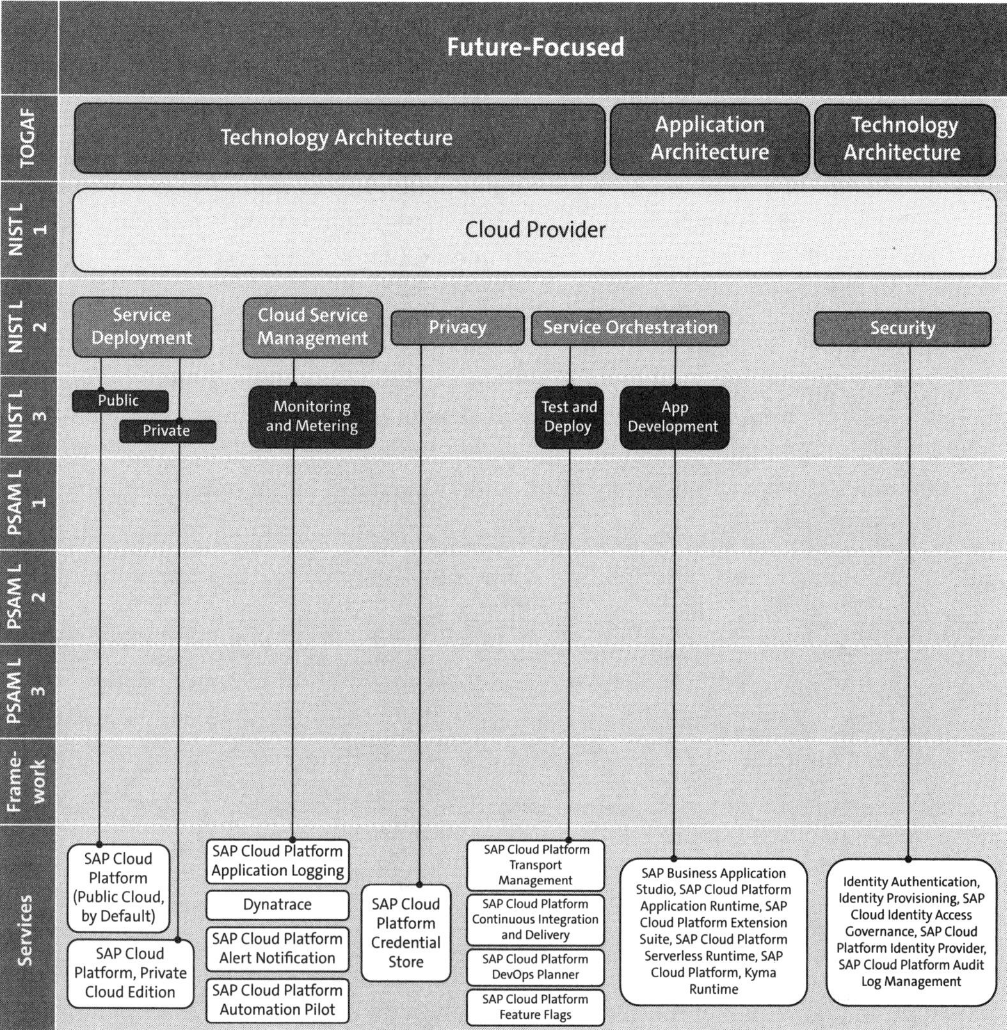

Figure 3.5 Future-Focused Stage in Platform Advisory Methodology

3.3.5   Trailblazers

The trailblazers stage is focused on remaining competitive in your industry and continuously innovating and transforming your organization. As noted previously, this stage builds on the previous ones.

Thus, to achieve continuous innovation and keep your application and system landscapes agile and flexible to adapt to changing business requirements, you need to build your organization's application and technology architecture based on all platform capabilities and methodologies covered thus far.

3.4   Data Architecture Principles

When we look at how technology can help a business, there is a strong consensus on the overall importance of data and analytics. With data volumes growing exponentially and an abundance of new intelligent technologies, organizations have countless opportunities to influence experiences and reimagine their business processes. To thrive in the digital economy, companies must use this data-driven innovation to respond quickly to the needs of their business and turn data into a valuable business asset.

Often, data is stored in silos within an organization; it's fragmented and scattered across multiple disparate systems over different deployment models, such as public and private clouds and on-premise systems, and can be both structured and unstructured. Organizations are seeking clear-cut guidance on data architecture best practices and the desired state for future-ready enterprises, as shown in Figure 3.6, which illustrates how to move from your current state to the target future state as a future-ready intelligent enterprise.

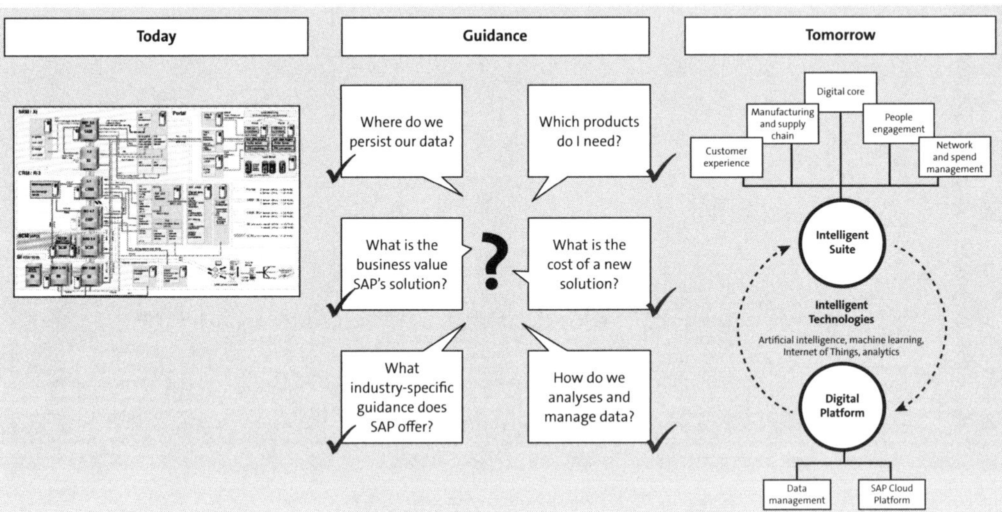

Figure 3.6 Data Architecture Desired State

SAP's Business Technology Platform provides database and data management services to provide enterprises with a level of modeling, governance, and protection.

For an enterprise to be ready to activate data, a concrete data architecture strategy is required that does the following:

- Maps and documents data sources and data flows
- Defines data value, assessing sources to determine data's purpose (see Figure 3.7)
- Defines data accuracy, standardizing the data format, cleansing the data, and matching complimentary data streams to derive meaningful insights

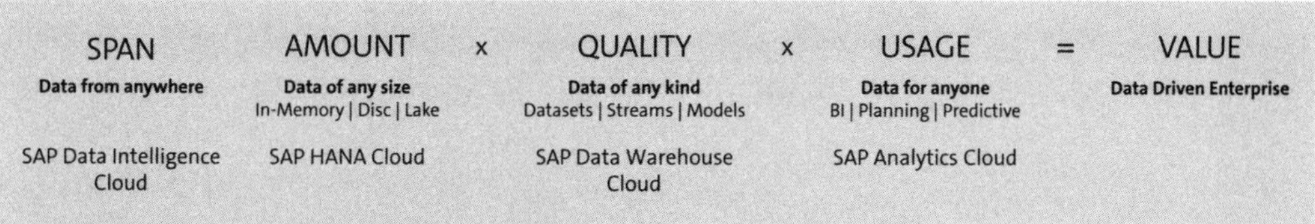

Figure 3.7 Business Technology Platform: Data Value

A data architecture pattern should define a standard set of products and tools an organization uses to manage data with the following elements:

- Data security
- Database and storage management
- Data migration
- Business intelligence and query modeling
- Data integration
- Master data
- Data quality
- Governance and stewardship
- Architecture, flows, and data modeling

The Business Technology Platform provides the capabilities listed in Table 3.1 to map all the data architecture patterns.

Data Pattern	Objective	Business Technology Platform Capabilities
Data governance and modeling	Capture data lifecycle and governance of data in landscape	- SAP PowerDesigner - SAP Data Intelligence - SAP Graph - SAP AI Business Services: Data Attribute Recommendation - SAP AI Business Services: Document Classification - SAP Data Warehouse Cloud
Data acquisition and orchestration	Data ingestion, pipelining, and orchestration of big data in complex landscapes	- SAP Data Intelligence - SAP HANA smart data integration (within SAP HANA Cloud) - SAP HANA smart data access (within SAP HANA Cloud)
Data quality	Capturing the right dataset	- SAP Data Quality Management - SAP Cloud Platform Data Enrichment

Table 3.1 Data Pattern Capability in SAP Business Technology Platform

Data Pattern	Objective	Business Technology Platform Capabilities
Data storage	Storage of data	■ SAP HANA Cloud ■ SAP Cloud Platform Document Management ■ SAP Cloud Platform Big Data Services ■ Object Store on SAP Cloud Platform ■ SAP Cloud Platform, SAP ASE service ■ PostgreSQL on SAP Cloud Platform, hyperscaler option ■ Redis on SAP Cloud Platform, hyperscaler option
Data exploration	Empowered analytics	■ SAP Analytics Cloud ■ SAP Data Warehouse Cloud
Master data management	Master data services	■ SAP Cloud Platform Master Data for business partners
Data migration	Enable smart data migration in the cloud	■ SAP Advanced Data Migration by Syniti
Data integration	Integration between systems	■ SAP Cloud Platform Integration Suite
Data security	Security, compliance and privacy of data	■ SAP Data Custodian ■ SAP Cloud Platform Data Retention Manager ■ SAP Cloud Platform Personal Data Manager

Table 3.1 Data Pattern Capability in SAP Business Technology Platform (Cont.)

3.5   Design Outcomes

As an outcome of this process, we recommend that every enterprise define its own set of best practices and enterprise architecture principles, consisting of the following process and steps:

■ The process of describing the desired future state of an organization's business process, technology, and information to best support the organization's business strategy

■ The definition of the steps required and the standards and guidelines to get from the current state to the desired future state

The architecture principles discussed in the following sections should be based on the platform capability model discussed in Section 3.3.

3.5.1 Technology Platform Principles

Those using a best-of-breed approach (i.e., selecting the best systems for each specialized function rather than a single integrated system) for a platform experience more challenges than single-vendor organizations. The most common overall challenges are integration issues, upgrade challenges, lack of internal skills, poor data governance, and inadequate partner support. Hence, the recommendation is to follow suggested technology platform principles:

- Choose a technology platform (a PaaS offering) from the dominant business application stack in your landscape.
- Have an aligned domain and data model.
- Have a cloud-native developer.
- The landscape should contain a minimal number of internal dependencies so that components can be separated easily in case of commercial breakups/divisions.

Note

If an enterprise is running its mission-critical applications, its digital core (e.g., SAP S/4HANA) system, or its primary source of truth from a specific vendor such as SAP, then it's recommended to choose the technology platform from the same vendor (in this case, SAP's Business Technology Platform via SAP Cloud Platform) to reduce TCD/TCO rather than choosing a third-party system.

3.5.2 Integration Principles

Building on the technology platform principles are your enterprise-wide integration development principles. The enterprise should start to map out which integrations are critical to control and rearchitect and should plan to implement the best process to consolidate, simplify, and integrate different technologies. The following architecture principles are supported by SAP Cloud Platform and should be used in the context of SAP application extension and integration, though they are generic and can also be applied for non-SAP applications and landscapes:

- Choose an integration platform as a service (iPaaS) of the dominant business application stack.
- Architect all integrations and set up an integration competency center (ICC) for governance if required.
- Avoid tightly coupled integration.
- Avoid multiple points of faliure (reduce multiple middleware vendors).
- Choose an integration solution that provides tools to cover all integration domains and styles.

- Buy rather than build to leverage prepackaged integration content.
- A common/shared API layer for all cloud products is desirable. This layer could also be used to expose APIs to external partners, vendors, and customers. For this, leverage API Management.

3.5.3   Development Principles

In addition to your design outcome principles, you need to define the development principles and guidelines in your organization while transforming your organization to make it future ready. Consider the following:

- Use a cloud-native development approach.
- Keep the core clean. For example, build side-by-side application extensions on SAP Cloud Platform rather than on SAP ERP/SAP S/4HANA.
- Reduce legacy enhancement modifications and move them into the PaaS. For example, move existing add-ons to SAP Cloud Platform.
- The landscape should contain a minimal number of internal dependencies so that components can be separated easily in case of commercial breakups/divisions.
- The landscape should support the separation of core solutions (both on-premise and in the cloud), while business-critical apps can be built on a cloud platform (e.g., SAP Cloud Platform).
- Separate the innovation layer to quickly build agile solutions on SAP Cloud Platform.

3.5.4   User Experience Principles

The success of your organization is heavily dependent on how engaged your employees and end customers are with your end user systems and how well they stay connected with your organization. Thus, it's very important to define user experience and design principles in your organization early to keep your end users excited and engaged. Consider the following:

- Cater for multichannel access to services and ensure seamless transition between channels.
- Ensure a consistent look and feel across channels.
- Implement adaptive content.
- Design for accessibility and design for mobile first.
- Offer frictionless access, anytime, anywhere, on any device.
- Offer a consumer-app-like user experience.

All these UX principles can be met via SAP Cloud Platform. Figure 3.8 shows an example of an as-is landscape for UX.

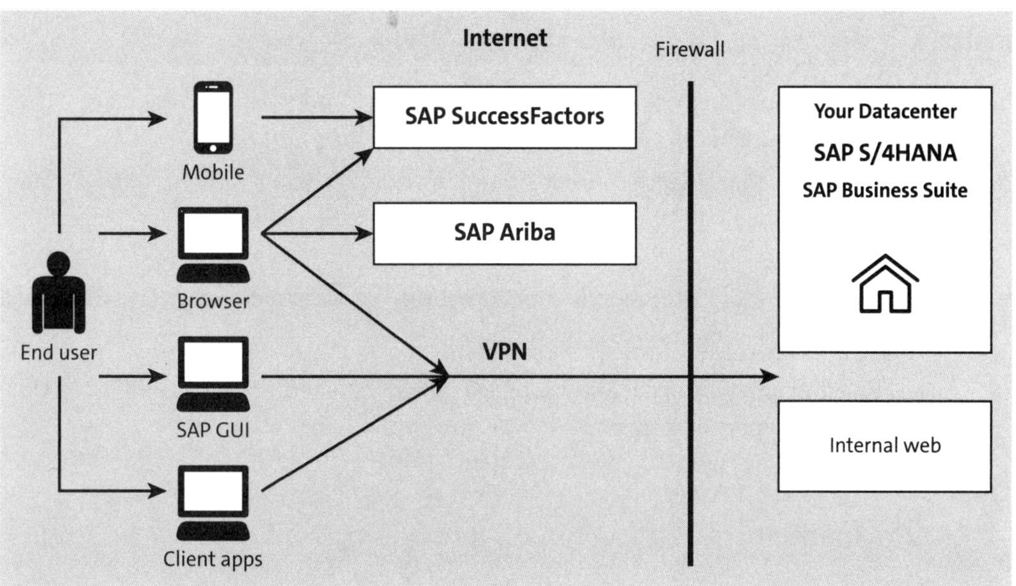

Figure 3.8 As-Is Landscape

Figure 3.9 shows the proposed landscape-to-be with SAP Cloud Platform.

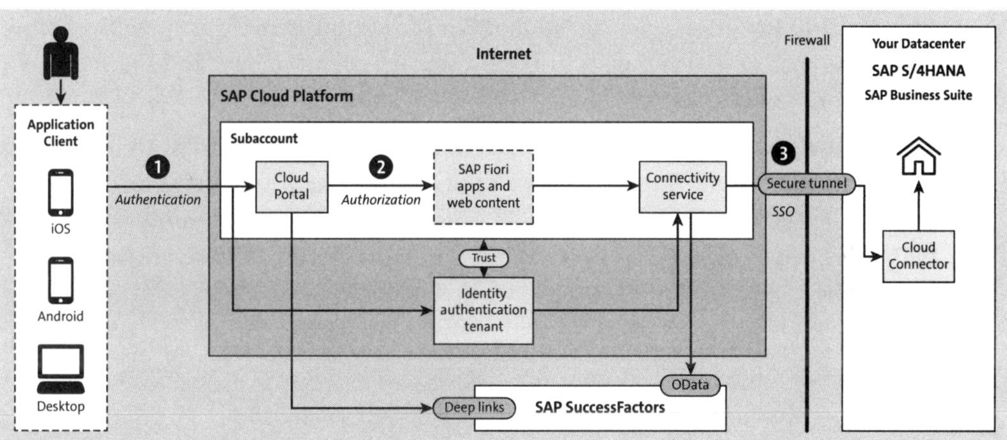

Figure 3.9 Landscape-To-Be

3.5.5   Database Requirement

Based on application requirements and use cases, an appropriate database and the respective schema should be selected for the scenario in consideration.

Various factors influence the selection—for example, if the storage is planned to be relational or is following NoSQL-based principles. There are four primary types of NoSQL databases, as follows:

1. A key-value store is a type of nonrelational database in which each value is associated with a specific key. These are fit for high performance and scalability.

2. Wide-column stores are dynamic, column-oriented nonrelational databases. They are also called column stores or extensible record stores.

3. Applications that require traversal between nodes or data points need *graph* databases. They facilitate storage of properties of each node, as well as the relationships between them.

4. A document store is a nonrelational database that stores data in JSON, XML, and other formats, such as document format.

Selecting the right type of database depends a lot on other criteria: availability, scalability, regional support, and so on.

3.6 Summary

In your journey to make your organization future ready, it's very important to have a good understand of your current landscape and define necessary design outcomes and principles.

In this chapter, we discussed the first step to map out the as-is landscape in your organization. SAP PowerDesigner can support you in this exercise to map your existing landscape, map the landscape to business processes, and understand the impact of architectural landscape changes.

We also discussed how your cloud architecture can be defined according to TOGAF and SAP Enterprise Architecture Framework standards. We also discussed the capabilities of SAP Cloud Platform, mapping them to TOGAF and NIST frameworks and various capabilities defined by these industry standards for the enterprise architecture transformation journey.

Because availability of data plays an important role in your transformation journey, we also spoke about various data architecture principles and tools in SAP Cloud Platform that play a pivotal part in your landscape transformation journey.

Finally, we discussed various design outcomes that define various principles and guidelines you need to consider early in your transformation process. Check off these design principles while you move toward your overarching goal of building a future-ready enterprise.

Chapter 4
Master Data Distribution

This chapter builds on the design outcomes discussed in Chapter 3 as part of your future-ready transformation journey. As you build on your path to make your systems, landscape, and business processes agile and future ready, one of your first considerations should be how you can enable a consistent, single view of master data across your organization. This chapter focuses on this aspect, on enabling availability of master data across the enterprise while managing legal compliance.

In the previous chapters, we've discussed the architectural guidance and principles for approaching your future-ready enterprise journey. This chapter will dig deeper into how you can approach the availability and facilitation of master data across your enterprise as you build upon the foundation from Chapter 3.

You need to facilitate the realization of a consistent view on master data within in your enterprise across cloud, on-premise, and all other third-party applications; this is key to achieving a scalable business model for future-ready use and to realizing an intelligent enterprise. Thus, it becomes an important consideration in your architectural guidance and design process to make available and distribute a consistent master data view across the organization. Master data is used across your enterprise, with updates and creation of different master data artifacts spread across different systems.

Thus, this chapter discusses this consideration and the different components available in SAP Cloud Platform to help you realize your goals. Specifically, we will discuss a number of use cases for master data distribution, dive into SAP Cloud Platform services you can use as part of your master data distribution strategy, and finish by discussing on managing compliance with retention and deletion policies and laws like GDPR.

4.1 Master Data Distribution Use Cases

Building a future-ready enterprise by transforming your organization with SAP S/4HANA requires you to share master data across multiple SAP and non-SAP systems, in the cloud and on premise.

Organizations should already have or plan to have a comprehensive master data management and governance policy in place to consolidate and centrally govern the master data lifecycle and ensure data quality and consistency. However, many organizations have distributed systems and landscapes, resulting in the need to distribute master data across many applications. Cloud systems add additional complexity. Such landscapes often result in data that is inconsistent and outdated, which causes issues in business processes downstream.

Hence an enterprise-wide master data management solution is needed for an organization to manage and govern data and its quality, as well as to synchronize/replicate the data with other systems. An enterprise-wide master data management landscape from SAP consists of SAP Master Data Governance and various SAP Cloud Platform services like SAP Cloud Platform Master Data Integration, SAP Data Quality Management, and SAP Cloud Platform Data Enrichment.

SAP Master Data Governance on SAP S/4HANA provides preconfigured, domain-specific master data governance and consolidation to centrally improve, govern, and distribute master data across your enterprise landscape. There are many key benefits of implementing master data governance solutions that are outside the scope of this book, but we will cover how such consistent master data can be made available across the enterprise for business processes and benefits to the same, while retaining a stable digital core system. Figure 4.1 shows an example of an enterprise-wide master data management and integration solution.

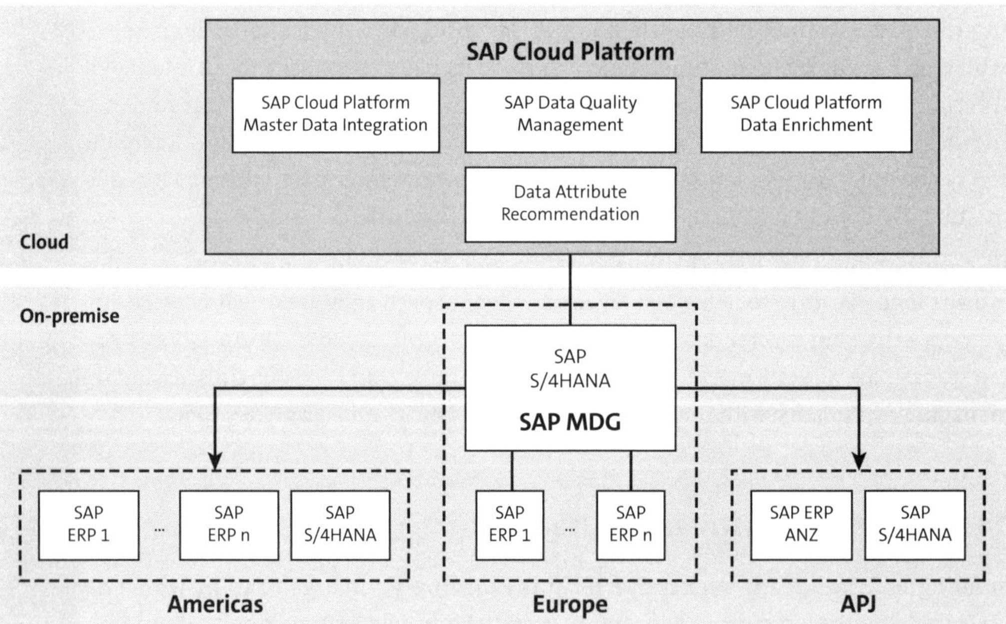

Figure 4.1 Master Data Management Solutions

Here are few scenarios and use cases for the usage of various master data services from SAP Cloud Platform in your landscape to improve your business process efficiency and transform your landscape:

- SAP Cloud Platform Master Data Integration services can act as central access layer for master data replication, sharing, and distribution. Figure 4.2 shows the future evolution of master data replication to a central place from a point-to-point integration method. This helps to facilitate a consistent single view of master data across your entire hybrid landscape, both SAP and non-SAP systems, as well as custom extensions on SAP Cloud Platform and applications from partners.

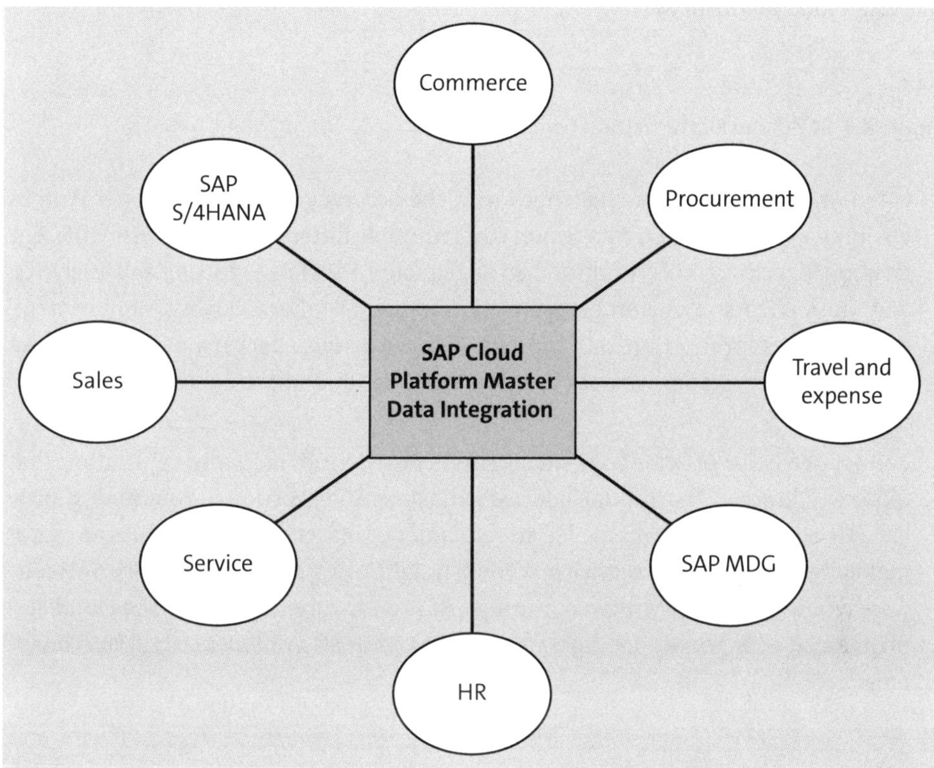

Figure 4.2 Evolution of Master Data Integration in Enterprise

- Another use case/scenario is the need to enrich the business partner, customer, and vendor master data with relevant data from third-party services. SAP Cloud Platform Data Enrichment offers a hub of multiple commercial data providers under a single umbrella from which you can consume the data of your choice without any further undue delay in integrating with these third-party providers. This service integrates with external data providers like Dun & Bradstreet and CDQ to get the master data and format it for enterprise applications. It then further interfaces with enterprise applications to pass required data to them. Figure 4.3 shows a high-level view of how this service integrates with an SAP S/4HANA system.

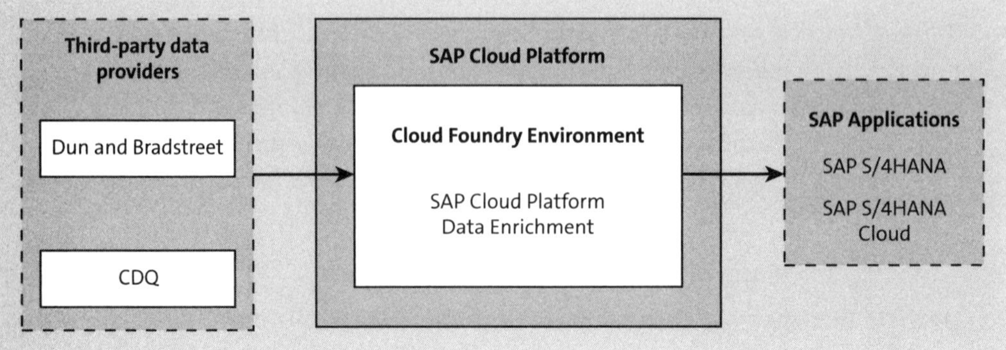

Figure 4.3 SAP Cloud Platform Data Enrichment

- Often organizations have challenges with the accuracy of location master data in business, especially when data comes from multiple different systems with different location identifiers. This results in bad data quality, impacts reporting and analytics, and often results in customer satisfaction issues. SAP Data Quality Management, microservices for location data can help eliminate these data maintenance issues, reduce IT costs, enable accurate location analytics, and improve customer satisfaction.

  The service helps provide high-quality and consistent data in any application that requires address entry through address validation and geocoding. You can instantly validate addresses, and misspelled street names or inaccurate postal codes are automatically corrected. This service provides straightforward integration with SAP software and with non-SAP software through REST APIs running both in the cloud and on premise, even while creating the data in SAP S/4HANA or SAP Master Data Governance.

Note

For more information about SAP Data Quality Management, microservices for location data, check out SAP API Business Hub. You can learn about this microservice at *http://s-prs.co/v515702*.

4.2   Developing a Master Data Distribution Strategy

The previous section discussed different use cases and scenarios for how organizations can have a consistent, single view of master data across the organization. We also saw use cases and benefits of ensuring location data quality for an organization and benefits of enriching business partner data from third-party services.

In this section, we'll look into various services for master data distribution in depth: SAP Cloud Platform Master Data for business partners; SAP Cloud Platform Data Enrichment; SAP Data Quality Management, microservices for location data; and Data Attribute Recommendation (part of SAP AI Business Services). This will show you how to get started and how your architecture and strategy can be developed while transforming your enterprise.

4.2.1   SAP Cloud Platform Master Data for Business Partners

SAP Cloud Platform Master Data for business partners allows for business partner data to be used in multiple business processes and integrated with any cloud or on-premise system. In the following sections, we'll discuss how to set up and configure this service in SAP Cloud Platform and also talk about various functionalities of the service, including role assignment.

Setup and Configuration

In the following sections, we'll first look at the prerequisites for setting up SAP Cloud Platform Master Data for business partners and then show you how to create a service instance.

Prerequisites

You already have a subaccount and space created in Cloud Foundry environment, so your first step will be to activate the service, using the following process:

1. Add the service from **Entitlement** section in your global account to the subaccount where you want to activate the service.
2. Inside the subaccount, navigate to **Subscriptions** and search for the **Master Data for Business Partner** tile.
3. Choose **Subscribe** to enable the service usage for your subaccount.
4. Before starting to use the service, you need to configure the appropriate authorizations for the user to use the service. Different role templates provided by the service are described in Table 4.1.

Role Template	Description
Master_Data_Admin_Template	Authorizes users to maintain necessary configurations for a business for maintenance of roles, authorization groups, number ranges, identification types, and industry sectors.

Table 4.1 Role Templates for SAP Cloud Platform Master Data for Business Partners

Role Template	Description
Master_Data_Specialist_Template	Authorizes users to search, view, and maintain business partners and view key mapping data. This role has an attribute for authorization groups that stipulates which business partner can be viewed/created at the instance level.
ExtensionDeveloper	Authorizes users to extend and activate models, but not to remove previously applied extensions.
ExtensionDeveloperUndeploy	Authorizes users to extend and activate models, as well as remove previously applied extensions.

Table 4.1 Role Templates for SAP Cloud Platform Master Data for Business Partners (Cont.)

5. Inside the subaccount, navigate to **Security • Role Collections** and choose **New Role Collections**. Enter a **Name** and optional **Description** and choose **Save**. Create different role collections for different authorization purposes.

6. Navigate back to **Subscriptions** in the subaccount and choose the **Master Data for Business Partner** tile. Choose **Manage Roles** in the service, click **+**, and add the predefined role from the application to the role collections previously created.

Note

You can also do this step as follows:

1. Inside the subaccount, navigate to **Security • Role Collections** and choose the role collection that you previously created.

2. Click **Add Role**, choose the **Application Identifier** of the business partner service, choose the **Role Template** and **Role**, and choose **Save**.

---

7. The next step is to add the role collection to the users. For this, navigate to **Security • Trust Configuration**, choose an **Identity Provider**, and enter **User** information (for the SAP ID service, this is the email address of the user).

8. Click **Assign Role Collection** and add the role collection to the user. If **Assign Role Collection** is inactive, click **Show Assignments** first.

Tip

We'll cover security concept details in Chapter 8.

---

9. Click the **Master Data for Business Partner** tile, then choose **Go to Application**. This will launch the application as shown in Figure 4.4.

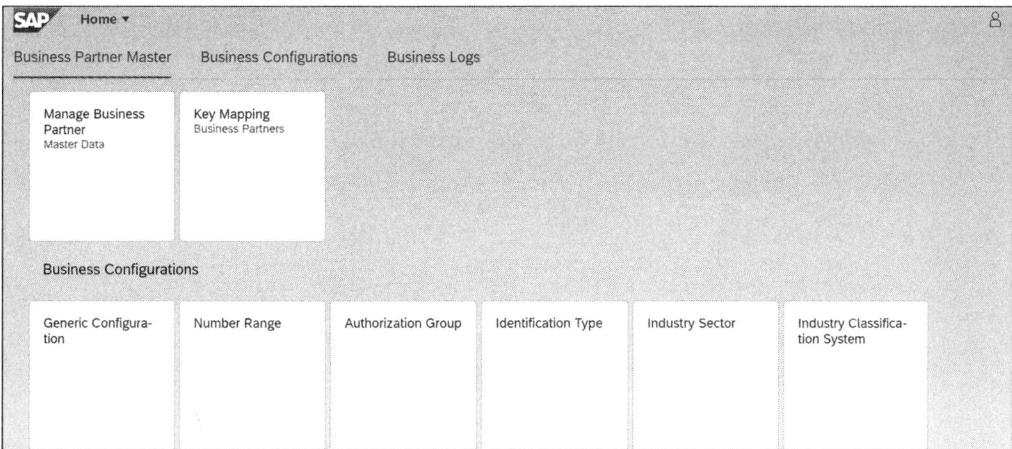

Figure 4.4 Master Data for Business Partner Home

The SAP Cloud Platform Master Data for business partners service offers the following apps to manage and access business partner master data:

- **Manage Business Partner**
  This app is used to create new business partner master data, update an existing business partner, and search for all maintained master data. You can perform following functions:
  - Create new business partner master data
  - Update existing business partner master data
  - View business partner master data
  - Select and search for business partner master data

  To access this app, you need access to the role defined in the Master Data Specialist template, as explained in Table 4.1.

- **Key Mapping**
  This app enables you to get business partner identification details, like ID, in the source backend system(s). You can search for the mappings based on business partner number, business system name, business partner GUID, or business partner external ID.

  This app will also help you to find the business partner number in the source system for the corresponding business partner number in SAP Cloud Platform Master Data for business partner.

  To access this app, you need access to the role defined in the Master Data Specialist template, as explained in Table 4.1.

- **Business Configuration**
  This app enables you to make various definitions and settings to meet your specific

requirements and make necessary adaptations as the business changes. This app allows you to perform following functions:

- Maintain and view configurations

- Add text or descriptions in other supported languages

To access this app, you need access to the role defined in the Master Data Admin template, as explained in Table 4.1.

- **Business Logs**

  This app enables you to monitor business logs while creating a business partner using this service. Here you can see business logs based on errors, warnings, and information and success messages, and even see logs based on business object ID or business object item ID.

  This app requires you to add the role identified in Table 4.2 to your role collection.

Role Template	Role Description
`Business_Process_Specialist_BL_AccessAll`	This role authorizes users to view business logs in the cloud application.

Table 4.2 Business Logging Role Template

Create a Service Instance

You can create a service instance that will provide endpoints to integrate SAP Cloud Platform Master Data for business partners with any applications. To begin, follow these steps:

1. Navigate to your subaccount and to the relevant space.

2. Inside the space, navigate to **Services • Service Marketplace** and select **SAP Business Partner Service**.

3. Navigate to **Instances** and choose **New Instance**. Go through the steps in the pop-up window, enter an **Instance Name**, and choose **Finish**.

> **Note**
>
> While creating the service instance, you can specify parameters in **Specify Parameters (Optional)** with different scopes available for the application. These scopes are defined here: *http://s-prs.co/v515703*.
>
> You also can provide additional configuration parameters to use data privacy capabilities.
>
> If your application needs to use the service with same features provided by SAP Cloud Platform Master Data for business partners, you can select **Application name** from the **Assign Application (Optional)** step or bind the service instance later to the application.

4. The next step is to create a service key for the instance that is created. For this, select the service instance that was created and navigate to **Service Keys**.

5. Choose **Create Service Key**, enter a **Name**, and click **Save**. A new service key is created that can be used to call the service from another application using OAuth 2.

> **Note**
>
> You need to note the client ID, client secret, URL, SOAP endpoint, and identity zone parameters from the service key for further configurations using this service.

Master Data Orchestration

In this section, we'll discuss how the master data created and managed through SAP Cloud Platform Master Data for business partners can be distributed and orchestrated among various providers and consumers, with SAP Cloud Platform as a central place for distribution. You can activate this service within the same subaccount as that of SAP Cloud Platform Master Data for business partners or activate it in a separate subaccount as a central place to distribute from. A possible scenario for a separate subaccount for orchestration service could be triggered by certain legal requirements, such as if a business partner needs to be stored in a particular geography but can be distributed from a central place. Data can be distributed directly to consuming applications or it can be first sent to SAP Master Data Governance for applying any governance rules in your organization and then distributed further from there. To begin, follow these steps:

1. Within your Cloud Foundry subaccount, navigate to **Subscriptions**, and choose the **Master Data integration (Orchestration)** tile.

2. Click the tile and choose **Subscribe** to subscribe to the application.

3. Before using the application, you need to configure required roles. Master data orchestration provides the role templates listed in Table 4.3.

Role Template	Description
`MasterDataOrchestrationAdminTemplate`	Authorizes users to be able to maintain, view, and search distribution models; maintain, view, and search object type owners; and maintain and view distribution status
`MasterDataOrchestrationDisplayTemplate`	Authorizes users to view distribution models, view object type owners, and view distribution status

Table 4.3 Master Data Orchestration Role Templates

4. Inside the subaccount, navigate to **Security • Role Collections** and choose **New Role Collections**. Enter a **Name** and optional **Description** and choose **Save**. Create different role collections for different authorization purposes if required, or choose one of the existing role collections previously created.

5. Navigate back to **Subscriptions** in your subaccount and choose the **Master Data Orchestration** tile. Choose **Manage Roles** in the service, click **+**, and add the predefined role from the application to role collections previous created.

6. The next step is to add the role collection to the users. For this, navigate to **Security • Trust Configuration**, choose an **Identity Provider**, and enter **User** information (for the SAP ID service, this is the email address of the user).

7. Click **Assign Role Collection** and add the role collection to the user. If **Assign Role Collection** is inactive, click **Show Assignments** first.

8. Click the **Master Data Orchestration** tile, then choose **Go to Application**. This will launch the application, as shown in Figure 4.5.

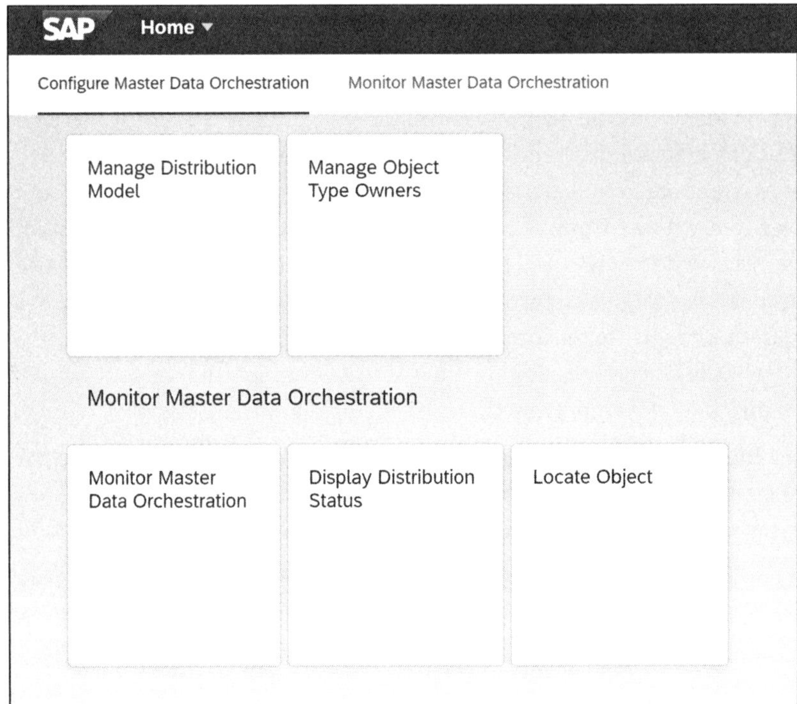

Figure 4.5 Master Data Orchestration Home Page

The master data orchestration service has the following apps to help distribute and orchestrate the master data for business partners:

- **Manage Distribution Model**
  This app enables you to define and modify the distribution model. This includes information like business object type, activation status, recurrence, and filtering, with which you can specify the master data that will be distributed.

You can also maintain the source and target systems, as well as at what point in time the model will be executed.

- **Manage Object Type Owners**
  This app allows you to maintain object type owners. This app provides following functionalities:
  - Definition of the object type owner
  - Selection of destinations

  The Manage Object Type Owners app will be called when creating new master data to find whether an object type owner exists and, if yes, whether it can be enriched with additional information or the planned changes can be executed. Manage Object Type Owners is the leading app in SAP Cloud Platform Master Data for business partners.

- **Display Distribution Status**
  This app is used to monitor the distribution status. It displays business logs for specific distribution triggers. Hence, you can use this app for following purposes:
  - Display information about the distribution status of models
  - Display error messages
  - Retrigger replication for failed distribution

- **Monitor Master Data Orchestration**
  With this app, you can visualize distribution models that have been activated once to monitor the distribution paths between source and target systems.

  This app provides the ability to view the distribution logs and offers error-resolution capabilities.

- **Locate Object**
  This app is used to obtain information about the systems a business object is known in, as well as details of the corresponding objects within these systems.

  Thus, you can track objects across the system landscape and receive information about an object's ID in other destinations, as well as display the most recent distribution status: started, success, or failed.

Distribution Model Maintenance

The distribution model is configured in the Manage Distribution Model app. The distribution specifies the master data to be replicated from the provider to the consumer. To create a distribution model, choose the Manage Distribution Model app from the **Configure Master Data Orchestration** home screen. Choose **Create** and a new screen will open, as shown in Figure 4.6.

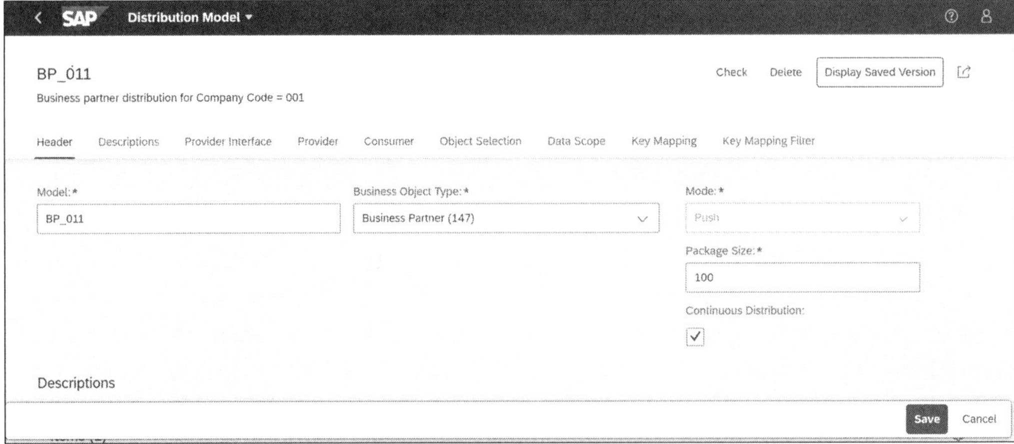

Figure 4.6 Maintain Distribution Model

Enter the following information:

- **Header**
  - Provide a **Model** name.
  - Select **Business Partner** for the **Business Object Type**.
  - Provide an appropriate **Package Size**.
  - Select the **Continuous Distribution** checkbox; otherwise, specify a **Recurrence Pattern**.

- **Descriptions**
  - Click **Create** and choose **Language**.
  - Under **Item**, maintain the description.

- **Provider Interface**
  - The provider interface API is based on the business object type.
  - Select the **Provider Interface** from the pop-up screen.

- **Provider**
  - Provider applications are all applications delivering data to the consumer applications. Thus, the provider is the recipient of the distribution model that specifies which data should be distributed to consumer. With this step, you establish connectivity between the master data orchestration tenant and provider application.
  - In your SAP Cloud Platform cockpit, navigate to **Connectivity • Destinations** and choose **New Destination**.
  - In the new destination configuration, enter the **Name**, **Type**, **URL**, **Proxy Type**, and **Authentication**.
  - Click **New Property** under **Additional Properties** and enter "MDOProvider" as the property with the value "true".

- Click **Create** and enter the **Cloud Platform Cockpit Destination** maintained previously.

- **Consumer**
  - The consumer is the application/system the data is replicated to based on the distribution model. Examples of consumer applications include SAP Cloud for Customer, SAP Ariba Cloud Integration Gateway, and SAP Master Data Governance.
  - In your SAP Cloud Platform cockpit, navigate to **Connectivity • Destinations** and choose **New Destination**.
  - In the new destination configuration, enter the **Name**, **Type**, **URL**, **Proxy Type**, and **Authentication**.
  - Click **New Property** under **Additional Properties** and enter "MDOConsumer" as the property with the value "true".
  - Click **Create** and enter the **Cloud Platform Cockpit Destination**.

Note

In a model, you can specify either one provider and one or several consumers, or one consumer and one or several providers, but not several consumers and several providers.

Tip

For more about destination configuration for the business partner service for distribution, see *http://s-prs.co/v515704*.

- **Object Selection**
  - This filter determines the business object instances that are replicated to consumer systems. The selection made in this section includes or excludes objects from the distribution, as shown in Figure 4.7.
  - If nothing is specified, all objects of the relevant business object type will be replicated.

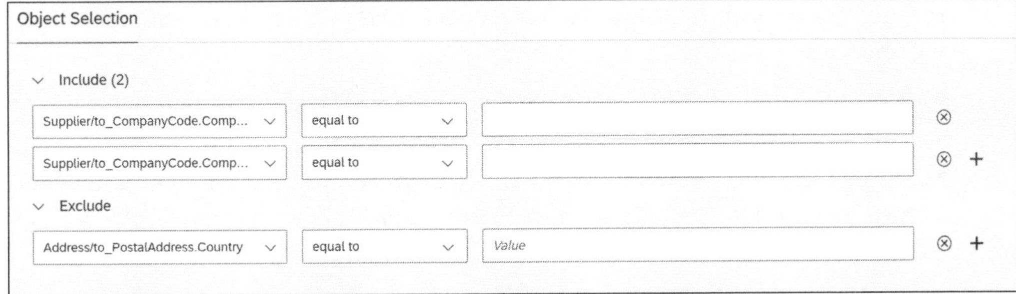

Figure 4.7 Object Selection in Master Data Distribution

- **Data Scope**
  - The data scope determines the data that will be replicated to the consumer systems. The filter specifies which segments will be included or excluded from the replication with this model. If nothing is specified, all data for the relevant object will be replicated. The attributes available in the selection depend on the implementation of the distribution service provider interface.
- **Key Mapping**
  - If no key mapping information is to be sent, do not set the **Distribute Key Mapping** checkbox.
  - If the **Distribute Key Mapping** checkbox is selected, the **Key Mapping Filter** is shown. Using this filter, you can restrict the key mapping information to the distribution by specifying **Business System**, **Business Object Type**, or **Object Identifier Type**.
  - You can specify the **Include** or **Exclude** option for **Key Mapping Filter**.

Once all of these settings have been made, click **Check** to check the consistency of the data in the distribution model. If all data is consistent, you can click **Save**. You can **Activate** the model, and the active model will trigger replication. For already existing data, distribution can be triggered manually.

> **Note**
>
> For more about establishing connectivity from SAP S/4HANA or SAP Master Data Governance to SAP Cloud Platform, visit *http://s-prs.co/v515705*.
>
> See also the documentation at the following URL for more on establishing a connection with SAP S/4HANA Cloud: *https://rapid.sap.com/bp/#/browse/scopeitems/1RO*.

4.2.2   SAP Cloud Platform Data Enrichment

SAP Cloud Platform Data Enrichment enables you to enrich your existing master data by consuming data from third-party data providers. The current third-party data providers supported are Dun & Bradstreet and CDQ. This eliminates the risk of errors in manual data creation. SAP Cloud Platform Data Enrichment integrates with the following solutions:

- Master data maintenance with SAP S/4HANA, for customer master data and supplier master data from SAP S/4HANA 1809 and on
- SAP Master Data Governance with SAP S/4HANA, for business partner master data, customer master data, and supplier master data applications from SAP S/4HANA 1809 and on
- SAP Master Data Governance, for business partner data, customer master data, and supplier master data applications starting from SAP Master Data Governance 9.2

- SAP S/4HANA Cloud, for customer master data and supplier master data from SAP S/4HANA Cloud 1805 release and on

In the following sections, we'll first look at the prerequisites for setting up SAP Cloud Platform Data Enrichment and then show you how to create a service instance.

Prerequisites

Your first step will be to purchase a data subscription plan from a provider at SAP App Center. Once purchased, you'll receive an email with the order number. Note this order number for later use. Depending on the provider you choose, you can find the configuration information at the following links:

- Dun & Bradstreet: *https://www.sapappcenter.com/en/product/display-0000023421_ live_v1#!editions*
- CDQ: *https://www.sapappcenter.com/en/product/display-0000031305_live_v1#! overview*

To activate the service, follow these steps:

1. Add the service from the **Entitlement** section in your global account to the subaccount where you want to activate the service.
2. Inside the subaccount, navigate to **Subscriptions** and search for the **SAP Cloud Platform Data Enrichment** tile.
3. Choose **Subscribe** to enable the service usage for your subaccount.
4. Before starting to use the service, you need to configure the appropriate authorizations for the user to use the service. Different role templates provided by the service are described in Table 4.4.

Role Templates	Description
Employee	Authorizes users to access the Data Search App only
MasterDataSteward	Authorizes users to access admin apps

Table 4.4 Role Templates for SAP Cloud Platform Data Enrichment

5. Inside the subaccount, navigate to **Security • Role Collections** and choose **New Role Collections**. Enter a **Name** and optional **Description**, then choose **Save**. Create different role collections for different authorization purposes if required, or choose one of the existing role collections previously created.
6. Navigate back to **Subscriptions** in the subaccount and choose the **SAP Cloud Platform Data Enrichment** tile. Choose **Manage Roles** in the service, click **+**, and add the predefined role from the application to the role collections previous created.

7. The next step is to add role collection to the users. For this, navigate to **Security • Trust Configuration**, choose an **Identity Provider**, and enter **User** information (for the SAP ID service, this is the email address of the user).

8. Click **Assign Role Collection** and add the role collection to the user. If **Assign Role Collection** is inactive, click **Show Assignments** first.

9. Click the **Master Data Orchestration** tile, then choose **Go to Application**. This will launch the application as shown in Figure 4.8.

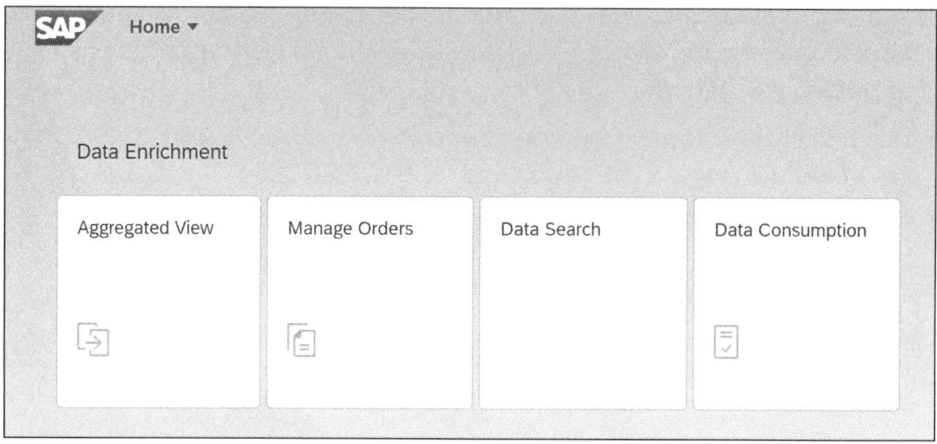

Figure 4.8 SAP Cloud Platform Data Enrichment Home

SAP Cloud Platform Data Enrichment offers the following applications:

- **Manage Orders**
  - This app allows you to link your data enrichment service with the third-party provider service purchased from SAP App Center.
  - Click this app, choose **Add**, and provide an **Order Number** and **Email**.
  - View the orders you have purchased via SAP App Center with information about the order number, data provider, name of the purchased product, and subscription status of the order.
  - To access this app, you need to add the MasterDataSteward role to the user.

- **Aggregated View**
  - This app enables you to define data views that serve as templates for data query and consumption.
  - You can choose required data points from a data provider for a specific domain that is applicable for your business.
  - You can select the data points for which you require updates and choose the frequency of updates required.
  - To access this app, you need to add the MasterDataSteward role to the user.

- **Data Consumption**
  - This app enables you to monitor the data consumption for your company's subscription to a product.
  - It gives a detailed view of data consumption from a data provider, dates on which the data is consumed, and the aggregated view used to consume the data.
  - To access the app, you need to add the `MasterDataSteward` role to the user.
- **Data Search**
  - This app enables you to search third-party data using search criteria specified after you have configured an aggregated view.
  - Once the matched data is available, you can get the details of the desired master data record in your integrated applications.
  - This can help reduce the cost of data consumption.

Create a Service Instance

You can create a service instance that will provide endpoints to integrate SAP Cloud Platform Data Enrichment with applications using the following steps:

1. Navigate to your subaccount and to the relevant space.
2. Inside the space, navigate to **Services • Service Marketplace** and select **Business Partner—Data Enrichment**.
3. Navigate to **Instances** and choose **New Instance**. Go through the steps in pop-up window, enter an **Instance Name**, and choose **Finish**.
4. The next step is to create a service key for the instance created. For this, select the service instance that was created and navigate to **Service Keys**.
5. Choose **Create Service Key**, enter a **Name**, and click **Save**.

> **Note**
>
> At this point, you should be sure to do the following:
>
> - Note the client ID, client secret, and URL from the service key for configurations of this service with other applications.
> - Use these parameters to configure integrations with other applications (e.g., SAP S/4HANA Master Data Maintenance). You can learn more about the configuration parameters and setup at *http://s-prs.co/v515706*.
>
> SAP Cloud Platform Data Enrichment supports mass enrichment of general data and addresses of business partners.

4.2.3   SAP Data Quality Management, Microservices for Location Data

In this section, we'll cover SAP Data Quality Management, microservices for location data, which provides the functionality for address cleansing with or without suggestion lists, geocoding with or without suggestion lists, and reverse geocoding. *Geocoding* assigns latitude and longitude coordinates to an address, while *reverse geocoding* translates latitude and longitude coordinates into address(es). With this service, users can eliminate operational challenges and infrastructure costs and it can be quickly deployed on many SAP applications with prebuilt integrations.

This solution uses reference data from global postal authorities for 240+ countries and territories to validate or correct addresses. Thus, you can quickly and easily correct, parse, standardize, validate, and enhance address data. The address cleansing, geocoding, and reverse geocoding capabilities of this service depend on the country in which an address is located.

Note

SAP Data Quality Management, microservices for location data is available in SAP API Business Hub: *http://s-prs.co/v515707*.

Before you begin, you'll need a subaccount in the SAP Cloud Platform Neo environment. Next, follow these steps:

1. Inside your Neo environment subaccount, navigate to **Services**.

2. Choose the **Data Quality Services** tile and click **Enable**.

3. Users who want to access SAP Data Quality Management, microservices for location data endpoints or the Configuration Editor are required to have the DQMICRO_ POWERUSER role. For this, navigate to **Authorizations • Groups** and, for an existing group, add the role to the group as shown in Figure 4.9. If there is no existing group, create a new group. We'll cover this configuration in more depth in Chapter 8.

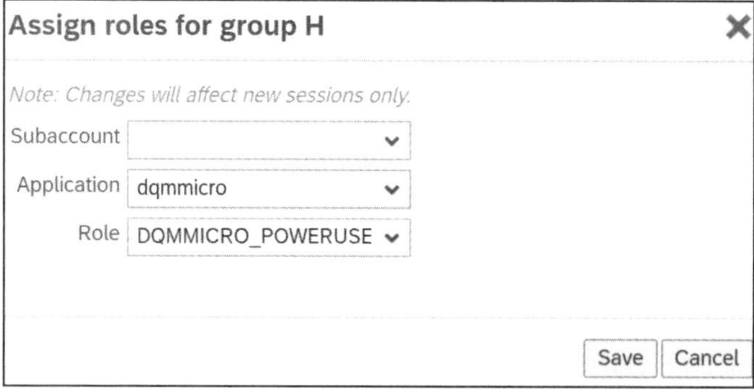

Figure 4.9 Authorization for SAP Data Quality Management, Microservices for Location Data

4. Click the **Data Quality Service** tile, then click **Application URL**. Note the available endpoint under the **Application URL** label. This URL is used for integration with other applications.

5. Click the **Data Quality Service** tile, then click **Configure Service**. This will open the configuration home screen as shown in Figure 4.10.

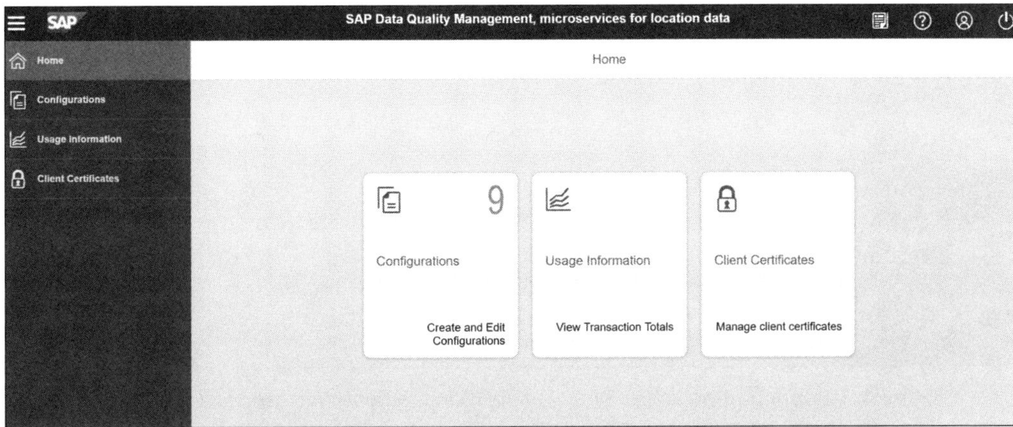

Figure 4.10 SAP Data Quality Management, Microservices for Location Data Configuration Screen

6. Choose the **Configurations** tile to create, edit, or view existing configurations. Configurations contain details about microservice location job settings. You can create a configuration for the following items:

   - A list of data model fields and how each field is mapped to a field in the addressInput object
   - The list of outputFields requested and how each field is mapped to a field in your data model
   - Your selection of attributes in the addressSettings object
   - Your selection of whether the suggestion list feature is enabled

> **Tip**
>
> If you use different fields or have different settings on a country-by-country basis, each of the preceding items may be saved differently per country in a configuration.

A sample configuration is as shown in Figure 4.11, as follows:

- **Data model fields**
  These are fields that exist in your application or database. The data model consists of input fields passed to microservices for location data for processing. Microservices for location data process this input data and produce a response.

- **Service fields**
  Microservices for location data output service fields processing can only happen after input data in your data model has been mapped to the appropriate input service fields.

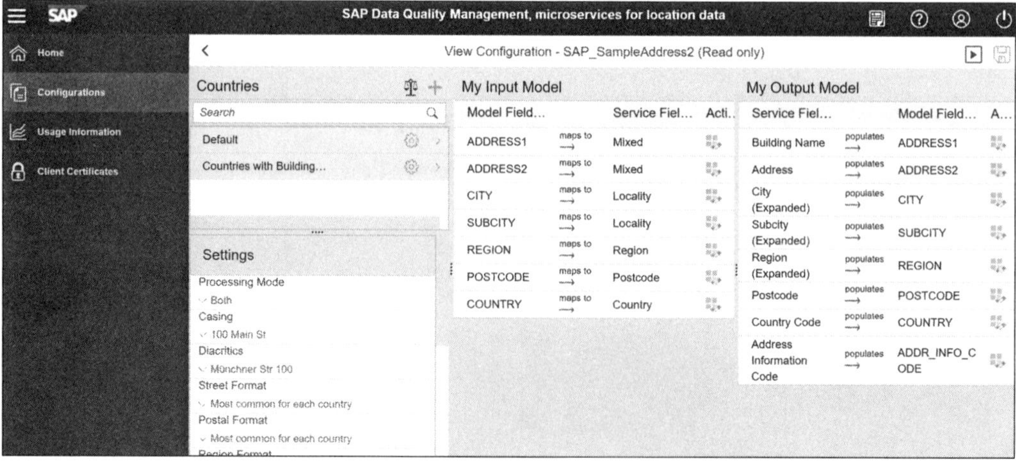

Figure 4.11 Configuration in SAP Data Quality Management, Microservices for Location Data

Note

Refer to the following URL to learn how to integrate microservices for location data using the application URL with various applications: *http://s-prs.co/v515708*.

4.2.4 Data Attribute Recommendation

Data Attribute Recommendation, part of SAP AI Business Services, is a machine learning service available on SAP Cloud Platform that helps master data specialists manage consistent master data with intelligent suggestions.

A significant use case for this service is to match point-of-sale information with product hierarchy information through trained models automatically before it's updated in the master data management system. Thus, it enriches missing attributes of the data record and results in saving of hours of manual work to maintain them, helping master data specialists to focus on more value-generating tasks. This service can be trained to your custom data model to deduce inferences from it.

Data Attribute Recommendation has the following three technical applications:

1. Data Manager, used for managing training data in the service. You upload training data to this application and manage its lifecycle, including validating, preprocessing, and deletion.

2. Model Manager, to manage and train the machine learning models, including model creation, activation/deactivation, and deletion.

3. Inference, responsible for deducing inferences; in other words, it helps classify datasets.

This service does not have a frontend user interface. The service activities, including uploading the training data, training the learning models, and classifying records, are performed through RESTful APIs, which are secured through HTTPS.

To perform the initial setup and configuration, follow these steps:

1. Navigate to your Cloud Foundry environment subaccount and to the relevant space.
2. Inside the space, navigate to **Services • Service Market Place** and select **Data Attribute Recommendation**.
3. Click the tile, navigate to **Instances**, and choose **New Instance**. Go through the steps in the pop-up window to create a new service instance.
4. Select the created service instance, navigate to **Service Keys**, choose **Create Service Key**, enter a **Name**, and choose **Save**.
5. Note down the contents of the **URL**, **CLIENTID**, **CLIENTSECRET** (inside the **UAA** section in the service key), and **URL** (outside the **UAA** section in the service key) fields for further configuration of Data Attribute Recommendation APIs.

> **Note**
>
> For further information on configuring the APIs, visit *http://s-prs.co/v515709*.
>
> The API details for Data Attribute Recommendation can be found in SAP API Business Hub at *http://s-prs.co/v515710*.

4.3   Compliance with Master Data Retention and Deletion

In the previous section, we looked at how different SAP Cloud Platform services can be used to provide a consistent, single view of master data for your landscape. However, you need to manage the data in the cloud carefully to ensure that you comply with all data retention rules.

Note that SAP Cloud Platform Master Data for business partners complies with the GDPR, with user consent available while creating data and with the SAP Cloud Platform Personal Data Manager to allow you to identify data subjects stored with the service. Another important aspect of compliance with regulation is the deletion of data stored in system. You must scale this operation correctly and be prepared to adapt your landscape, systems, and applications for any future legal regulatory changes. For this, in this section, we'll look at the SAP Cloud Platform Data Retention Manager service that can help you manage this process.

SAP Cloud Platform Data Retention Manager helps your applications running in SAP Cloud Platform, Cloud Foundry environment to comply with data protection and privacy rules and manage data residence and deletion activities based on set business purposes.

Figure 4.12 shows the complexity of defining data residence and retention policies, which need to be managed across many business applications for multiple business processes and datasets to meet preset conditions.

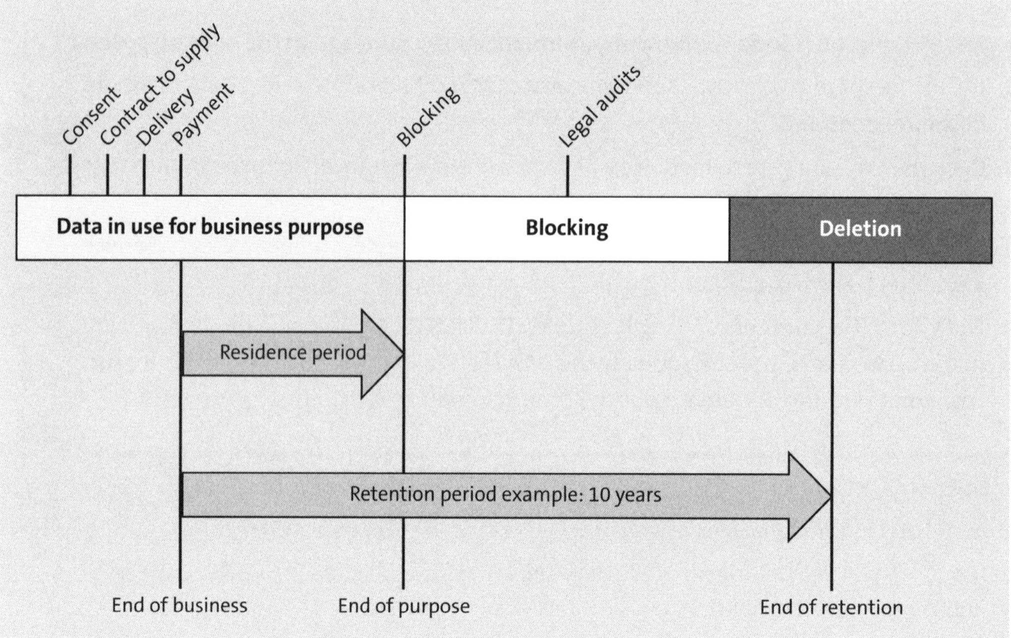

Figure 4.12 Data Residency and Retention Lifecycle

To configure this service for users, follow these steps:

1. Add the service from the **Entitlement** section in your global account to the subaccount where you want to activate the service.

2. Inside the subaccount, navigate to **Subscriptions** and search for the **Data Retention Manager** tile.

3. Choose **Subscribe** to enable the service usage for your subaccount.

4. Before starting to use the service, you need to configure the appropriate authorizations for the user to use the service. The role templates provided by the service are described in Table 4.5.

Role Template	Description
DataProtectionOfficer	Authorizes users to manage (create, edit, and delete) business purposes for different applications, roles, and legal entities; and to export data
DPPSpecialist	Authorizes users to identify data subjects up for deletion and send them for deletion to applications

Table 4.5 Role Templates for Data Retention Manager

Role Template	Description
Administrator	Administrator role

Table 4.5 Role Templates for Data Retention Manager (Cont.)

> **Note**
>
> In the context of SAP Cloud Platform Master Data for business partners, this scenario is applicable if the master data service is the leading system for managing the master data. If SAP Cloud Platform Master Data for business partners is not the leading system, the deletion would be triggered from the source system.

5. Inside the subaccount, navigate to **Security** • **Role Collections** and choose **New Role Collections**. Enter a **Name** and optional **Description** and choose **Save**. Create different role collections for different authorization purposes if required, or you can choose one of the existing role collections previously created.

6. Navigate back to **Subscriptions** in the subaccount and choose the **SAP Cloud Platform Data Enrichment** tile. Choose **Manage Roles** in the service, click **+**, and add the predefined role from the application to the role collections previous created.

7. The next step is to add the role collection to the users. For this, navigate to **Security** • **Trust Configuration**, choose an **Identity Provider**, and enter **User** information (for the SAP ID service, this is the email address of the user).

8. Click **Assign Role Collection** and add the role collection to the user. If **Assign Role Collection** is inactive, click **Show Assignments** first.

9. Click the **Data Retention Manager** tile, then select **Go to Application**. This will launch the application as shown in Figure 4.13.

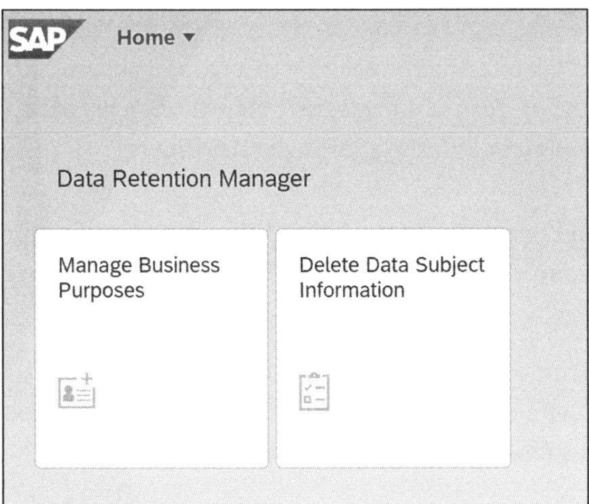

Figure 4.13 SAP Cloud Platform Data Retention Manager Home

Before going into detail about the SAP Cloud Platform Data Retention Manager applications, let's first look at terminology used within this service:

- **Business purpose**
  Name of the purpose for which data privacy rules need to be defined. A purpose is an end-to-end business scenario for which data residency and retention is addressed based on legal or contractual regulations.

- **Application group**
  Unique name of the application(s) for which data residency and retention rules need to be defined.

- **Data subject role**
  This role is an entity identifiable directly or indirectly in a business process who forms the base for defining data privacy and protection rules—for example, a customer or employee.

- **Legal entity**
  Legal entity associated with the data subject role. For example, a country associated with a customer could be a legal entity.

- **Legal ground**
  Legal grounds are rules defined by the consuming application at a more granular level associated with the data subject role. These rules set and define the residence and deletion criteria for data—for example, the completion date for a sales order or posting of financial documents.

- **Data subject**
  A data subject is a combination of the data subject role, legal entity, and legal ground. This field is not visible in the UI, but it's used to define a list of data subject roles, legal entities, and legal grounds as parameters for the SAP Cloud Platform Data Retention Manager service instance in a JSON array format while binding it to consuming applications.

  Thus, the data subject encapsulates all information relating to a natural person who can be identified directly or indirectly—for example, a customer who can be identified directly based on name or location or indirectly through sales orders.

- **Reference dates**
  The date from which residence and retention rules need to be applied. You specify reference dates per application group, data subject role, and legal ground combination.

Figure 4.14 outlines the relationships among these terms. This is important for you to understand in order to design and define different residence and retention rules for business purposes based on the data subject role and legal entity.

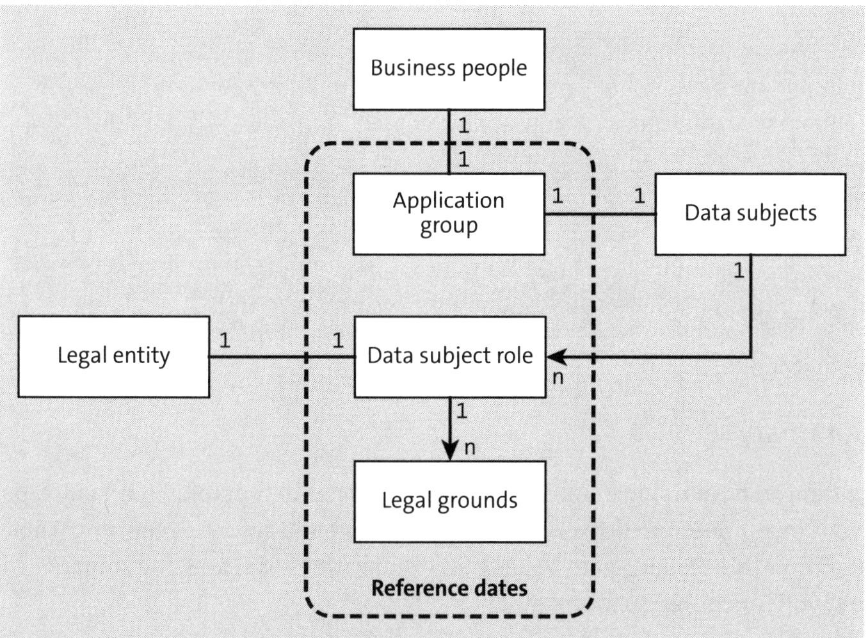

Figure 4.14 Relationships in SAP Cloud Platform Data Retention Manager Terms

SAP Cloud Platform Data Retention Manager has following applications:

- **Manage Business Purposes**
  This app is used for following purposes:
  - Create a new business purpose for application groups, data subject roles, and legal entities.
  - Edit a business purpose.
  - Set or edit defined legal grounds and define residence and retention periods for each business purpose.
  - Activate business purposes.
  - Set reference dates based on application group, data subject role, and legal grounds. This is an important attribute in the deletion process.

  This app requires a role with the `DataProtectionOfficer` role template assigned to the user.

- **Delete Data Subject Information**
  This app is used for following purposes:
  - Schedule and identify data subjects that need to be deleted for the combination of application group and data subject role.
  - Send a deletion request to consuming application.

  This app requires a role with the `DPPSpecialist` role template assigned to the end user.

> **Note**
>
> As noted, you can also use/configure SAP Cloud Platform Data Retention Manager to manage your data residence and retention periods for any custom SAP Cloud Platform, Cloud Foundry environment applications.
>
> For this configuration, you need to create a service instance of the SAP Cloud Platform Data Retention Manager service by providing the data subjects, data subject role, legal entity, and legal grounds parameters per the consuming application in JSON format. You can read more about this configuration at *http://s-prs.co/v515711.*

4.4   Summary

It's important to have a single, consistent view of master data across your landscape and in your organization to drive customer satisfaction and accurate reporting. Thus, we covered how this challenge can be addressed and various use cases and scenarios to help make your organization future ready.

We covered different SAP Cloud Platform services that help distribute business partner master data, enrich business partner data with third-party services, and cleanse address data using geocoding/reverse geocoding features. We saw how these services help keep data clean by reducing manual entry errors and getting rid of inconsistencies, and how this data can be distributed to other systems.

We know data privacy regulations are very important to consider while dealing with data based on different business process and legal reasons. Hence, SAP Cloud Platform Data Retention Manager helps the data protection officer manage this process, thus allowing your organization to remain compliant with these regulations and adapt your business applications and landscape easily to any future legal requirements.

Chapter 5
Process and Data Integration

A critical element in your enterprise transformation journey to become future ready is designing your enterprise architecture process and data integration based on current business requirements, with the flexibility to adapt quickly to future business cases. In this chapter, we'll build on various design considerations for your process and data integration and look at various capabilities that you can leverage.

As you move into your implementation phase with SAP S/4HANA or consider a move to SAP S/4HANA, an important aspect to consider is how to design and architect your integration of SAP S/4HANA with multiple other systems. Integrations are crucial to your end-to-end business outcomes, and a lot of thought is put into designing your landscape accordingly.

As organizations are looking for new, intuitive, and engaging user experiences through mobile devices, including a need for access outside of corporate network for both on-premise and cloud applications, enterprise architects have to devise integration strategies for cloud and on-premise systems and develop APIs to present data appropriately via UIs. In addition to this, organizations are looking to improve their end user experience for employees and external users (vendors, suppliers) by exposing their APIs, which needs to be governed and managed. A heterogenous landscape in an organization with systems from different vendors increases the manifold complexity, with different systems having different connectivity and data model requirements. Thus, integrations become a major obstacle in many organizations in their quest to become future ready.

All these integration requirements bring different process and data movement requirements, which expand connectivity to different devices. These requirements follow different integration patterns that are best suited for particular business and integration requirements.

Hence, organizations need integration platforms that offer the following qualities:

- **Secure connectivity**
  A secure way to connect and expose APIs from an organization's on-premise systems outside of the corporate network.

- **Flexible and scalable**
  Unlimited flexibility and scalability to meet your future business expansion needs.

- **Support different integration patterns**

  Different integration requirements have different architectural patterns to be followed. You need a solution that supports these integration best practices.

- **Nondisruptive continuous innovation**

  Cloud services offer continuous innovation capabilities, and you need a platform that offers these in a nondisruptive way so that your critical business processes are not impacted.

- **Manage APIs**

  The ability to manage, govern, operate, secure, and publish your organization's API for the end user experience.

- **Open connectivity**

  The ability to connect to different third-party applications in a heterogenous landscape and govern these connections across the enterprise in a scalable and easy manner.

In the following sections, we'll look at different integration use cases and patterns; connecting securely to your backend systems; and various capabilities of SAP Cloud Platform Integration Suite to meet your enterprise-wide integration requirements for application-to-application (A2A), business-to-business (B2B), API management, and connectivity with third-party systems. We'll also look at how event-driven patterns can be realized using SAP Cloud Platform Enterprise Messaging service with your SAP and non-SAP systems in your landscape. A critical integration requirement in your landscape centers on data exchange and learning and identifying patterns from data. For this, we'll also look at how you can use SAP Data Intelligence to meet your data integration and orchestration needs and apply machine learning capabilities. Lastly, we'll also talk about using the SAP Internet of Things service for your "things" integration needs. Thus, in this chapter we'll look at various services and capabilities offered by SAP Cloud Platform to help you meet your enterprise integration requirements and explain how you can select the right integration tool for your enterprise in your digital transformation journey while making yourself future ready.

5.1   Integration Use Cases and SAP Integration Solution Advisory Methodology

In this section, we'll investigate various integration patterns in a heterogenous landscape. We'll also look at the best practices guidance published by SAP in the form of the SAP Integration Solution Advisory Methodology, which can guide you through the process of designing your landscape and integrations.

You can expect (some) of the following system types in a diverse system landscape:

- SAP ERP
- SAP S/4HANA

- SAP S/4HANA Cloud
- Various SAP and non-SAP SaaS applications
- Data lake and data mart solutions like SAP HANA, SAP Analytics Cloud, SAP Data Warehouse Cloud, and SAP BW/4HANA
- Internet of Things applications/edge devices
- Mobile and other engaging portals for user experience
- Data and process exchange with external vendors and suppliers
- Data and process exchange with governments for legal requirements

Interactions with these different systems would involve different integration requirements and needs, and the SAP Integration Solution Advisory Methodology provides a guided journey to defining these integration patterns. The methodology identifies various integration domains, as shown in Figure 5.1.

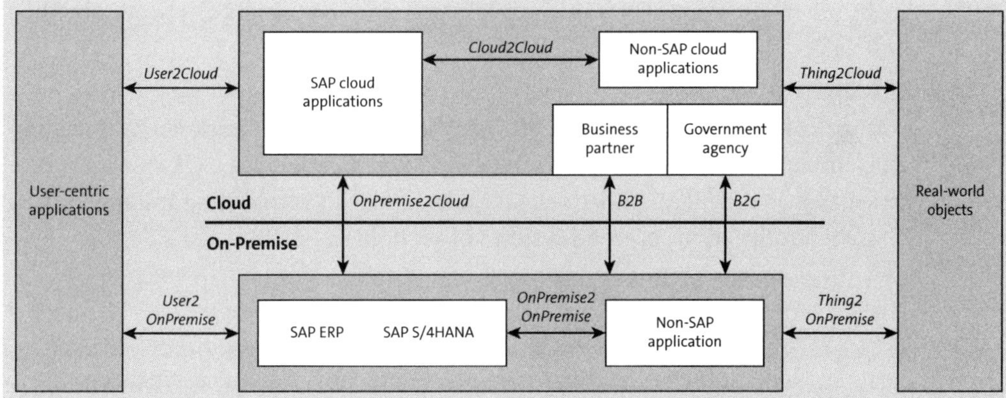

Figure 5.1 Integration Domains of SAP Integration Solution Advisory Methodology

These integration domains result in four types of integration patterns and use cases from an integration architecture perspective based on business requirements and scenarios. These integration styles result in different integration patterns that organizations need to apply while developing an integration strategy, which we'll cover in the following section.

5.1.1   Process Invocation Use Case Patterns

Process invocation is a business scenario involving the chaining of processes between different applications. For example, in a lead-to-cash scenario, a quote created in SAP Customer Experience triggers a sales order to be created in SAP S/4HANA. Some specific requirements for process invocation are as follows:

- Message orchestration
- Transactional process integrity

- Guaranteed message delivery

- Business-to-business (B2B) and business-to-government (B2G) support

Under process invocation, we have the following use case patterns:

- **Application to application**
  In an application-to-application (A2A) use case pattern, two applications are chained together to complete a business process by exchanging transactional data.

- **Business to business**
  In a business-to-business use case pattern, two business partners are chained together to complete a business process by exchanging transactional data. This use case happens through standard B2B protocols and standards like UN/EDIFACT or ANSI X.12.

- **Business to government**
  A business-to-government use case pattern mostly originates from legal requirements for businesses to communicate with a government in a format set by that government. An example for this use case pattern is a business electronically filing its taxes.

- **Message orchestration**
  This use case pattern includes stateful exchange of messages between multiple services or applications by applying service-oriented architecture (SOA)/business process management (BPM) principles. An example for this is collecting and aggregating messages from various backend systems or services.

5.1.2   Data Movement Use Case Patterns

The data movement business scenario involves data movement and synchronization of data between different applications. Two examples of this are as follows:

1. A data mart or data warehouse scenario for reporting

2. Initial load of data into business applications

Some specific requirements for data movement are as follows:

- Data orchestration

- Big data processing

- Data quality services

- Complex transformation from multiple data sources

In data movement, use case patterns involve moving data between different applications. These patterns are as follows:

- **Data warehouse/data mart**
  This use case pattern includes extracting, transforming, and loading data from multiple data sources to a data warehouse or database for data mart scenarios.

- **Data migration/initial data load**
  In this use case pattern, you extract, transform and load data from multiple data sources to a new business application. An example scenario is when your organization implements a new business application, an important step of which is loading the initial data while setting up the system.

- **Data quality management**
  Load data from multiple sources for cleansing, matching, or consolidation. This scenario includes master data address cleansing from different applications to have consistent master data.

- **Data virtualization**
  In this use case pattern, data from multiple different systems is accessed and federated from a single place, without actual physical data movement. An example for this use case for this is virtual data access in SAP HANA Cloud to remote data sources.

- **Data orchestration**
  This use case pattern includes orchestrating data from different applications, IoT devices, data warehouses, data lakes, and other business applications to analyze and corelate data in order to identify patterns, outliers, and the like and make meaningful sense of the data. This is a typical data science requirement, and data treated through orchestration can be further used for training machine learning models.

5.1.3   User Consumption

The user consumption business scenario involves consumption of data from different applications in a user interface like a portal or mobile device. Examples of user consumption are as follows:

- A user interface or mobile application showing data from SAP S/4HANA, SAP Customer Experience, or SAP SuccessFactors in a single user interface

- A chatbot for employees that offers data from an organization's different applications

User consumption is triggered by a user event and has the following requirements:

- End user orchestration
- Online/offline support
- Device management
- Application management

In a user consumption or user-centric integration style, use case patterns are centered on presenting data to end users to improve end user efficiency and engagement. These use case patterns are as follows:

- **User interface integration**
  End users of applications appreciate having access to data from multiple sources in an easily consumable way. User interface integration looks to solve the challenges

associated with presenting data from different applications in a single UI. An example of this scenario is SAP Cloud Platform Launchpad, which provides a single entry point for SAP S/4HANA, SAP SuccessFactors, or custom applications.

- **Mobile integration**
  End users of business applications look to consume backend application data from mobile applications, leveraging native mobile capabilities, including when they're offline. We cover these integration patterns in detail in Chapter 7.

- **Chatbot integration**
  In this use case pattern, chatbots are created for including data from different applications. These chatbots are further embedded within different communication channels and integrated with different business applications to trigger a transaction or retrieve further information.

5.1.4 Internet of Things Integration

The Internet of Things business scenario involves consumption of data from various *things* inside business applications and in the context of business process. For example, you might need real-time data capture from things to monitor the temperature, humidity, and so on in a shipping container.

Some specific requirements for IoT integrations are as follows:

- Thing management
- Edge processing and intelligence
- High stream data processing
- IoT protocol support

Use case patterns include bringing data from things to business applications in the context of business transactions and the ability to apply intelligence at the edge for use cases in which there is limited connectivity. The following use case patterns apply under this integration style:

- **Thing to application**
  Integration of thing data with an application to trigger a business process.

- **Thing to analytics**
  Integration of thing data for real-time monitoring and analytics and to detect outliers.

- **Thing to data lake**
  Collection of thing data into a data lake to correlate and derive meaningful value from the data—for example, using machine learning. This use case pattern can go in hand with the data orchestration integration pattern.

- **Thing to thing**
  Integration and communication between two or more things, like robots.

Now that we've looked at different integration styles applicable in the context of the landscape-to-be definition/transformation, in the following sections we'll look at various SAP Cloud Platform capabilities and how they can help you map and transform your organization's integration landscape to become future ready.

5.2   SAP Cloud Platform Connectivity and Cloud Connector

SAP Cloud Platform Connectivity allows applications running on SAP Cloud Platform to connect to and securely access remote services that either run via the internet or on-premise. This service allows you to do the following:

- Specify configurations at the subaccount level to establish connectivity to applications through destinations

- Make connections to on-premise systems through the cloud connector, which we will cover ahead

- Establish a secure tunnel from your on-premise network to an application on SAP Cloud Platform

SAP Cloud Platform Connectivity supports the following:

- HTTP(S) to connect between your cloud application and other services via the internet or in on-premise systems

- RFC to connect and invoke ABAP function modules in backend systems

- TCP to connect to on-premise systems via TCP-based protocols using a SOCKS5 proxy

- Creation of service channels to connect with SAP HANA databases in Cloud Foundry

In the Cloud Foundry environment, SAP Cloud Platform Connectivity provides the following two services:

1. **SAP Cloud Platform Connectivity service**
   Provides a standard HTTP connectivity proxy that can be used to access on-premise system:
   - You use this service to set up your Cloud Foundry application connection to on-premise systems via HTTP or RFC.
   - Establish a service channel to the SAP HANA database in SAP Cloud Platform from your on-premise system.

2. **SAP Cloud Platform Destination service**
   You can use SAP Cloud Platform Destination for configuring the following scenarios:
   - To connect your application to any other application accessible via the internet. This scenario can be configured without the connectivity service.
   - To retrieve and store technical information about destinations that are required to consume the connectivity service. Setting this up is optional, but recommended.

Normally, exposing APIs securely from an on-premise system for inbound access involves opening up ports in the firewall and using reverse proxies in DMZ. This process takes a long time to review with an organization's IT security team, thus slowing down the innovation speed and flexibility.

The cloud connector offers a fast and easy way to establish a connection with your on-premise system without the need to open up additional ports and configure your firewall. Thus, the cloud connector provides the following functionality:

- Offers a secure way to establish connectivity from SAP Cloud Platform to on-premise systems, without needing to configure your on-premise firewall to allow external access

- Provides fine-grained control over on-premise systems and resources that can be exposed and accessed by cloud applications

- Supports HTTP and RFC protocols to access ABAP systems by invoking function modules

- Lets you propagate the identities of cloud users to the on-premise system in a secure way

The cloud connector is a small, on-premise agent that can be configured in the DMZ of your landscape, which acts as a reverse invoke proxy between the network and SAP Cloud Platform. After installation and configuration, the cloud connector establishes a secure tunnel encrypted with TLS to SAP Cloud Platform. Figure 5.2 demonstrates the connectivity settings and flow.

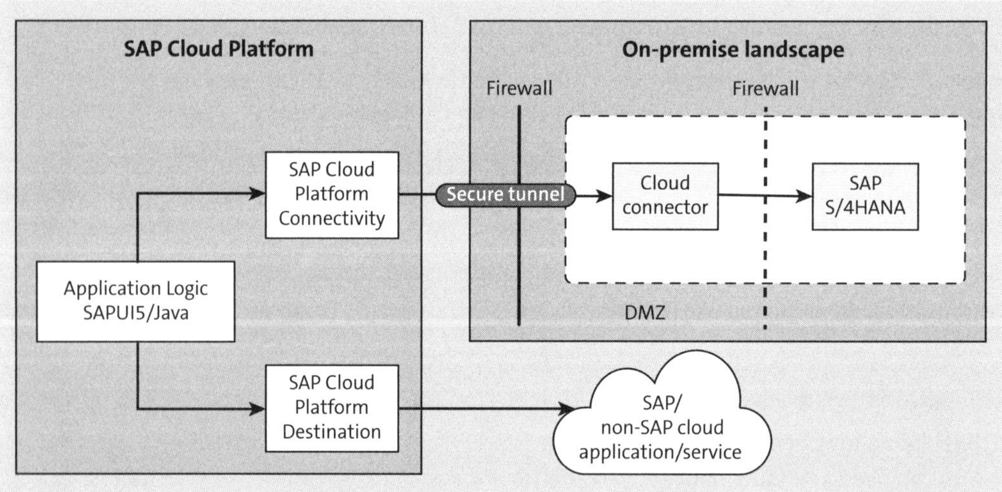

Figure 5.2 SAP Cloud Platform Connectivity and SAP Cloud Platform Destination Service Configuration

We'll now look at how to configure and set up the cloud connector. First, there are a number of prerequisites that you must meet:

- The cloud connector is installed in the on-premise landscape of your organization. SAP recommends meeting hardware prerequisites in the physical or virtual machine in terms of CPU, memory, and free disk space.
- You must have downloaded the cloud connector installation archive.
- JDK 7.0 or 8.0 must be installed, and the JAVA_HOME environment variable must be set.
- Your user must have subaccount administrator rights in the SAP Cloud Platform subaccount in which you want to configure the cloud connector.

To start your configuration, follow these steps:

1. Navigate inside your global account and click on the **i** icon in your subaccount. This opens a pop-up as shown in Figure 5.3. Note the content of the **ID** field; you'll use it later.

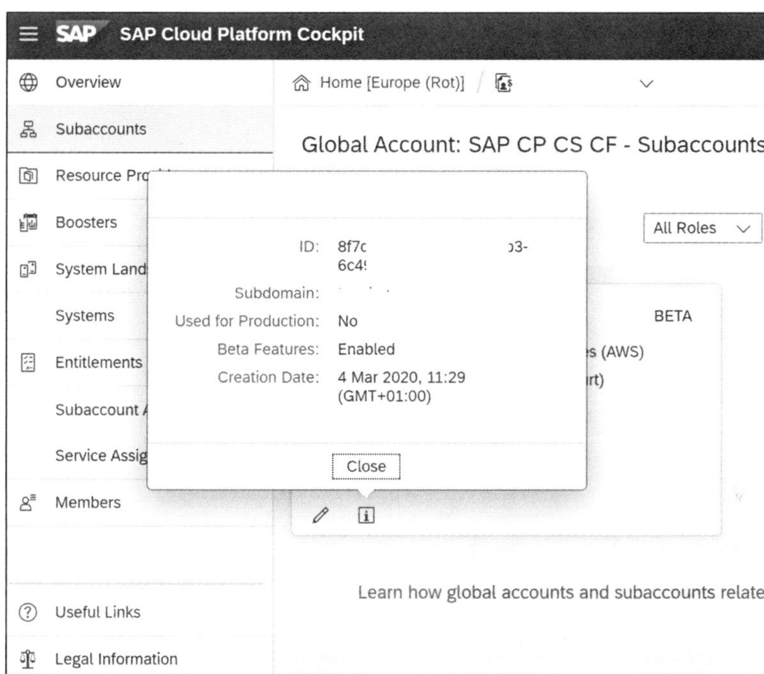

Figure 5.3 ID of Subaccount

2. Extract the cloud connector archive file, click the installation file, and navigate through the steps to install the cloud connector in the system.
3. Once installed, you can access the cloud connector at *https://<hostname>:<port>*, where *hostname* refers to the machine and *port* is the port specified during installation. The initial user is set to *Administrator* and initial password to *manage*.
4. On initial login, you'll be asked to change the initial password.
5. In the home screen of the cloud connector, click on **+Add Subaccount** and provide the following information:

- **Region**
  Data center location of the subaccount in which you want to connect the cloud connector

- **Subaccount**
  ID of the subaccount noted above

- **Display Name** (optional)
  Provide an optional display name

- **Subaccount User**
  Provide the user name of the user in the subaccount with appropriate roles

- **Password**
  Password for the user provided above, used to login to SAP Cloud Platform

- **Location ID** (optional)
  Provide an optional location ID to differentiate the location of cloud connector

- **HTTPS Proxy** information
  Provide the HTTPS proxy info (if any exists)

6. Click **Save.** A secure connection is established between SAP Cloud Platform and the cloud connector to enable secure connectivity, as shown in Figure 5.4.

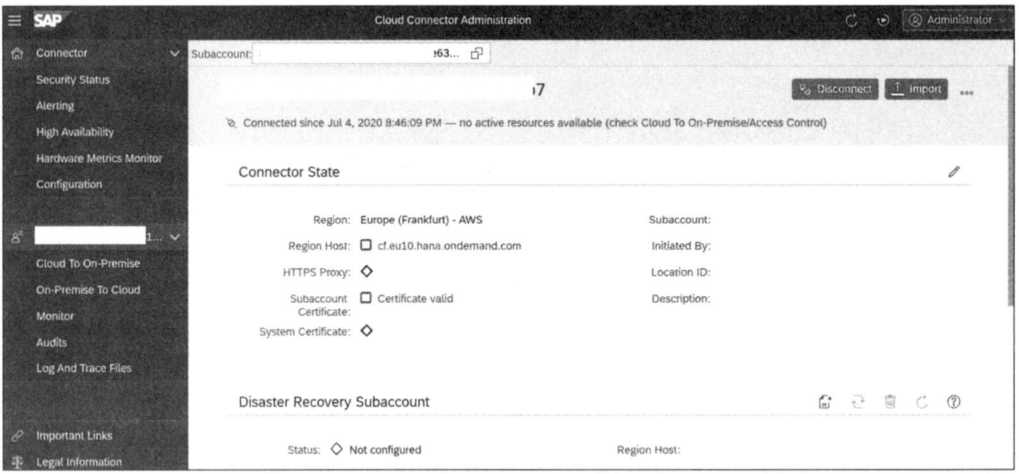

Figure 5.4 Cloud Connector Administration

7. Navigate inside your subaccount to **Connectivity • Cloud Connectors**. You'll see that the cloud connector is connected.

8. Now configure your backend system to trust the cloud connector. The cloud connector uses a system certificate for all HTTPS requests to the backend system. This means that the system certificate in the cloud connector needs to be trusted by backend systems to which the cloud connector is supposed to connect.

9. To set the system certificate, navigate to **Configuration • On Premise** in the **Cloud Connector Administration** screen.

10. Choose **System Certificate**. You can import a certificate or create and import a self-signed certificate. The public key used for establishing trust with a server can be exported using the **Export** button.

11. To set up mutual authentication between the cloud connector and the ABAP backend system connected via RFC, you need to configure secure network communications (SNC) for the cloud connector. The prerequisite for this connectivity is that the ABAP system be configured for SNC. For this, navigate to **Configuration • On Premise** and then navigate to the **SNC** section.

12. Enter **Library Name**, **My Name**, and **Quality of Protection**, then choose **Save**. **Library Name** is the location of the SNC library, **My Name** is the SNC name that identifies the cloud connector, and **Quality of Protection** is the level of protection that is required for connectivity to the ABAP system.

13. Now establish and expose connectivity to the backend system. For this, navigate to **Cloud to On-Premise • Access Control** to define connectivity to the backend ABAP system.

14. Define a **Virtual Host** and **Virtual Port** mapped to your internal system host name and port. Virtual host and port information is used in SAP Cloud Platform while configuring destinations.

15. Define the services path that should be accessible from the cloud.

> **Note**
>
> As previously mentioned, an important feature of the cloud connector is user ID propagation, in which the user identity in the cloud is propagated to the backend system in secure way. You can read about this configuration at *http://s-prs.co/v515719*.

5.3   SAP Cloud Platform Integration Suite

Intelligent enterprise processes like hire-to-retire, lead-to-cash, procure-to-pay, and design-to-operate involve multiple applications, which include on-premise, SaaS, and custom applications from SAP and non-SAP sources.

SAP Cloud Platform Integration Suite helps simplifies building these end-to-end processes through harmonized APIs, prepackaged integrations, and more than 150 open connectors to third-party applications. It helps accelerate enterprise integration and thus supports your organization's quick transition to an intelligent enterprise. It helps organizations integrate anything (applications, process, data, things, suppliers, vendors) anywhere (on-premise, cloud) and manage their APIs to share information internally with employees and external vendors. Thus, SAP Cloud Platform Integration Suite helps orchestrate end-to-end integration scenarios in your organization across a complex heterogenous landscape in a cohesive way.

SAP Cloud Platform Integration Suite is an open and modern solution built from the ground up with enterprise-grade capabilities such as security and high availability, and it's available across multiple hyperscaler data centers across the world, thus supporting your organization's transformation and helping your landscape become future ready.

In the following sections, we'll first look at how to set up SAP Cloud Platform Integration Suite itself, and then we'll look at various capabilities provided by SAP Cloud Platform Integration Suite: Cloud Integration, API Management, Open Connectors, and Integration Advisor.

> **Note**
>
> Cloud Integration, API Management, Open Connectors, and Integration Advisor are not the only products and functionalities that can be found within the SAP Cloud Platform Integration Suite umbrella; they are just the ones discussed in this section.

5.3.1   Configuration

To perform the initial set up of SAP Cloud Platform Integration Suite, follow these steps:

1. In your Cloud Foundry environment subaccount, navigate to **Subscriptions**.

2. Select the **Integration Suite** tile and choose **Subscribe**. Your subaccount is now subscribed to the SAP Cloud Platform Integration Suite.

3. Assign a role to a user to enable provisioning of capabilities in SAP Cloud Platform Integration Suite.

4. Inside your subaccount, navigate to **Security • Trust Configuration**.

5. Select the SAP ID service, enter your email ID, and choose **Show Assignments**. Choose **Assign Role Collection** and assign the Integration_Provisioner role to your user ID.

6. Navigate back to **Subscriptions** in the Cloud Foundry subaccount and choose the **Integration Suite** tile, then choose **Go to Application**.

7. In the pop-up screen, select the capabilities that you want to activate while provisioning the tenant. You have the following capabilities to choose from while activating the tenants, as shown in Figure 5.5:

   - **Design, Develop and Operate Integration Scenarios**
     Capability to enable end-to-end process integration through a transactional exchange of data between cloud and on-premise applications

   - **Design, Develop and Manage APIs**
     Capability to let you manage, govern, and publish APIs in a secure and scalable environment

   - **Extend Non-SAP Connectivity**
     Capability that allows you to seamlessly connect and integrate with a variety of non-SAP cloud applications

- **Implement Interfaces and Mappings**
  Capability that enables you to develop B2B interfaces and integrations via crowd intelligence and machine learning

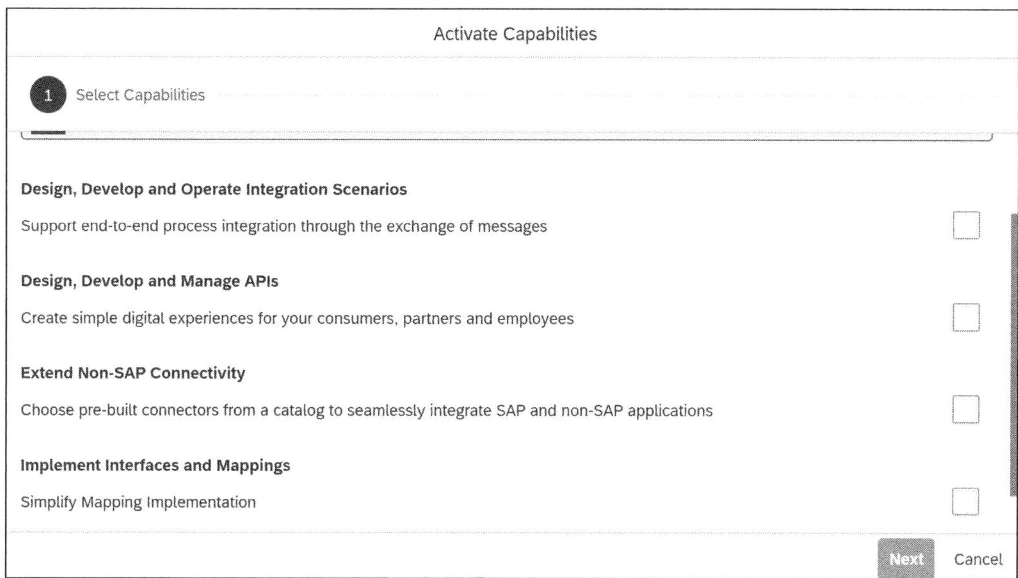

Figure 5.5 SAP Cloud Platform Integration Suite Capabilities

8. Click **Next** and choose the required features for the capabilities selected, then choose **Activate**. This completes the tenant provisioning and takes you to the tenant provisioning home screen, shown in Figure 5.6.

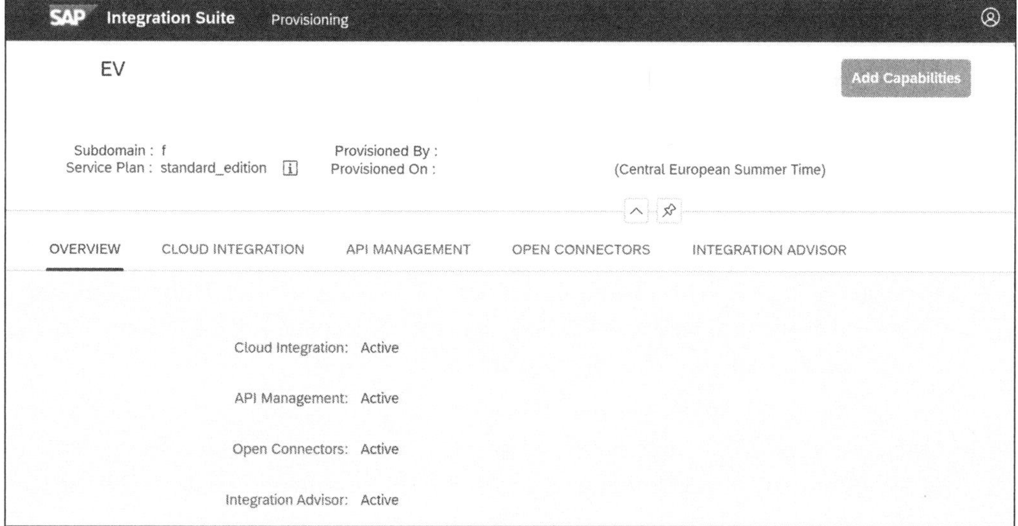

Figure 5.6 SAP Cloud Platform Integration Suite Provisioning Home Screen

9. If you haven't activated any capabilities during this process, you can always go back to the SAP Cloud Platform Integration Suite home page, shown in Figure 5.7, and click **Add Capabilities**.

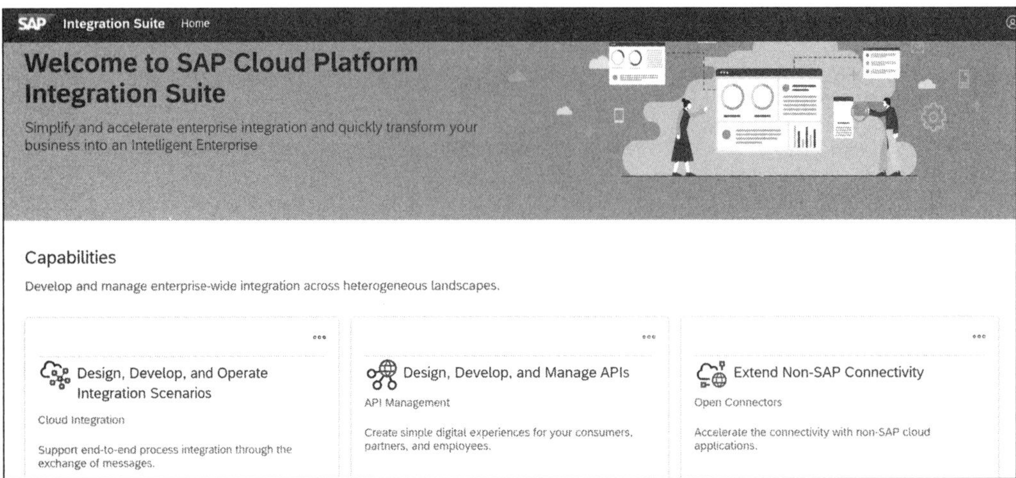

Figure 5.7 SAP Cloud Platform Integration Suite Home Page

5.3.2   Cloud Integration

The Cloud Integration capability within SAP Cloud Platform Integration Suite allows you to develop cloud and hybrid integrations supporting A2A, B2G, and B2B patterns. The artifacts developed using Cloud Integration are called *integration flows*. Cloud Integration follows Business Process Model and Notation (BPMN) capabilities to develop integration flows. It supports 40+ different orchestration capabilities to enable integration flow development.

Cloud Integration comes with prepackaged integration contents and scenarios for SAP solutions. These prepackaged integration contents and scenarios can help you start setting up integrations for your business transformation easily and can be customized using a web-based development interface for cloud integration. Cloud Integration provides out-of-the-box prebuilt adapters that support connectivity with IDocs, OData, SAP SuccessFactors, SOAP/HTTPS, Lightweight Directory Access Protocol (LDAP), Java Message Service (JMS), Java Database Connectivity (JDBC), Advanced Message Queuing Protocol, and more, as well as SAP OEM adapters for Salesforce, Microsoft Dynamics 365 CRM, SugarCRM, and Amazon Web Services.

You can use Cloud Integration and SAP's on-premise integration platform, SAP Process Orchestration (SAP PO), in a seamless, integrated way. You can use Cloud Integration as your central discovery and design tool while developing integration flows and use SAP PO to run scenarios to connect on-premise to on-premise applications.

To set up Cloud Integration, you must first have met the following prerequisites:

- You already have activated and set up SAP Cloud Platform Integration Suite as described earlier.
- You have assigned **Process Integration Runtime** to your Cloud Foundry subaccount from the **Entitlement** section in your global account.

The first step to access Cloud Integration is to provide users with required rights and accesses to allow them to administrate and develop integration flows. This requires assigning roles to users.

Table 5.1 provides a few important roles in the Cloud Foundry environment that you need to assign to users to give them the right roles to start development with Cloud Integration.

Role Template	Description
`AuthGroup_ Administrator`	This role allows users to do the following: - Deploy integration content - Deploy security content - Monitor integration flows and their status - Delete messages from transient data store in integration flow This role and role template is already added to a role collection: `PI_Administrator`.
`AuthGroup_ IntegrationDeveloper`	This role allows users to do the following: - Deploy integration content - Monitor integration flows and their status. This role provides required rights to access the web development interface of Cloud Integration. This role and role template is already added to a role collection: `PI_Integration_Developer`.
`AuthGroup_ ReadOnly`	Role allows users to do the following: - Monitor messages - Access the read-only view of the data store. This role and role template is already added to a role collection: `PI_Read_Only`.
`AuthGroup_ BusinessExpert`	This role allows users to do the following: - Read message payloads and attachments - Monitor integration flows and status. This role is provided to users who need to access message payloads. This role and role template is already added to a role collection: `PI_Business_Expert`.

Table 5.1 Role Collection for Cloud Integration in Cloud Foundry

You can either create a new role collection and add the mentioned roles from the role templates in Table 5.1 to existing role collections, or assign the predefined role collection to a user ID directly. To add a Cloud Integration role to an existing role collection, follow these steps:

1. Navigate inside your Cloud Foundry subaccount to **Security • Role Collections** to create new role collection. Navigate inside the created role collection and click **Add Role**.

2. Select **it*** for the **Application Identifier**, then select role templates from Table 5.1 to add the corresponding roles.

3. Navigate to **Security • Trust Configuration**, select the trust provider and provide the user ID, and click **Show Assignments**. Click **Assign Role Collection** and add a predefined role collection or custom role collection.

4. Navigate back to the SAP Cloud Platform Integration Suite home screen and select **Web UI URL for Cloud Integration**. This will open the Cloud Integration home screen, as shown in Figure 5.8.

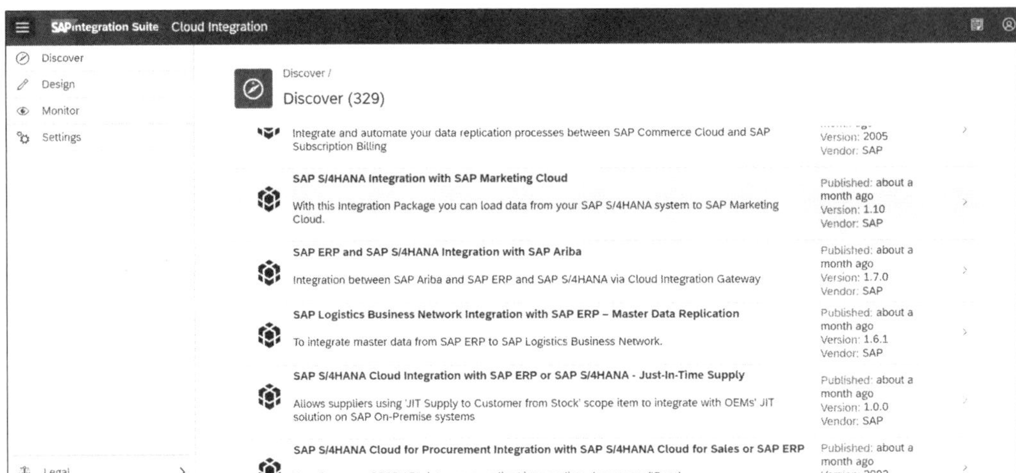

Figure 5.8 SAP Cloud Platform Integration Suite, Cloud Integration Home

The home screen has four sections:

- **Discover**
  The **Discover** section allows you to discover and browse published prepackaged integration content. You can select the prepackaged integration content and click **Copy** to copy the content to your integration tenant.

- **Design**
  The copied integration package from the **Discover** section will be visible in the **Design** section. This section allows you to create/edit/configure flows and packages, as well as perform version management of the flows, as shown in Figure 5.9. This is the web-based design interface for Cloud Integration. Once your integration flow design is

ready, you can run the integration flow simulation to test it before you deploy. To deploy the integration flow, click **Deploy**.

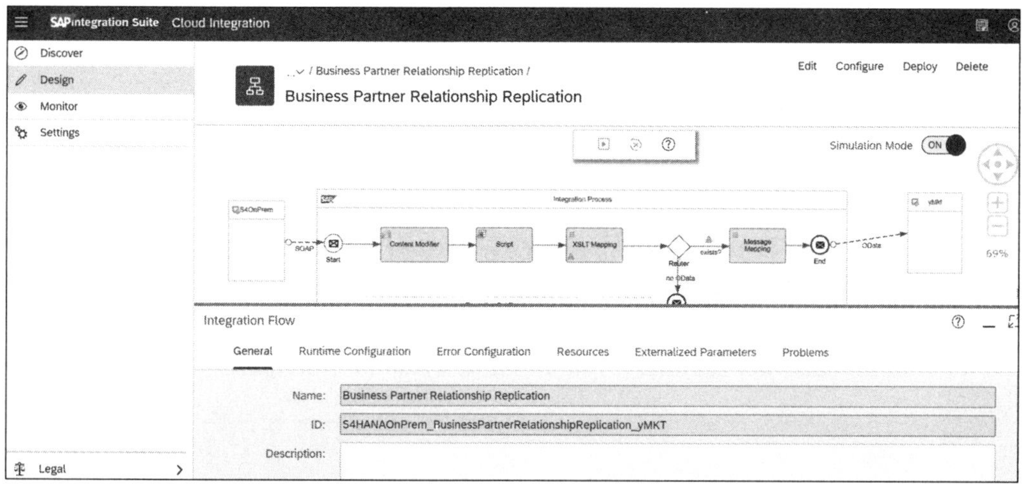

Figure 5.9 SAP Cloud Platform Integration Suite, Cloud Integration Web-Based Integration Flow Development

- **Monitor**

  The **Monitor** section allows you to monitor the deployment status of a deployed integration flow. You can create flexible monitoring tiles for your needs, such as to monitor the status of a particular integration flow only, and view detailed logs that provide a visual representation of the message processing status.

  If your user has administrator rights, you can use this view to deploy security contents.

- **Settings**

  The **Settings** view allows you to maintain product profiles, the enterprise service repository from which message mappings can be imported, and custom tags.

To access the endpoints of the deployed integration flow, you need to create an instance of process integration runtime as follows:

1. Navigate inside **Spaces** in your Cloud Foundry subaccount to **Services • Service Marketplace**.

2. Select the **Process Integration Runtime** tile and navigate to **Instances**.

3. Click **New Instance** and, on the **New Instance Creation** screen, select **integration-flow** as the service plan and click **Next**. The service plan API provides access to SAP Cloud Platform Integration's public APIs.

4. On the **Specify Parameter** screen, enter the code shown in Listing 5.1. This role authorizes the sender system to call a tenant and allows you to process messages on the tenant. Click **Next** until you reach the screen to provide an instance name.

```
{
    "roles":[
        "ESBMessaging.send"
        ]
}
```

Listing 5.1 Specify Parameter in Cloud Integration Service Instance Creation

5. On the last screen, provide an **Instance Name** and click **Finish** to create a new service instance.

6. Select the created service instance and navigate to **Service Keys**. Select **Create Service Key**.

7. Provide a **Name** and click **Save**. This creates the service key; the `clientid` and `clientsecret` in the key are used to call the integration flow instance.

> **Note**
>
> Prepackaged integration flows for SAP S/4HANA can be found in SAP API Business Hub at *https://api.sap.com/themes/S4HANA*.
>
> The SAP API Business Hub is hub where all APIs from SAP are published.

5.3.3   API Management

APIs have become the primary facilitator for ensuring digital transformation and have become the foundation to make your organization future ready. APIs enable modern application development and open integrations. Thus, to cite a few examples, APIs have enabled organizations to do the following:

- Collaborate with partners, vendors, and suppliers to increase transparency and build closer business relationships.
- Empower your developers to build modern applications, including mobile apps.
- Enable new revenue streams by monetizing your data and digital assets.
- Empower and enable your ecosystem to engage in modern ways.
- Securely integrate with other applications in your landscape, enabling an end-to-end business process.

Thus, APIs have facilitated increasing the engagement index for your business among your employees, vendors/suppliers, and ecosystem and help you gain better insight into these streams to gauge and understand their needs. This makes it clear that you need to have one solution that can help you manage the full lifecycle of your APIs and provide analytics on top of it to understand the most consumed data and processes. API Management provides such an ability, providing a single experience for managing and monitoring APIs in your landscape, and it's enriched with real-time analytics.

API Management also provides advanced capabilities to secure your APIs and selective data exposure based on user roles, thus protecting your APIs against any security incidents.

API Management capabilities in SAP Cloud Platform Integration Suite help you manage your APIs securely across your organization. It's also closely integrated with SAP applications and landscapes, which enables you to create APIs and define policies for your SAP systems easily by providing reusable building blocks. The features provided by API Management help to create low-level APIs without requiring new complex development on the backend, thus helping to modernize and empower your landscape for further integration and collaboration. This helps to create a consistent user experience for your stakeholders by abstracting the underlying system complexities.

Further, APIs created and exposed through API Management can be easily accessed from other services on SAP Cloud Platform, like SAP Cloud Platform Mobile Services and SAP Web IDE, which enable you to quickly create applications on top of these APIs.

The following are the key components of API Management:

- **API portal**

  This is a web-based API modeling environment. It helps secure, govern, monitor, analyze, and monetize API access. It enables discovery of SAP and select partner APIs and policy templates from SAP API Business Hub.

  An API admin persona in your organization responsible for defining organization-wide API governance and security policies will interact with the API portal.

- **Developer portal**

  The developer portal enables you to publish your organization's APIs to a developer community. This provides a turnkey developer portal with features for developers to discover, explore, test, experience and onboard (via self-service) APIs.

  Any application developer who wants to consume these APIs to develop applications or integrations will engage with the developer portal.

- **API designer**

  The API designer is used to design and develop APIs based on open standards. It supports code generation for rapid API development. APIs can expose your backend on-premise systems, database, and cloud services in your landscape. The API developer in your organization responsible for creating APIs will interact with the API designer.

- **API gateway**

  The API gateway is a component of API Management responsible for providing enterprise-grade API security, like handling authentication, requesting API access, and so on, with end-to-end user propagation. API gateway also provides API monitoring and analytics.

Before you begin to set up API Management, you must have activated the ability to design, develop, and manage APIs in SAP Cloud Platform Integration Suite. Enabling this capability creates different role collections in your SAP Cloud Platform subaccount for the API portal and developer portal capabilities.

Predefined role collections relevant for the API portal are shown in Table 5.2.

Role Collection	Description
`APIManagement.SelfService.Administrator`	Provides access to users for onboarding of the API portal and access to its settings page Assign this role to the user responsible for onboarding of API Management
`APIPortal.Administrator`	Provides access to users to API portal user interface and services
`APIPortal.Service.CatalogIntegration`	Establishes a connection from the developer portal to the API portal
`APIPortal.Guest`	Provides access to users to the API portal in read-only mode Enables viewing all APIs, policies, API providers, and analytics

Table 5.2 API Portal Predefined Roles

Table 5.3 provides predefined role collections for the developer portal.

Role Collection	Description
`AuthGroup.SelfService.Admin`	Assign this role during onboarding of the developer portal for access to it Assign this role to the user onboarding the developer portal
`AuthGroup.API.ApplicationDeveloper`	Lets users to access the developer portal; create, update, and delete applications; view analytics information on usage; and view and download bills for subscribed applications
`AuthGroup.API.Admin`	Provides access to users to manage the application developer's access to the portal; manage roles for users by adding/deleting roles; create, update, and delete applications; create custom attributes for applications; and provide an app key and secret while creating and updating an application
`AuthGroup.Content.Admin`	Provide access to users to create and update categories
`AuthGroup.ContentAuthor`	Provides access to users to publish content to the developer portal and establish connections from the API portal to the developer portal

Table 5.3 Developer Portal Role Collections

Role Collection	Description
`AuthGroup.Site.Admin`	Provides access to users to configure updates and make changes to the portal, like changing the logo, name, or description

Table 5.3 Developer Portal Role Collections (Cont.)

To enable access to the API portal and developer portal for API Management, you must first onboard the API portal and developer portal. For this step, you need to assign an onboarding role collection as identified in Table 5.2 and Table 5.3 to the user to provide the required access rights. To proceed with enabling access, follow these steps:

1. Navigate in your Cloud Foundry subaccount to **Security • Trust Configurations**.

2. Select the configured identity provider and enter a user ID, and click **Show Assignments**. Click **Assign Role Collection** to assign the role collection for the API portal and developer portal.

3. From the SAP Cloud Platform Integration Suite home screen, click the **Design, Develop and Manage APIs** tile, or from the SAP Cloud Platform Integration Suite **Provisioning** screen, navigate to **API Management • API Portal** and click the URL, which launches the screen to configure the API portal service.

4. In the **Account** section, select the **Account Type**. The following options are available:

   - **Nonproduction**
     Account type for non-business-critical activities, integrating test systems, testing new scenarios, performance testing, and sandbox activities

   - **Production**
     Account type for business-critical usage, integrating production systems, and productive APIs

5. In the **Virtual Host** section, enter a **Host Alias**. The host alias will appear in your API proxy as follows: *https://{VirtualHost}.apimanagement.hana.ondemand.com*.

6. Click **Set Up** and a confirmation screen will appear. Click **Confirm** on this screen.

7. A new screen appears, which shows the progress of onboarding the API portal service.

8. If you have assigned other API portal roles to the user before, log out of the API portal and log back in for changes to be reflected. If not, assign other required roles to the user as explained previously and then log in to the API portal using the user. On login, the API portal is displayed as shown in Figure 5.10.

Note

You can also add other role authorizations to the user at this point.

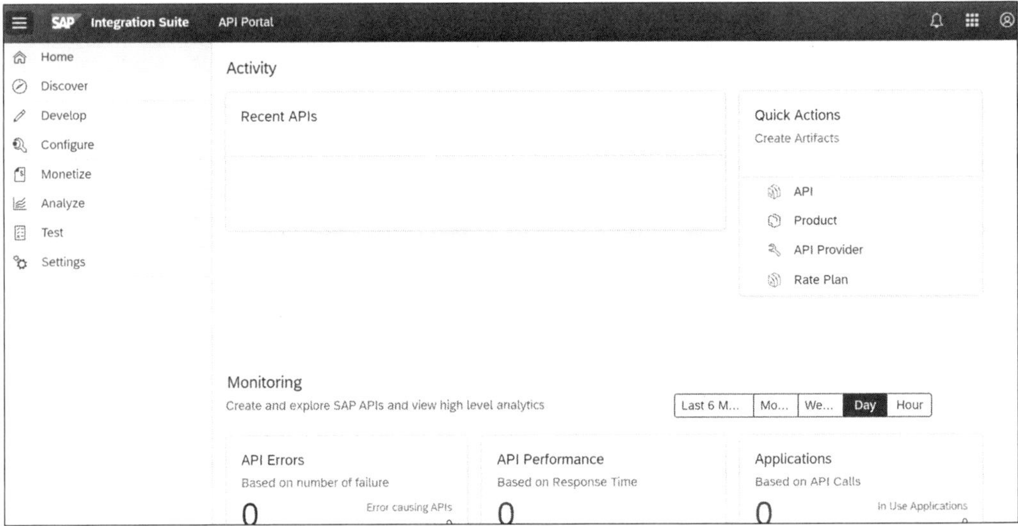

Figure 5.10 SAP Cloud Platform Integration Suite API Portal Home Screen

Next, you need to configure API Management's developer portal. Before this, ensure that the onboarding role for the developer portal is assigned to a user, as follows:

1. Navigate in the SAP Cloud Platform Integration Suite provisioning page to **API Management • Developer Portal**. Click the URL and enter the user login credentials. After successful login, wait for some time until the developer portal is successfully set up.

2. After a successful setup, log out and log in again to the developer portal for all changes to be reflected.

3. Click the **Register** button to onboard another user as an application developer.

4. In **Register** pop-up screen, select the country of origin and choose **OK**.

5. This completes the registration; log out and log back in for changes to be reflected for the application developer.

6. Click **Manage** to make changes to **Company Logo**, **Color Scheme**, **Manage Home Page Name and Description**, and **Manage Users**.

5.3.4   Open Connectors

If your organization landscape has applications and systems from multiple different vendors, these systems have different connectivity layers and requirements, making integration across these different systems quite often a complex task to realize end-to-end business value.

SAP Cloud Platform Integration Suite's Open Connectors capability helps to simplify this task by providing capabilities to create a unified API layer and standards-based integrations across these systems. Thus, it helps to create a single place from which you can set up, manage, and govern these connections, which can be reused across your environment, across different integrations and APIs.

Open Connectors already comes with 150+ common third-party integration connections and catalogs. In addition, for the systems not in the list, Open Connectors provides the ability to create and manage connectivity with minimal effort.

Before you use Open Connectors, you must have already met the following prerequisites:

- You have activated the extend non-SAP connectivity capability as part of SAP Cloud Platform Integration Suite.

- Your user is assigned to the `IntegrationSuite_OpenConnectors_Access` role collection. This role is assigned to your user automatically if you have provisioned this service.

To access Open Connectors, click the **Non-SAP Connectivity** tile from the home page of SAP Cloud Platform Integration Suite, or navigate inside your Cloud Integration **Provisioning** screen to Open Connectors and click the URL. This will open the Open Connectors home page, as shown in Figure 5.11.

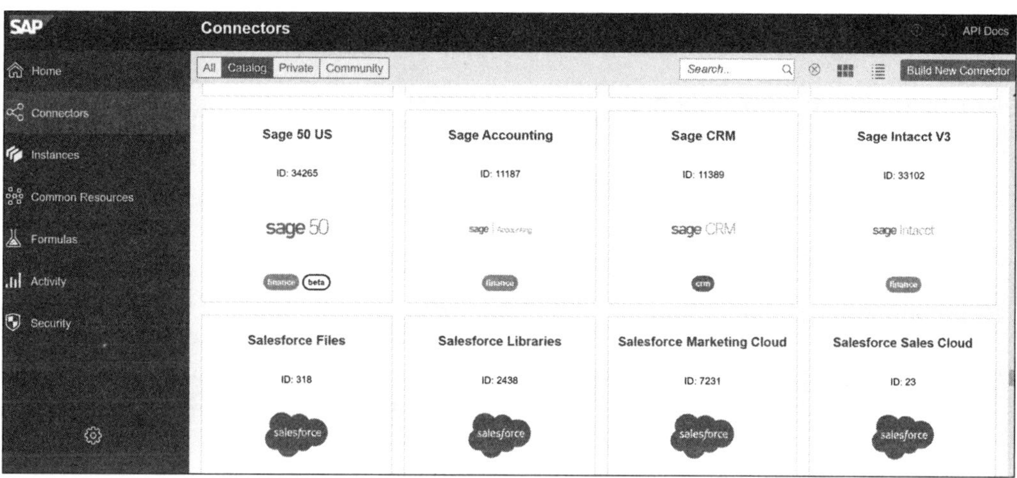

Figure 5.11 SAP Cloud Platform Integration Suite, Open Connectors

The following are the components in Open Connectors and the activities you can perform:

- **Connectors**
  Provides a list of prebuilt connectors; over 150 of them are displayed here. You can search for required connector here. You can also build your own new connectors with features including authentication, pagination, and errors with automatic generation of API documentation in the OpenAPI specification. Created connectors are published to private connector catalogs.

- **Instances**
  Lets you manage your authenticated connections to API providers (connector instances) and executed workflows (formula instances), which we'll cover ahead.

- **Common Resources**
  Create a uniform API wrapper around multiple API providers. For example, you can build a common resource like contacts, which could be mapped to different third-party systems like Salesforce Sales Cloud, HubSpot, and so on, resulting in a one-to-many model that scales with integration. You can build new resources from scratch or build from a JSON object. Thus, this helps to develop and map data models in order to extend prebuilt connectors.

- **Formulas**
  Formulas are user-defined workflows that have a trigger (incoming event, API request, timer, etc.); when triggered, they begin executing a series of steps across different applications. This helps to move complex logic from your application.

- **Activity**
  Shows various logs and metrics. You can filter logs and metrics based on certain connectors, accounts, and connector instances.

- **Security**
  Manage security settings, including setting up trusted identity providers, accounts, and users. You can add additional users (S-User ID/P-User ID) from here to provide them with access.

5.3.5   Integration Advisor

Organizations need to connect with other business partners in your landscape. Inherent complexity exists in such integrations as different business partners use different industry standards like UN/EDIFACT, IDocs, and ASC X12 to communicate. Each standard means new interfaces that need to be created, and doing so manually is an extremely time-consuming job.

SAP Cloud Platform Integration Suite's Integration Advisor capability helps simplify the creation and management of such interfaces for your organization. Integration Advisor uses crowd intelligence with the help of the library of type systems as a starting point and applies machine learning capabilities on top of it to create new message implementation guidelines. This creates mapping guideline to map the standards in your organization to standards used by your business partners, and automatically generates integration runtime artifacts that can be used in Cloud Integration. It also generates the documentation for created integration guidelines and mappings, which reduces the human effort needed for creating, maintaining, and updating them on a regular basis.

The libraries are collections of message templates provided by agencies that maintain the required B2B standards. For example, type system ASC X12 is maintained by ANSI

X12 and IDocs are developed and maintained by SAP. As mentioned before, this can act as a good starting point and message implementation guidelines can be further customized for your scenario-specific requirements based on industry standards. Message implementation guidelines are then used in mapping guidelines to create mappings for developing integration flows to communicate with business partners.

Using this approach can speed the development of B2B/B2G integrations, and the contents you have generated and adjusted can be shared with other users of the applications, which helps to improve and enhance the crowd intelligence capabilities of Integration Advisor over time.

Before you set up Integration Advisor, you must have activated the implement interfaces and mappings capability in SAP Cloud Platform Integration Suite. Then, you can proceed as follows:

1. Inside your Cloud Foundry environment subaccount, navigate to **Entitlements • Configure Entitlements** and click **Add Service Plans**.

2. From the pop-up screen, select a service plan, **SAP Integration Advisor B2B Library**, and click **Add <X> Service Plans**.

3. Choose **Save**.

4. Now, provide access to Integration Advisor for your user. Navigate in your Cloud Foundry environment subaccount to **Security • Trust Configurations**.

5. Select the SAP ID service, enter your email ID, and click **Show Assignments**.

6. Choose **Assign Role Collection** and add the `iadv-content-developer` role.

7. From the SAP Cloud Platform Integration Suite home page, click the **Implement Interface and Mappings** tile; or from the SAP Cloud Platform Integration Suite **Provisioning** page, navigate to **Integration Advisor** and click the URL. Figure 5.12 shows the home page of Integration Advisor within SAP Cloud Platform Integration Suite.

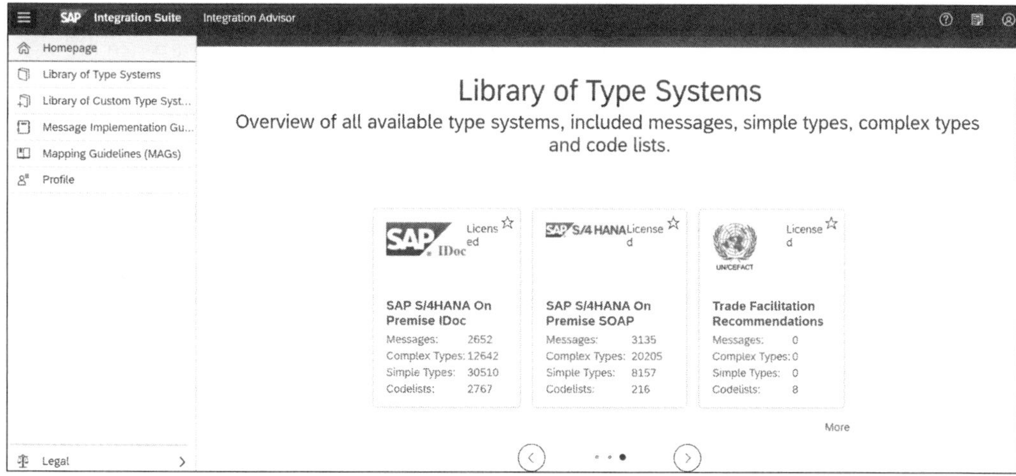

Figure 5.12 Integration Advisor Capability in SAP Cloud Platform Integration Suite

5.4   SAP Cloud Platform Enterprise Messaging

As previously mentioned, enterprise landscapes often have multiple applications and systems that need to be connected together. The integration of processes involves exchanging messages across systems and applications. These messages can be exchanged in two different ways:

1. Directly sent to receiving systems

2. Sent to a message broker, from which they're sent to the receiving system or picked up by the receiving system

Sending the messages from the sender system to a message broker system allows you to decouple the message processing from the sender system. The sender system would send the message to the broker and continue further with the processing; the responsibility to deliver the message to the receiving system is on the broker.

Using this pattern, applications can send events notifying the change in state of their business objects and publish these to message brokers, from which these events can trigger or complete the business process in one or more receiving business applications. Thus, it makes your application design on the sending side easier as it takes the load away from them to manage while they scale, eventually improving the performance, and allows an architecture pattern to loosely couple with receiving applications and allow them to receive these messages asynchronously.

SAP Cloud Platform Enterprise Messaging is a cloud-based offering that allows applications to communicate asynchronously. The events published to SAP Cloud Platform Enterprise Messaging service can be subscribed to by different receiving applications interested in this data independently, thus creating a loosely coupled architecture. This service enables asynchronous (nonblocking) communication of events between applications, services, and systems. Such an architecture allows your organization to scale your landscape, making it more agile and set for a future-ready transformation.

In the following sections, we'll first introduce SAP Cloud Platform Enterprise Messaging in general and then show how to perform the necessary setup.

5.4.1   Product Overview

SAP Cloud Platform Enterprise Messaging supports standard messaging protocols and asynchronous communication in a secure, reliable manner with high throughput and low latency. Let's look some important terminology in the context of this service:

- **Queues**
    - Publishing apps send their messages/events to a queue that a receiving application can subscribe to in order to receive the messages.
    - Messages are stored in the queue until consumed by a client.

- There is typically a 1:1 relationship between a queue and a client.

- This creates a point-to-point messaging pattern model.

- **Topics**

  - Publishing app scan send their messages/events to a topic. One or more different receiving applications/clients can pick up messages from a topic. This is used when each message needs to be sent to several receivers, and there is a 1:N relationship between a topic and clients.

  - Messages sent to receivers are not stored, so receiving systems need to be active.

  - Event notifications from the SAP S/4HANA system can only be sent to a topic.

  - This enables a publish-subscribe messaging pattern.

- **Queue subscriptions**

  - Queues and topics can be combined. The messages sent to topics are directly sent to a queue.

  - This is used when you need to store the events sent by SAP S/4HANA to a topic to be picked up by a receiving system that may not be active all the time. In this case, receiving systems point to the queue.

  - This enables a publish-subscribe pattern with message storage until a client picks it up, as shown in Figure 5.13.

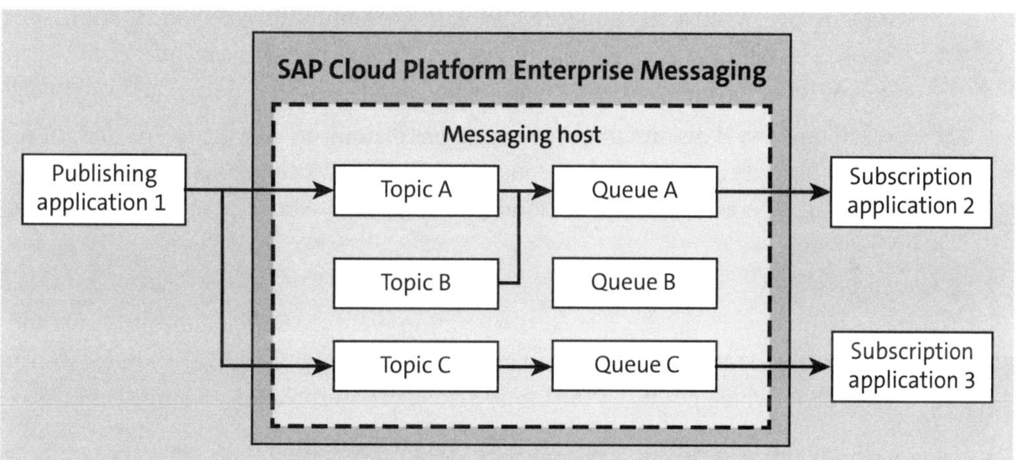

Figure 5.13 SAP Cloud Platform Enterprise Messaging Publish-Subscribe Pattern

SAP Cloud Platform Enterprise Messaging supports the following messaging protocols to communicate between sending and receiving applications:

- **Advanced Message Queuing Protocol 1.0 over WebSocket**
  Advanced Message Queuing Protocol is an open-source protocol used for messaging between applications. The recommendation is to use this protocol for messaging between applications running on Cloud Foundry.

- **Message Queuing Telemetry Transport (MQTT) 3.1.1 over WebSocket**
  MQTT is a lightweight protocol designed specifically for low-bandwidth, high-latency, or unreliable devices. The recommendation is to use this protocol to communicate with applications not running in the cloud.

- **HTTP RESTful APIs for messaging**
  Enterprise messaging provides REST APIs for messaging to send and receive messages.

> **Note**
>
> SAP Cloud Platform Enterprise Messaging supports standard libraries for Node.js and Java developers to communicate with the service. This helps to reduce the development effort to connect to and communicate with this service while developing applications.
>
> You can get more information about these libraries at *http://s-prs.co/v515720*.

SAP Cloud Platform Enterprise Messaging supports out of the box a capability to receive events and to enable event look up and event discovery from the following SAP solutions, thus enabling different event sources to publish events to the service:

- **SAP S/4HANA Cloud**
  Event discovery and activation for SAP S/4HANA Cloud with SAP Cloud Platform Enterprise Messaging is available out of the box. The connectivity established between the services is through the MQTT protocol.

  In Chapter 9, we'll discuss the configurations to easily enable, configure, and set up this connectivity automatically through SAP Cloud Platform's extension capabilities. The events thus available in SAP Cloud Platform can also be used to develop modern extensions.

- **SAP S/4HANA**
  SAP S/4HANA systems can enabled for sending events to SAP Cloud Platform Enterprise Messaging through SAP Cloud Platform Connectivity and SAP Cloud Platform Destination, which points to SAP Cloud Platform Enterprise Messaging via MQTT.

> **Note**
>
> You can read more about enabling this for SAP S/4HANA on-premise at *http://s-prs.co/v515721*.

- **Other SAP applications**
  SAP Cloud Platform Enterprise Messaging also supports event enablement and discovery with other SAP applications like SAP Customer Experience, SAP SuccessFactors, and SAP ERP.

Thus, this service provides a single event-enablement platform for your integrations to enable end-to-end business scenarios and develop event-driven smart extensions for your business needs. In addition, SAP Cloud Platform Enterprise Messaging supports CloudEvents, a specification for describing event data in a common way. CloudEvents is supported by multiple third-party applications and systems, which enables a consistent integration between these systems.

Using SAP Cloud Platform Enterprise Messaging, you can enable multiple use cases in your organization, like events triggering SAP Cloud Platform Integration Suite Integration Flows (iFlows) to trigger and complete business processes, triggering serverless functions to further extend processes (explained further in Chapter 9), thus making your landscape ready for future-ready changes and uncovering new business models that adapt to your business process in a quick and agile way.

5.4.2 Configuration

To start, you must have assigned SAP Cloud Platform Enterprise Messaging to your Cloud Foundry environment subaccount. To begin your initial setup, follow these steps:

1. Navigate inside your Cloud Foundry environment subaccount to **Space** and once there, select **Services • Service Marketplace.**

2. Select the **Enterprise Messaging** tile and navigate to **Instances.** Click **New Instance.**

3. In the New Instance Creation wizard, select a service plan and choose **Next.**

4. On the next screen, **Specify Parameters (Optional),** provide a JSON file as defined in Listing 5.2 and click **Next.**

```
{
    "options": {
        "management": true,
        "messagingrest": true,
        "messaging": true
    },
    "rules": {
        "topicRules": {},
        "queueRules": {}
    },
    "emname": "<your messaging client name>",
    "namespace": "<yourorgname>/<yourmessageclientname>/<uniqueID>",
}
```

Listing 5.2 SAP Cloud Platform Enterprise Messaging Service Instance Creation in JSON Format

5.   Skip the next step and in last step, provide an **Instance Name** as same as defined in emname in the JSON file defined in Listing 5.2, and click **Finish**. This creates a service instance of SAP Cloud Platform Enterprise Messaging.

> **Note**
>
> These steps can be automated using SAP Cloud Platform Extension Suite for SAP S/4HANA Cloud, which is explained in Chapter 9.

6.   Navigate inside your Cloud Foundry environment subaccount to **Subscriptions**.

7.   Click the **Enterprise Messaging** tile and choose **Subscribe**. Wait until the application is subscribed.

8.   Assign roles to users to access this service. Table 5.4 provides the default roles provided by this service.

Role	Description
ManageRole	This SAP Cloud Platform Enterprise Messaging administrator role provides users access to create, edit, delete, and view queues, rules, service descriptors, webhook subscriptions, message clients, and event channel groups.
ReadRole	Provides users access to view queues, rules, service descriptors, webhook subscriptions, message clients, event channel groups, and explore events.
TestRole	Provides users access to test sending and receiving messages through SAP Cloud Platform Enterprise Messaging.

Table 5.4 Roles in SAP Cloud Platform Enterprise Messaging

9.   If you have existing role collections created in SAP Cloud Platform, add the required roles to your existing role collections. Otherwise, you can create them from your Cloud Foundry environment subaccount by navigating to **Security • Role Collections**.

10. Add the role collection to your user. Navigate to **Security Trust Configuration**. Select **SAP ID Service** and enter your user ID. Click **Show Assignments** followed by **Assign Role Collection** to add a role collection to your user ID.

11. Navigate back to **Subscriptions**, choose the **Enterprise Messaging Service** tile, and click **Go to Applications**. Log in with your user ID and password, which will launch the home screen for SAP Cloud Platform Enterprise Messaging, as shown in Figure 5.14. Here you can see the message client that was created during the service instance creation process.

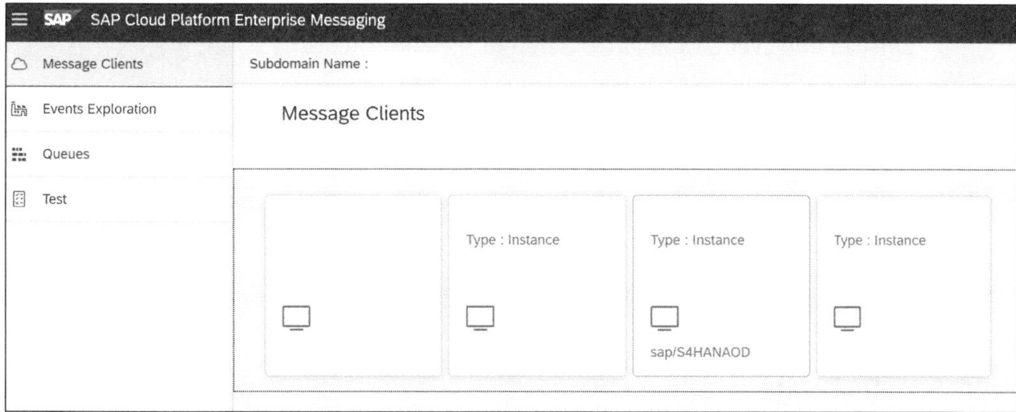

Figure 5.14 SAP Cloud Platform Enterprise Messaging Home

On the home screen, you can see following elements:

- **Message Clients**

  A message client is created when you create a service instance of SAP Cloud Platform Enterprise Messaging from your space as explained earlier. Message clients allow you to connect to SAP Cloud Platform Enterprise Messaging to send and receive message using unique credentials.

  You can have multiple message clients created and define queues or topics inside message clients via which you can send and receive messages. You can define connections between different message clients using the unique credentials defined in the service descriptor.

  Inside the message client, you'll find the following:

  - **Rules**

    View connection rules for queues or topics with permissions to subscribe or publish as defined during service instance creation.

  - **Queues**

    Manage queue and topics creation. You can manage queue subscription from here and handle delete queues.

  - **Webhooks**

    You can create a webhook to a specific queue from here. The messages received in a queue can be pushed to webhooks.

  - **Events**

    You can view the event catalogs for a specific message client.

  - **Service Descriptor**

    You can view the service descriptor file used to create the service instance here.

  - **Test**

    You can test publishing or consuming messages.

- **Events Exploration**
  You can find event catalogs for different message clients here.

- **Queues**
  You can view various queue details from this selection, including the number of messages, unacknowledged messages, and size of messages.

- **Test**
  You can use this view to publish or consume messages from queues for the selected message client.

5.5   SAP Data Intelligence

Complex enterprise landscapes can also include multiple data warehouses, data lakes, databases, SAP and non-SAP on-premise or cloud applications, and IoT systems. A very common requirement in an application landscape is to manage the data that flows through these systems. The complexity is multidimensional because of various facets of data, which could be structured, semistructured, or unstructured, like images, video or audio files, and geospatial data. Different systems also have different semantics for data and may require data in batches or in real time.

Organizations today understand that data is the new currency that accelerates business innovations. Analyzing data, understanding patterns, and driving intelligence through data to uncover new business innovations has become very critical for organizations to drive data-driven intelligence and innovations across enterprises. This has made the data scientist's job a demanding profession to analyze and drive intelligence via machine learning models. Many organizations deploy different disparate tools and processes to achieve this scenario.

Thus, organizations face a complex challenge to manage and govern these various data sources across various transactional and analytical applications and, more importantly, to manage, scale, and deliver various data-driven machine learning models to give purpose to data.

SAP Data Intelligence provides a single platform to address these challenges. You can use this service to manage various data sources and execute ETL/ELT processes, as well as provide functionality to scale and manage your machine learning models in one place in your organization, and at same time managing the data quality. SAP Data Intelligence brings the following advantages to your landscape:

- **Govern and manage metadata and architecture across disparate data sources**
  - You can use SAP Data Intelligence to discover your data landscape and its interconnections and connect to data sources regardless of type.
  - You can create reusable custom connectors with ease and extend to even more data sources with Open Connectors.

- You can further refine your data, apply data transformations, and curate so everyone can find what they need by creating reusable data pipelines for data orchestration.

- All this can be done using an intuitive GUI, replacing cumbersome coding and custom applications.

- **Govern, manage, and scale machine learning models**
  - Provides the capability to create, manage, and productize machine learning models for the various data orchestration pipelines in a single place.

  - Data scientists can use tools in SAP Data Intelligence—like JupyterLab using Python/R, libraries like PAL and APL, and open-source libraries like TensorFlow—to create and experiment with various data models. The production-ready data models can be promoted to production with a few clicks, thus providing a central place to manage all your machine learning models.

Before you begin, you must have SAP Data Intelligence as an entitlements for your Cloud Foundry environment subaccount. To start your initial setup, follow these steps:

1. Navigate to **Spaces** inside the Cloud Foundry environment subaccount.

2. Inside **Spaces**, navigate to **Services • Service Marketplace** and click the **SAP Data Intelligence** tile.

3. Navigate to **Instances** and choose **New Instance**.

4. In the New Instance Creation wizard, select a service plan and choose **Next**.

5. On the next screen, provide your **Username, Password, Number of Nodes, Warm Storage Size**, and (optionally) **VPC CIDR IP Address Range** and click **Next**.

> **Note**
>
> The default CIDR IP range value is 10.0.0.0/16. If your network can't use this value, you can specify a CIDR block value between a /22 netmask and a /16 netmask (both inclusive).

6. Provide an instance name and click **Finish**. It takes some time for the SAP Data Intelligence service instance to be created.

7. Once the instance is created, click **Actions • Open Dashboard** and login using the username and password you provided during instance creation. This will launch the SAP Data Intelligence launchpad, as shown in Figure 5.15.

> **Note**
>
> For more on the various supported connections provided by SAP Data Intelligence, visit *http://s-prs.co/v515722*.

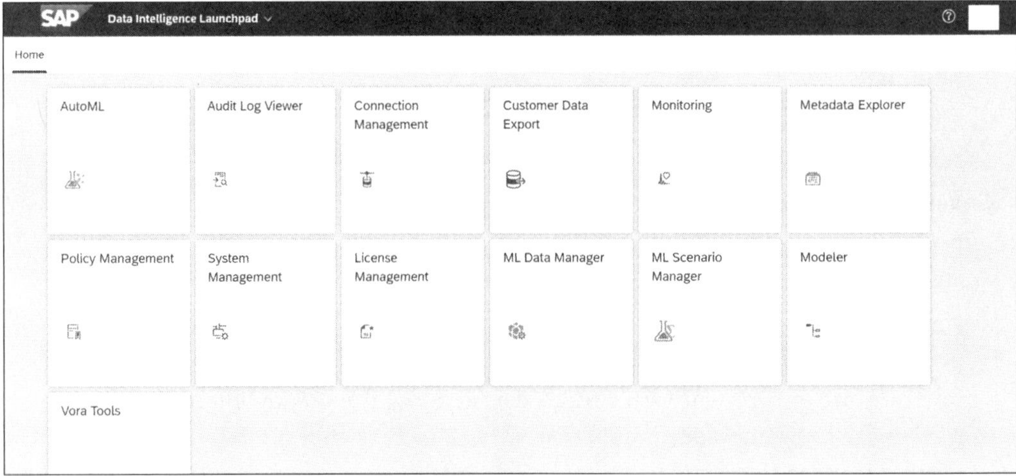

Figure 5.15 SAP Data Intelligence Launchpad

5.6   SAP Internet of Things

SAP Internet of Things enables your organization to develop, customize, and operate IoT applications. It helps to onboard and connect devices securely using different IoT protocols. The service also supports managing the device lifecycle from onboarding to decommissioning. SAP Internet of Things includes following components:

- **Cockpit**
  The cockpit is the user interface of the service. It provides access to administration, user management, device management, and data visualization.

- **Gateway cloud**
  The gateway cloud provides the capability to connect bidirectionally with devices and sensors online. You can receive measures sent by devices to SAP Cloud Platform and send commands to devices from the cloud. It supports MQTT and REST interfaces.

- **Edge platform**
  The edge platform provides capability to connect with devices on the edge. This service acts as a connector between your on-premise landscape and services in the cloud and helps to manage the transactional integrity of business applications. The edge platform supports different protocol adapters like HTTPS REST, MQTT, FILE, SNMP, Modbus, CoAP, OPC UA, and Sigfox to connect with devices, as well as providing an SDK that allows users to build new gateway adapters.

  You can download the IoT edge platform from SAP Software Center.

- **Messaging**
  This capability handles all incoming device data streams and device messages.

- **Message processing**
  This capability lets you define how data derived from devices needs to be processed.

You can store data in a SQL database or forward it to an IoT service, a Kafka cluster, or an HTTP endpoint.

- **Device management**
  This capability provides functionality for the management of the lifecycle of IoT devices.

  This includes defining a format for messages from devices and commands, configuration of protocols to be used, and providing credentials to access the service.

Before you begin, you must have configured the SAP Internet of Things service from **Entitlements** in your Cloud Foundry subaccount. To start your initial setup, follow these steps:

1. Navigate to **Spaces** and then select **Services • Service Marketplace** and choose the **Internet of Things** tile.
2. Navigate to **Instances • Create New Instance**.
3. In the **Create New Instance** wizard, provide a service plan and click **Next** until you reach the last screen to provide an instance name, then click **Finish**.
4. Click the service instance created and navigate to **Service Keys • Create Service Key**.
5. In the **Create Service Key** wizard, provide a name and click **Save**. A service key is created with an initial username and password. Note the initial username and password.
6. Navigate to **Referencing Apps • Open Dashboard**. This opens the dashboard for the IoT service. Log in with the initial username and password from previous step.
7. In the **New Administration Cockpit** screen, you can create additional users and IoT tenants and assign IoT tenants to users.
8. Once a tenant is assigned to your user, log in with your user to display the tenant assigned. Click the tenant to launch the IoT tenant home page, as shown in Figure 5.16. You can further configure the service for the tenant from here.

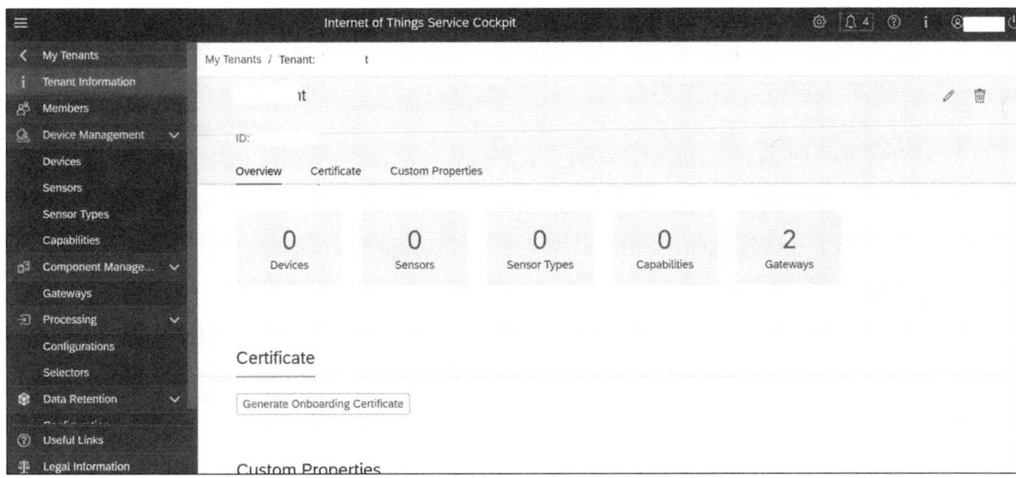

Figure 5.16 SAP Internet of Things Tenant Home

5.7   Defining Integration Pattern Design

To put the previously discussed integration patterns into practice, your organization needs an approach and methodology to follow, with decisions to be made step by step via checklists.

The following are step-by-step design consideration decisions you need to think about in your organization while defining and designing the right integration pattern to make your enterprise future ready:

1. **Expose and secure APIs and business events**
   This process includes setting up your landscape and business applications first to enable integration. The first step in cloud integration is to secure your backend system APIs to be accessed securely from integration tools. This includes two steps:

   - **Expose your backend application API securely for your on-premise applications**
     You first need to securely expose your on-premise backend system APIs to enable integration. SAP Cloud Platform services that could be used for this are as follows:
     - SAP Cloud Platform Connectivity's cloud connector
     - SAP Cloud Platform, serverless runtime, OData provisioning capability (optional; discussed in Chapter 9, Section 9.3.2)

   - **Expose your business events from SAP S/4HANA on-premise**
     For event-based integration, you need to enable your events to integrate with other systems and applications. SAP Cloud Platform services that could be used for this are as follows:
     - SAP Cloud Platform Connectivity's cloud connector
     - SAP Cloud Platform Enterprise Messaging
     - SAP Cloud Platform Extension Suite (for SAP S/4HANA Cloud, which helps automate this process; discussed in Chapter 9)

2. **Manage, govern, and secure APIs**
   It's good practice to define policies to manage, govern, and secure APIs across the organization consistently while opening them up further for integration and usage by external users. This is applicable for your cloud APIs, on-premise APIs, or custom APIs. SAP Cloud Platform services that could be used for this are as follows:
   - SAP Cloud Platform Integration Suite, API Management
   - SAP Cloud Platform Connectivity's cloud connector (for on-premise connectivity)

3. **Define integration styles and data exchange**
   You need to define integration styles for the integrations to and from your SAP S/4HANA system (on-premise or cloud). By now, you've defined how to expose your data from the backend system. During this process, you'll also define and develop a landscape-to-be diagram that includes the following information for each system you integrate with:

- Integration domain (process invocation, data movement, user integration, or thing integration)

- Transactional data or master data exchange

- Availability of APIs or business events for transactional data and master data

- Number of different systems in the landscape which SAP S/4HANA is integrated with and which it sends the same data to

- Availability of standard out-of-the-box integration content from SAP published on SAP API Business Hub

- Schedule or frequency of integration

- Application development/extensibility requirement for data exposed from the core system

Based on these criteria, Figure 5.17 shows the decision-making process guidelines that you can follow while designing your future-ready integration.

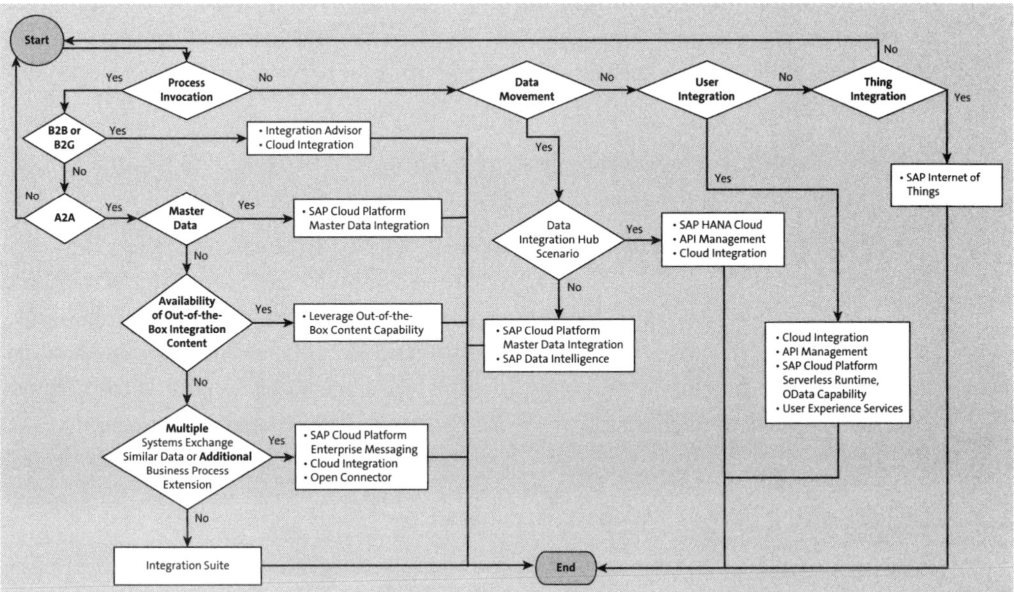

Figure 5.17 Integration Style Design Pattern Decision-Making

Note

It's possible for more than one integration style to be part of the decision-making process in one end-to-end integration scenario.

5.8   Integration with Legacy Applications

Many organizations have other message brokers (hyperscalers like Kafka and RabbitMQ) in their landscapes and may also have applications hosted in the Kubernetes hyperscaler service. For realizing an end-to-end business process in an organizational landscape, you need to connect your SAP applications and SAP Cloud Platform services with services or applications hosted in non-SAP hyperscalers. SAP Cloud Platform services provide the capability to integrate with these services and applications to realize end-to-end business processes. The following use cases are a few examples of such scenarios:

- Your organization has applications hosted in a Kubernetes cluster. You have a use case in which you need to create, manage and govern APIs to modernize this application.

- Your organization already has a message broker streaming event and messages from other applications. You need to be able to connect non-SAP data from these message brokers with messages and events from SAP applications.

- Your organization has a central audit logging storage system, which is used for access and governance in your landscape. You need to write logs from applications running in your Cloud Foundry-based SAP Cloud Platform environment to this central logging system.

Your organization will have multiple such scenarios and use cases, and in this section, we'll look at various capabilities in SAP Cloud Platform to realize them.

SAP Cloud Platform provides the ability to create service instances for applications running outside of SAP Cloud Platform and bind them to applications running on SAP Cloud Platform. This allows a single place to manage service instance creation and to share services and service instances between different environments and application bindings from a central place. In the following sections, we'll see various options to integrate third-party applications more natively with your SAP Cloud Platform applications.

5.8.1   Creating User-Provided Service Instances

User-provided service instances enable you to use services that are not available in SAP Cloud Platform Marketplace with your applications running in your Cloud Foundry-based SAP Cloud Platform environment. Before you begin, be aware that applications need to be accessible from the network. You must have the URL, port, and authentication parameters to access the application.

Follow these steps to activate this scenario:

1. Navigate inside **Spaces** and select **Service Instances**.
2. Select **Create Instance** and in the dropdown, select **Create User-Provided Service Instance**
3. In the **New Instance Creation** wizard, provide an **Instance Name**, **System Logs Drain URL**, **Route Service URL**, and credentials in the JSON format. Click on **Save**.

4. Now, you can bind this service instance to your applications like other normal service instances created for services from the services marketplace.

This is an easy way to stream logs from your applications running in your Cloud Foundry environment to your central audit logging system.

5.8.2   Service Management

Service management allows you to consume SAP Cloud Platform services in any connected runtime environment. It supports services that implement the Open Service Broker API and thus can be consumed natively from an OSBAPI-enabled platform like Cloud Foundry or Kubernetes. Service management is tightly integrated with SAP Cloud Platform services and enforces service access rules and quotas.

To consume SAP Cloud Platform services in your Kubernetes cluster, the cluster must be registered in service management within the context of a subaccount. Once registered and set up, you would be able to see and create a service instance of an SAP Cloud Platform service available for Kubernetes consumption.

Follow these steps:

1. You need the Service Management Control (SMCTL) command-line tool to manage platforms, service instances, and service bindings using service management. Download the latest release of SMCTL at *https://github.com/Peripli/service-manager-cli/releases/latest*.

2. Download the right version for your OS. Rename *smctl-<os name>* to *smctl.exe* and install the executable.

3. Add the `smctl` executable path to your system environment variable under `<PATH>`.

4. Inside your Cloud Foundry environment subaccount, navigate to **Subscriptions** and choose the **Service Management** tile.

5. Click **Service Management** and choose **Subscribe**.

6. Provide your user with required roles. Service management has the subaccount service administrator role collection.

7. Navigate to **Security • Trust Configuration** and choose the SAP ID service. Enter your user ID and click **Show Assignments**, followed by **Assign Role Collection**. Assign the subaccount service administrator role collection to your user.

8. You can log in to service management by executing the command in Listing 5.3.

```
smctl login -a https://service-manager.cfapps.<region_ dc>.hana.ondemand.com
--param subdomain=<subdomain>
//<region_dc> parameter example would be eu10 for AWS Europe Frankfurt based
subaccount
```

Listing 5.3 Service Management Login Command

Note

Further configuration within the context of your Kubernetes-based environment is explained at *http://s-prs.co/v515723*.

Once configured within your Kubernetes cluster, you can see the available SAP Cloud Platform services through the marketplace, as shown in Figure 5.18.

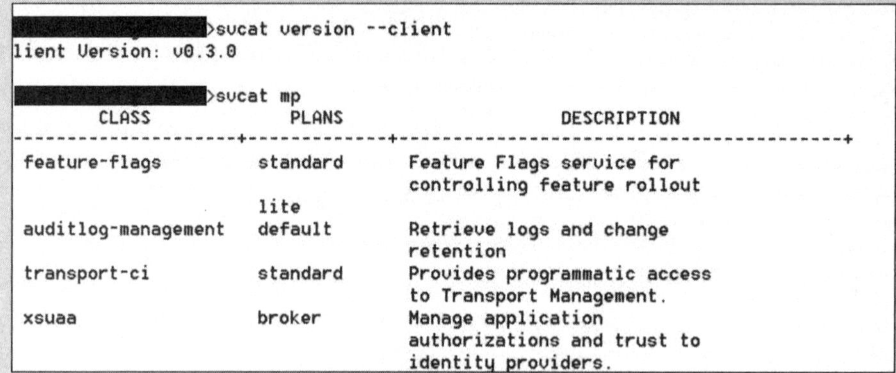

Figure 5.18 Sample Service Management Marketplace from Kubernetes Cluster

---

Note

At time of writing this book, not all SAP Cloud Platform services are available for integration through service management.

5.8.3   Other Integration Methods

There are other ways to integrate your existing services or applications with SAP Cloud Platform services. The following are a few examples, to explain on how SAP Cloud Platform can integrate with other capabilities to enable end-to-end business processes in your organization:

- We will cover integration with some hyperscaler services through open-source service brokers in Chapter 9, Section 9.4.2.

- Connectivity is available through the Open Connectors capability in SAP Cloud Platform Integration Suite for most third-party applications.

- Cloud Integration supports the Advanced Message Queuing Protocol sender and receiver adapter for integration with messaging systems in your organizational landscape. Thus, it enables you to use SAP-standard content and connectivity along with other messaging systems.

Integration scenarios are complex and need to operate in a heterogenous environment with different applications and systems. These are a few scenarios and use cases to help you understand the capabilities and design your architecture based on your organizational landscape. The openness of SAP Cloud Platform helps you operate in a heterogenous environment, while offering comprehensive capabilities to best take advantage of integrating with your business applications, bringing forth business value by helping you with quick time to value, reducing overall implementation and maintenance costs, and offering complete agility and flexibility.

5.9   Summary

In this chapter, we saw how various capabilities in SAP Cloud Platform Integration can be leveraged to meet your complex integration requirements in a highly heterogenous landscape. We saw how various capabilities can be used to suit various integration styles in your organization.

We covered the SAP Cloud Platform Connectivity service and its cloud connector, which can be used to expose your SAP S/4HANA on-premise service quickly to the cloud without complex network security requirements.

We also discussed how these exposed APIs from on-premise or cloud systems can be managed and governed using API Management, which can be used to integrate with other applications using the Cloud Integration capability, and how easily you can connect and manage third-party application connectivity using Open Connectors. We also discussed how the Integration Advisor capability can simplify highly complex B2B integration requirements using a crowd-learning capability.

We also covered how various event-driven architecture patterns can be established using SAP Cloud Platform Enterprise Messaging and how SAP Data Intelligence is used to orchestrate various data pipelines and make meaning out of data using machine learning, providing a single, consistent place to manage and govern your data and machine learning models.

We also discussed SAP Internet of Things, which enables your organization to connect business applications with IoT devices and sensors while managing transactional integrity.

Finally, we covered how openness in SAP Cloud Platform can be leveraged to integrate with legacy applications and services running in your landscape.

Thus, we covered in extensive detail how various integration capabilities in SAP Cloud Platform can help you meet, scale, and integrate your requirements in a very complex and multidimensional landscape.

Chapter 6

Business Process Automation and Optimization

Efficient business processes are key pillars of a successful enterprise. Technologies like robotic process automation, intelligent business process management, chat bots, and artificial intelligence services play a holistic role together in solving enterprise-wide business problems by automating key processes and transforming the enterprise.

Automation is one of key facets for building a future-ready enterprise. Automation paves a path for increasing the focus on high-value, knowledge-based work by humans and reduction of repetitive, non-value-generating tasks.

Enterprise process automation of critical business processes is key to improving innovation by opening space to focus on areas where human analytical capabilities can take over to work on innovation and improvement of those processes.

The intelligent business services are built on a microservices concept; they can talk to each other and serve to transform a large business process by solving one piece at a time via one service. This way, a complex business process can be automated and transformed seamlessly.

In the following sections, we'll walk through what it means to build automations into your business process and discuss two tools that can be used for automation: SAP Intelligent Robotic Process Automation (SAP Intelligent RPA) and SAP Intelligent Business Process Management (SAP Intelligent BPM). We'll end this chapter by providing some uses cases and a decision matrix for automation.

6.1 Build Efficient Automations in Business Processes

This section focuses on the importance of automation in the context of future-ready enterprises and how an enterprise can automate its business processes. Organizations have realized the value they can derive from automation to give a better experience to their employees and customers by helping their stakeholders focus on key process outcomes rather than mundane tasks that can be done by digital assistants.

One key aspect of automation is to ensure that processes and automation platforms are scalable and can cater to the increasing demands of an organization. The automation of processes should be looked at holistically based on how they can work on top of an orchestration platform like SAP Cloud Platform to provide a complete tool set, covering the repository, monitoring, scheduling, logging, and security.

Organizations should also work on an extension strategy to ensure they can utilize pre-built API bots for things like additional fields and can modify the predelivery code to handle the additional fields and validations. The strategy must focus on the criticality of business processes and the best way to automate or optimize them. The precursor to this stage is to have a good process understanding, which can be achieved by deploying process-mining tools and gaining visibility into these processes for deep insights.

In the following sections, we'll introduce the services and products SAP Cloud Platform provides for automating business processes. We'll also discuss scenarios delivered by SAP for automation and provide some additional introductory information on robotic process automation (RPA) and artificial intelligence (AI).

6.1.1   Services and Products on SAP Cloud Platform

Various services available as part of SAP Intelligent BPM play a role in process optimization, transformation, and automation, as shown in Figure 6.1.

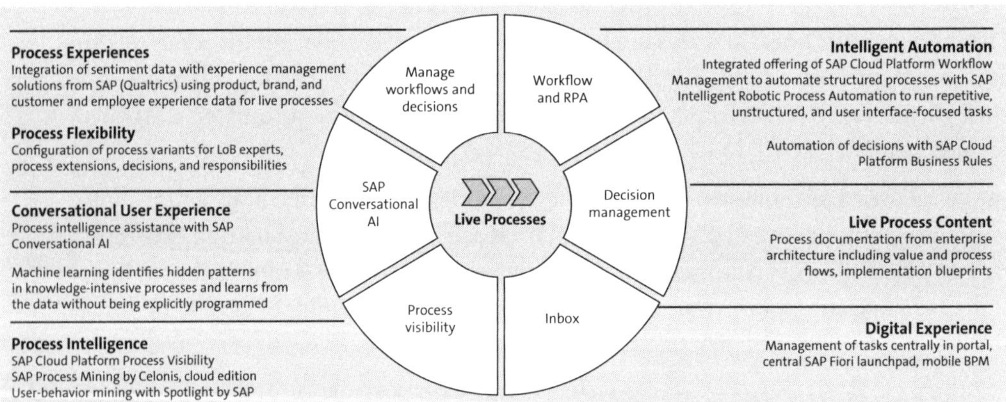

Figure 6.1 Services Available as Part of SAP Intelligent Business Process Management

The architecture approach is to ensure that enterprises effectively use these technologies in the following contexts, as shown in Figure 6.2:

- **Interact**
  SAP Conversational AI should be used for automation of all processes for which information can be provided via chatbots.

- **Design**

  The ability to create process variants from existing process is a great tool for hands-on business users to manage and administer their processes in a more flexible way, leading to better utilization and faster implementation of process changes.

- **Optimize**

  Machine learning should be used for optimizing executions and interactions based on model trained using existing process data.

- **Execute**

  Execution of tasks and processes should be done by an RPA engine.

- **Think**

  Machine learning services can be utilized for learning based on past outcomes and improve execution of a process.

- **Logging**

  Qualtrics is integrated with these processes to manage the logging of experiences during the execution of these processes.

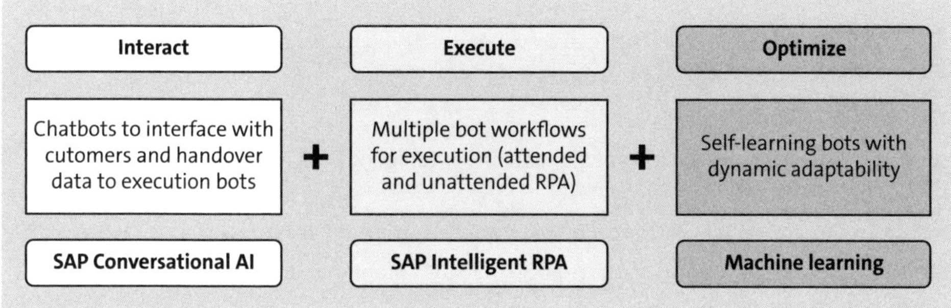

Figure 6.2 Scenario-Based Usage of Services

Enterprises can utilize SAP Cloud Platform Process Visibility for a two-fold advantage: helping business owners monitor and track processes and helping them get insights into potential areas of improvement by seeing the tasks that are laggards in the context of an enterprise process.

SAP Intelligent BPM and SAP Intelligent RPA bring all these features together under an integrated umbrella to manage a future-ready enterprise.

Many enterprise processes are complex and often involve unstructured data like images to automate a process. Artificial intelligence plays a key role in inferring the relevant information from these processes and automating processes completely to get the desired outcomes.

SAP Intelligent RPA leverages technology for automating business processes or creating a virtual workforce (robot) governed by business logic and structured inputs. This technology can capture and interpret applications for processing a transaction, manipulating data, triggering responses, and communicating with other digital systems.

SAP Intelligent RPA and SAP Intelligent BPM can connect to various systems, including SAP and non-SAP ones, and work in conjunction to digitize, automate, pull insights, and improve processes (see Figure 6.3).

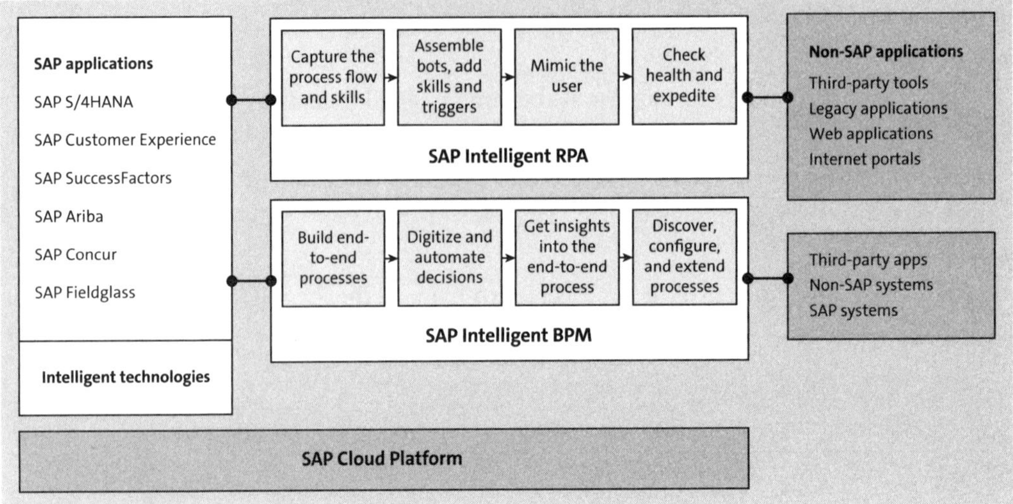

Figure 6.3 Steps and Systems Involved with SAP Intelligent BPM and SAP Intelligent RPA

SAP AI Business Services helps an organization to manage innovations and bring value to enterprise processes and scenarios. Some such business services are as follows:

- Service Ticket Intelligence
- Document Classification
- Document Information Extraction
- Data Attribute Recommendation

These services can be utilized across industry and LOB scenarios to ensure faster implementation. SAP AI Business Services provides strategic machine learning capabilities that automate and optimize processes and enrich the customer experience across the intelligent suite. They are provided as reusable services for SAP Cloud Platform customers. These services can be consumed from any application and can add value via machine learning (see Figure 6.4).

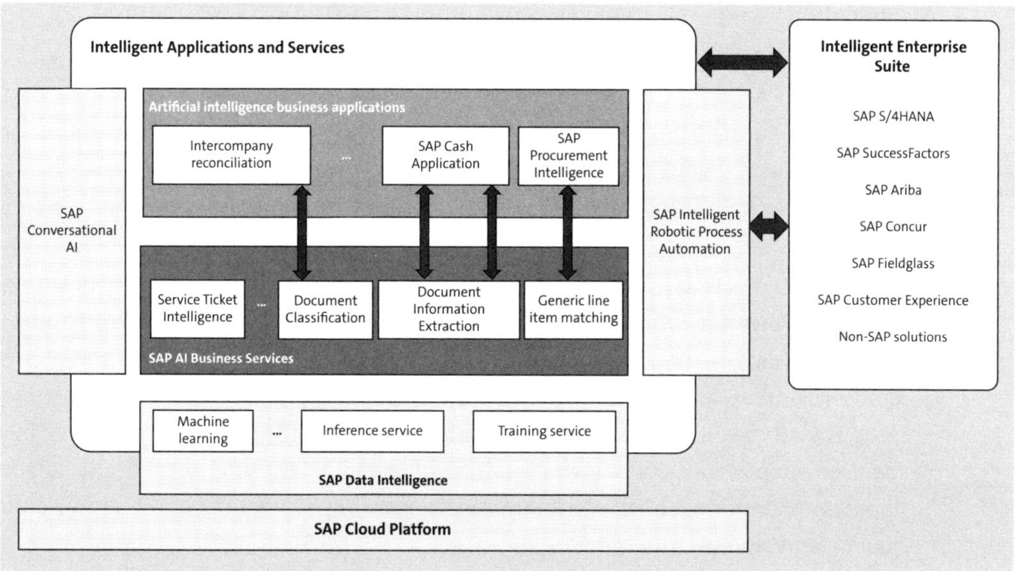

Figure 6.4 SAP AI Business Services

6.1.2   SAP-Delivered Scenarios

As a strategic initiative, SAP delivers SAP Intelligent RPA content for most common and critical scenarios for organizations, enabling faster automation and easily managing version changes for SAP standard processes. This helps to onboard innovations from SAP by adopting standard predelivered bots.

Let's take one business process as an example in this chapter and see how it can be automated and optimized using various tools. We'll consider a sales inquiry example. There are several ways a sales inquiry can be done for an organization: via a web form, a chat option, manually, via mail, and so on. Keeping track of all channels and ensuring that all prospects are catered to in the most effective way is a challenging task for an organization.

With new content provided by SAP, enterprises can easily scale-out automation to new processes. For already automated processes, with end user access based on SAP S/4HANA, the implementation team can focus on the flow of business processes and achieve automation without much need to reinvent security.

From deployment and component perspectives, SAP Intelligent RPA is a hybrid solution made up of the following components:

- On-premise Desktop Studio to design the automation processes
- Cloud Factory to orchestrate the automation processes
- On-premise Desktop Agent (can be multiple) to execute the automation processes

As you move through this chapter, keep in mind the two types of robotic process automation:

- **Attended RPA**

    Because end users have already set up access, it's easy to set up an attended bot for such users as a digital assistant option. This can improve innovation by opening bandwidth to help employees and customers focus on impactful areas.

    Attended RPA (previously known as robotic desktop automation) runs on the desktop. The robot acts as a software assistant for the human user and interacts with a desktop application while respecting applicable business logic. It reads an application window's contents, identifies the fields containing useful data, copies them into another window, and launches a transaction, for example. While performing these tasks, the robot will, if required, hand control back to the user for decisions requiring the user's experience and knowledge. The robot can carry out checks on the data it handles to provide additional guarantees regarding regulatory compliance, data quality, and results.

- **Unattended RPA**

    The other side of coin is the unattended bot. These bots run without any human interaction. The common scenario for such unattended bots, which are executed from a server, is to execute processes that do not involve any interactions and update the log for any further monitoring. These bots act as digital workers that are independent and can be triggered based on an event via an API, or as scheduled at certain times for execution of a task.

This separation (or opposition) between *attended RPA* and *unattended RPA* no longer applies when you take a closer look at company expectations, which ultimately are simple: improve the efficiency of the most commonly used business processes, with quick ROI, thereby accelerating the company's digital transformation with a measurable effect on the working conditions of its employees and improvement of the customer experience.

In the context of our sample process, automation of scenarios using SAP Conversational AI and responding to emails can be done by an unattended worker, although for complex inquiries a robot that assists a salesperson in developing a proposal or processing an order can be built to pull information from various sources.

RPA bots can make business processes more efficient and employees more productive by taking advantage of structured data stored in various databases and application silos. But what about the increasing number of business scenarios involving unstructured data such as images, text, and speech, which often comes from mobile devices?

For these types of specialized cognitive tasks, the answer is artificial intelligence. AI has advanced rapidly in recent years. Thanks to the huge amounts of data now available to train models, machine learning and deep learning algorithms have reached very high levels of confidence, beating human cognition in many cases.

In the context of our example sales inquiry scenario, process visibility can help lead creators to validate inquiries based on machine learning algorithms and categorize them into various priority buckets.

Providing RPA bots with AI capabilities can deliver speed and efficiency—and it doesn't stop there. Within a few years, AI could help to bring self-learning capabilities to RPA bots. By understanding what a human user is doing, bots could replicate some tasks and even adapt themselves to minor changes in their working environment, such as handling exceptions or updates in the applications they interact with.

In the context of our example process, via machine learning and natural language processing (NLP), SAP Conversational AI can help break down sales inquiries (in multiple languages) into their components and then send those results for validation to a sales representative.

6.2   Robotic Process Automation

SAP Intelligent RPA is an orchestration platform to provide a complete tool set covering the repository, monitoring, scheduling, logging, and security. You can extend the prebuilt API bots with additional fields and modify the predelivery code to handle additional fields. Automation helps in the following ways:

- Operate round the clock. Bots can be triggered based on customer needs or can be scheduled to do a task. This can help multinational corporations who operate around the globe to save any lead time caused by regular 9–5 work schedules.
- Perform with great precision in accordance with bot design time tasks.
- Highly scalable (can cater to peak loads/volume, such as for seasonal requirements).
- Provide detailed operation logs for audit and traceability purposes.
- Require no physical office space for execution.
- Minimize impacts on existing systems by using an existing UI.

Most of an enterprise's mission-critical processes need an RPA solution with intelligent capabilities to cater to a complex flow of steps involving unstructured data and document processing. That's why it's important to ensure your RPA platform is intelligent. This is key to ensuring that on your future-ready enterprise path, the platform will offer capabilities to ensure complex scenarios can be addressed.

SAP Intelligent RPA can be subscribed to in SAP Cloud Platform. As a customer, you need to have an entitlement to subscribe to the service. Once you subscribe, you can build and deploy various bots. Various components of SAP Intelligent RPA are as follows (see Figure 6.5):

- Desktop Studio
- Cloud Factory
- Desktop Agent

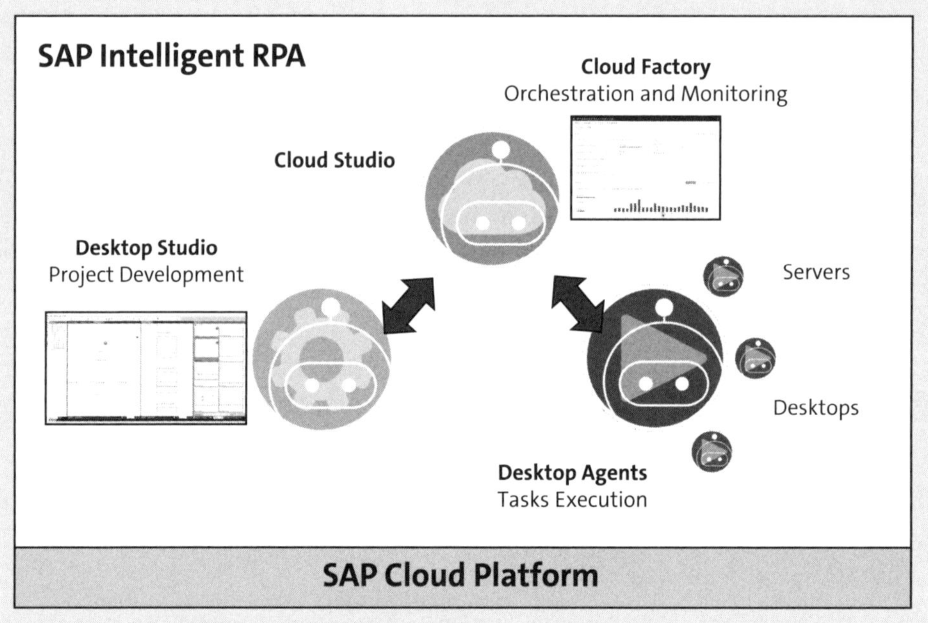

Figure 6.5 SAP Intelligent RPA Components

We'll look at the most important components in the following sections.

6.2.1   Cloud Factory

SAP Intelligent RPA is available in the SAP Cloud Platform cockpit based on service entitlement. The service can be found in the **Service Marketplace** section of the cockpit. After activation of the service, as shown in Figure 6.6, the Cloud Factory can be accessed from Cloud Foundry and packages can be deployed in the cloud.

The factory is meant for management, orchestration, and monitoring of deployments. A project-based view of bots is available from this tool. The main page of the Cloud Factory is shown in Figure 6.7.

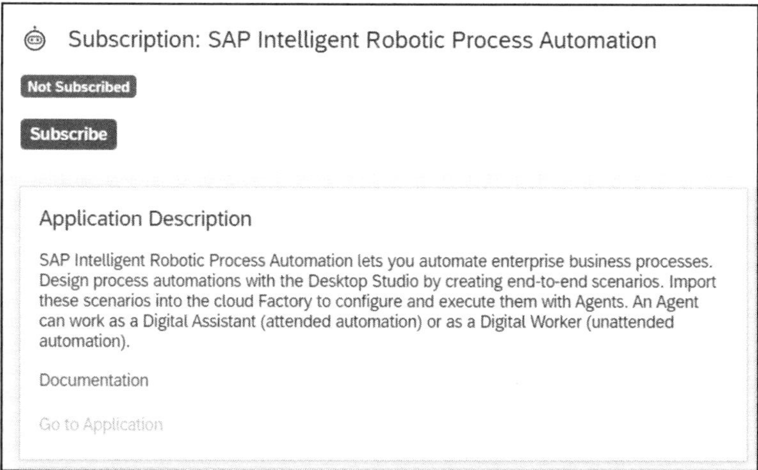

Figure 6.6 SAP Intelligent RPA Service in Cockpit

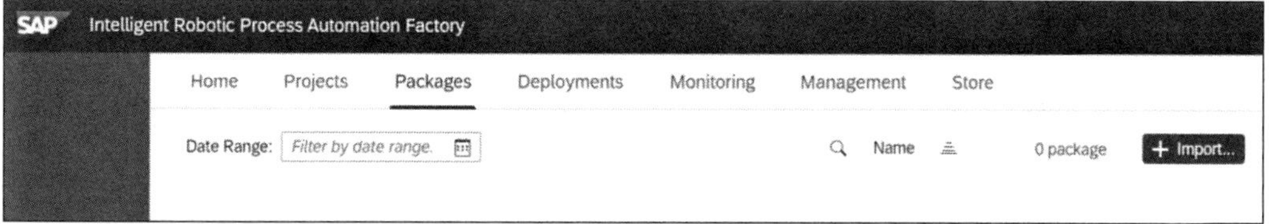

Figure 6.7 SAP Intelligent RPA Automation Factory

For building a new bot, a new project can be created and either a desktop package or files can be imported (see Figure 6.8). A new project can be created from the **Project** section of the Cloud Factory.

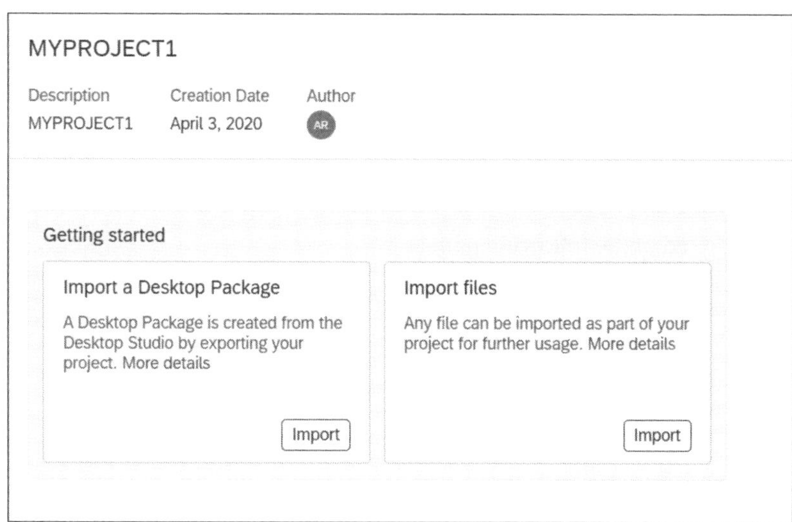

Figure 6.8 SAP Intelligent RPA Project

Once bots are deployed, their statuses can be found from the monitoring view of the Cloud Factory (see Figure 6.9).

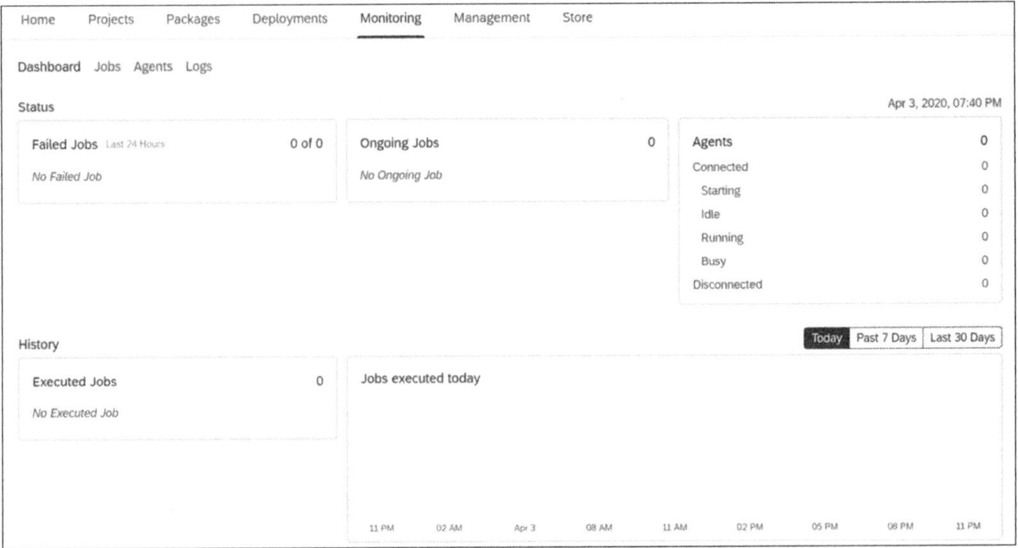

Figure 6.9 SAP Intelligent RPA Jobs Monitoring

6.2.2   Desktop Studio

The SAP Intelligent RPA Desktop Studio is the desktop-based client/tool for creating automations. It helps with the following tasks:

- Recording applications that your bot will interact with to perform automation
- Design of workflows and scenarios with steps the RPA bot needs to perform
- Testing and debugging your scenarios

You can download the SAP Intelligent RPA desktop components from *https://tools. hana.ondemand.com/#cloud*.

After installation, when the studio is launched, it opens the main window shown in Figure 6.10.

This window is comprised of the following panels:

❶ The menu bar

❷ The toolbars

❸ The perspective selector

❹ The current perspective panel

❺ The status bar

Once the development is completed, you can export it as a project.

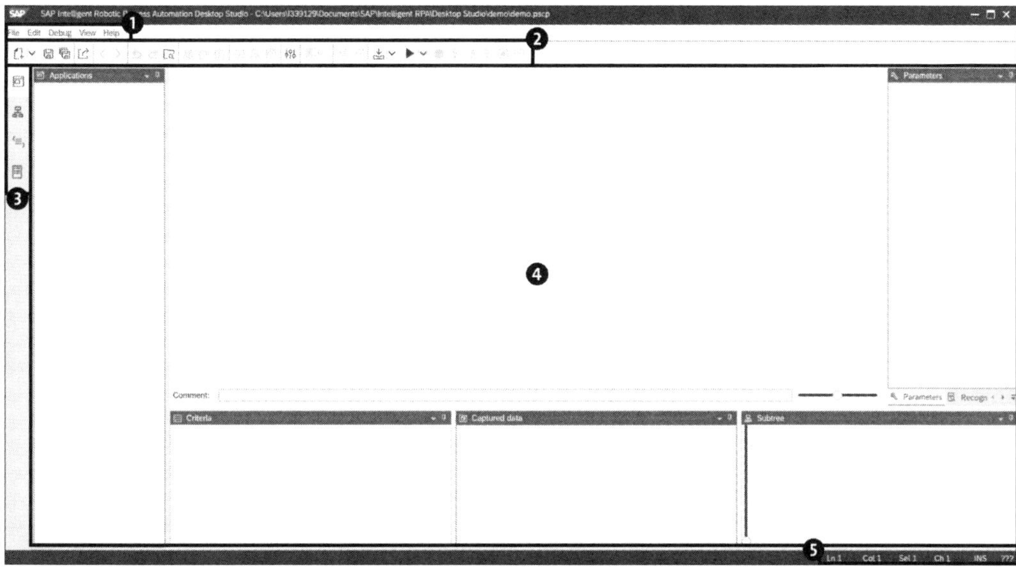

Figure 6.10 SAP Intelligent RPA Desktop Studio

6.2.3   Desktop Agent

The SAP Intelligent RPA Desktop Agent is installed on your workstation for attended bots to act as digital assistants or on a virtual machine for autonomous unattended bots as digital workers. It runs automation jobs that replicate human actions and perform mundane, repetitive tasks. You connect your agent to the tenant on which it should execute the automation job and set its trigger mode.

Another key aspect of a robust RPA platform is to ensure that it can cater to attended and unattended modes in executing business processes. Because RPA acts as the nuts and bolts of an execution engine for automated business processes, the categorization of processes and mode of execution plays a key role in successful outcomes. Figure 6.11 demonstrates various RPA modes.

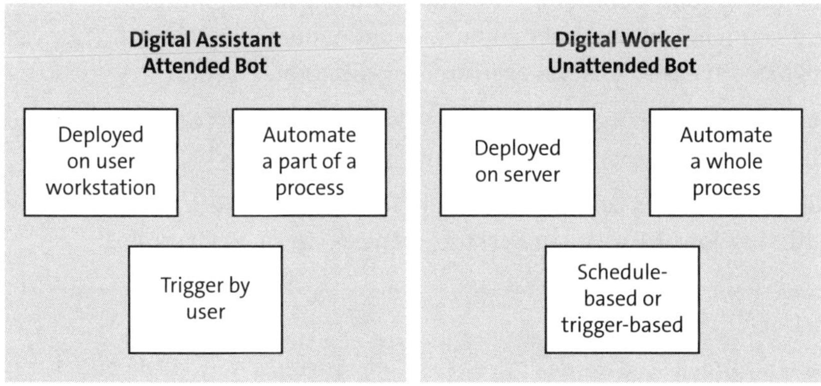

Figure 6.11 RPA Execution Modes

Some processes can be automated from end to end by using robots installed in server clusters. Known as unattended RPA processes, autonomous software robots work on their own at the heart of the information system or in the cloud (see Figure 6.11).

They can connect to databases to retrieve data and information, perform tasks, orchestrate end-to-end processes that produce new data, and create new entities in other applications using their own APIs.

These robots complete their tasks without human intervention. However, they remain under human control: processes are monitored to ensure they are executed properly, and a "robot supervisor" will identify and correct any anomalies or problems. Because they're installed on servers and are therefore inside the information system, unattended RPA bots require some infrastructure. And because they act directly on application data, they need to use APIs, which requires programming work. As a result, unattended RPA projects are likely to be more complex and take a little longer.

With cross capabilities of RPA and machine learning, enterprises can handle unstructured data scenarios and end-to-end automation.

To derive value from automation scenarios, it's critical to start with effective identification of potential cases, which can help in enterprise transformation by focusing on highly critical repetitive tasks and establishing an end-to-end operational process for the same, from security to monitoring to logging process outcomes.

Intelligent enterprises achieve greater cost efficiencies by relying on resource-intensive processes. They can gain a stronger market posture and reputation by focusing personnel on high-value processes rather than on those requisite but repetitive low-value processes that until now have resisted attempts at automation.

SAP Intelligent RPA provides the tools and technologies required to help your organization optimize both costs and personnel allocations. Intelligent bots created using the services can execute repetitive manual processes quickly, consistently, and without error. They can scale in step with the needs of your enterprise, which can help you respond proactively to emerging opportunities and evolving customer needs.

And these are not the only benefits you can gain from SAP Intelligent RPA. Intelligent bots can help you reinvent processes and drive continuous improvement. They can help you make smarter decisions and overcome organizational barriers.

In short, they can help your organization move ever further along the path of becoming an intelligent enterprise.

You can utilize predelivered content and enable SAP Intelligent RPA content for SAP S/4HANA with the Cloud Factory and Desktop Agent, as shown in Figure 6.12.

Reference Link

All details of the content and the deployment files are stored and can be referenced at *http://s-prs.co/v515746.*

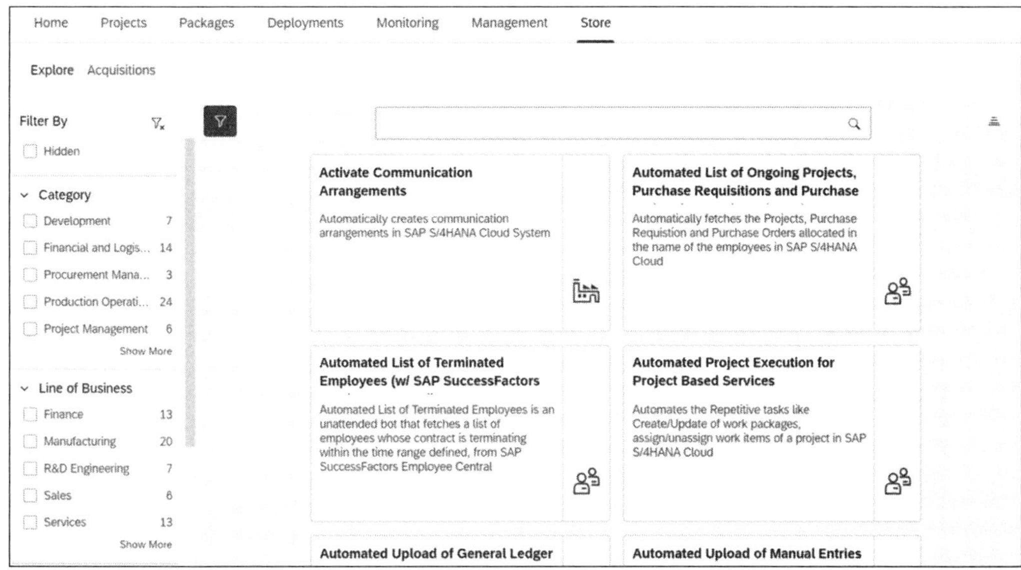

Figure 6.12 SAP Bot Store

6.3   Business Process Visibility and Management

SAP Intelligent BPM is a bundle of services that includes the following:

- SAP Cloud Platform Process Visibility
- SAP Cloud Platform Workflow Management
- SAP Cloud Platform Business Rules
- SAP Conversational AI
- SAP CoPilot
- SAP Intelligent RPA

In this section, we'll just focus on the first product offerings. SAP Cloud Platform Process Visibility enables an enterprise to automate structured business processes, manage decision logic, and gain end-to-end process visibility. SAP Cloud Platform Process Visibility will support end users to achieve process excellence, process transparency, and process transformation by providing one view of the process no matter where it runs (SAP or non-SAP cloud or on-premise applications). It provides actionable insights on running processes and identifying bottlenecks.

SAP Cloud Platform Workflow Management helps to you to build, run, and manage simple and complex workflows spanning various organizations and applications. With standard applications and forms and custom-built interfaces, users can participate in decision-making processes and data entry.

In the following sections, we'll go into more detail about SAP Cloud Platform Process Visibility, its integration with SAP Cloud Platform Workflow Management, and its use cases.

6.3.1   Activating SAP Cloud Platform Workflow Management and SAP Cloud Platform Process Visibility

You can activate SAP Cloud Platform Workflow Management by accessing your SAP Cloud Platform account's cockpit, as shown in Figure 6.13.

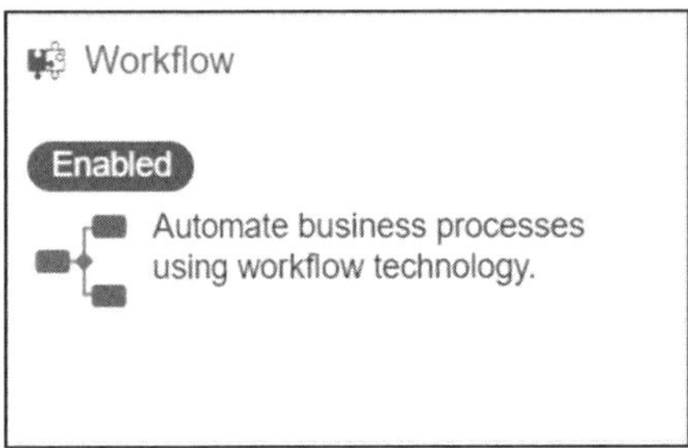

Figure 6.13 SAP Cloud Platform Workflow Management Service

You need to enable SAP Web IDE full stack and then assign the necessary role to access the service. This can be done in the SAP Cloud Platform cockpit at the subaccount or space level.

To use the Workflow Editor, you need to first activate the extension in the SAP Web IDE in the **Extensions** section, as shown in Figure 6.14.

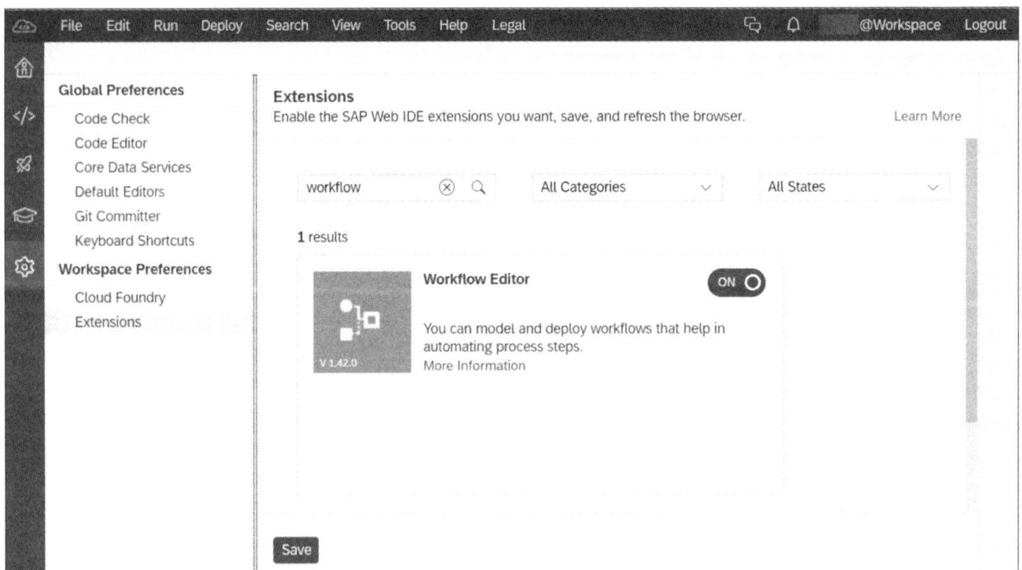

Figure 6.14 Workflow Setup

Developers can access SAP Web IDE and SAP Intelligent BPM templates to create a new workflow project, as shown in Figure 6.15.

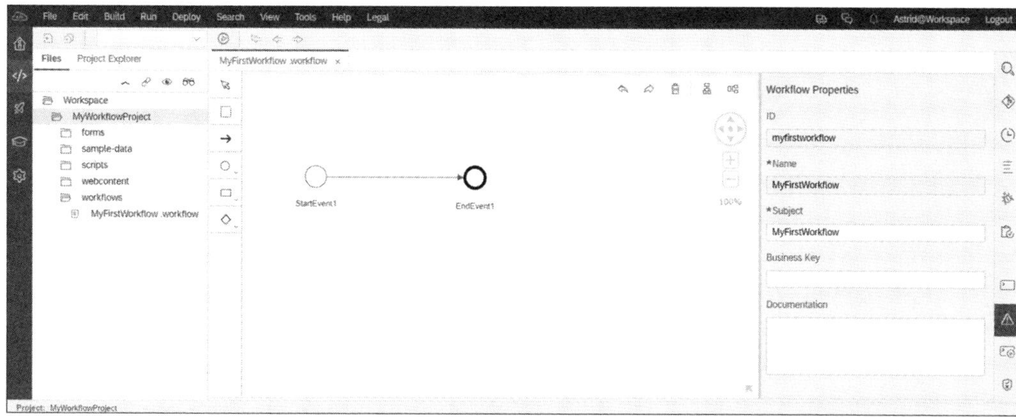

Figure 6.15 Workflow Editor Project

To activate SAP Cloud Platform Process Visibility in the SAP Cloud Platform cockpit on Cloud Foundry, you have to create and activate an instance of SAP Cloud Platform Process Visibility (see Figure 6.16).

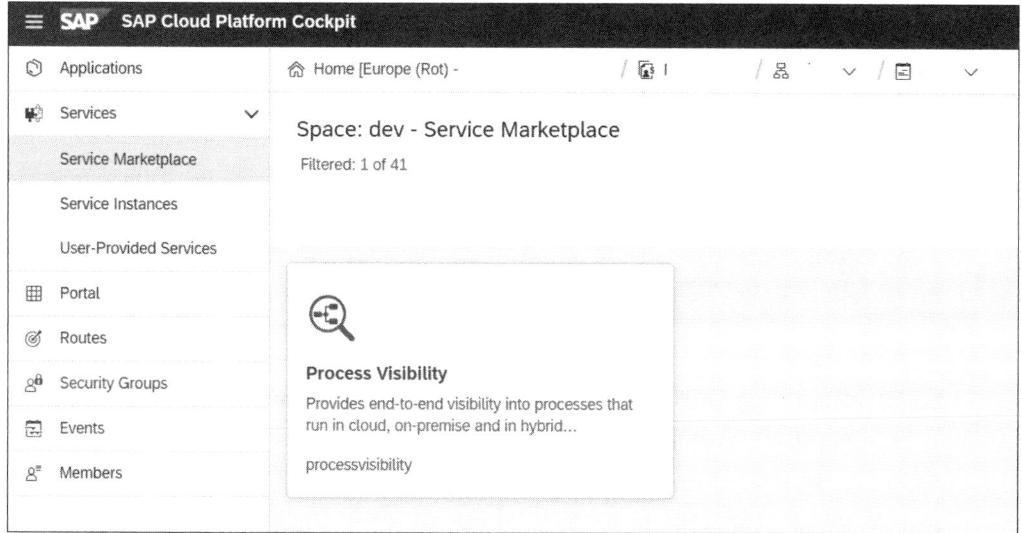

Figure 6.16 Process Visibility

A new instance for SAP Cloud Platform Process Visibility can be created in your Cloud Foundry space (see Figure 6.17).

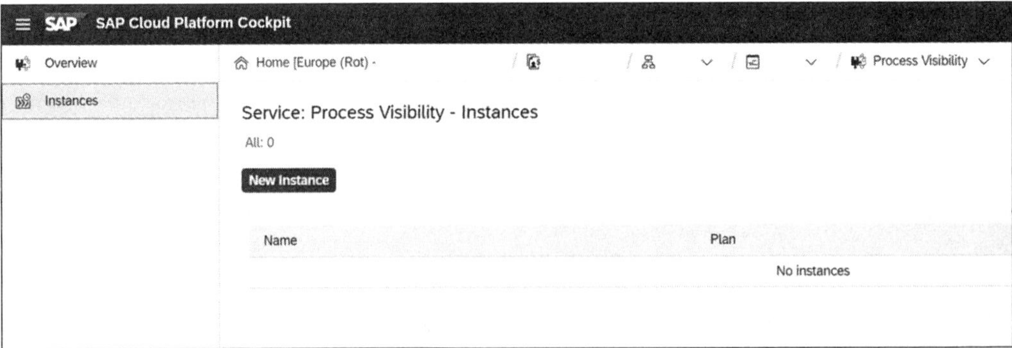

Figure 6.17 Activation of Instance in Cloud Foundry Space

The appropriate service plan needs to be selected to allocate the resources, as shown in Figure 6.18.

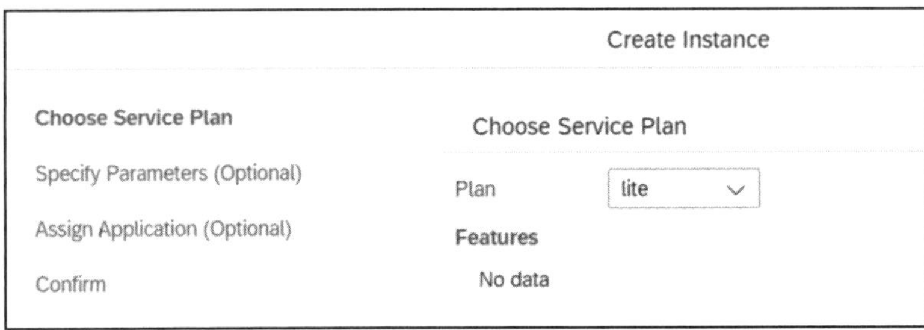

Figure 6.18 Service Plan Selection

Note

More details on services can be found at *https://help.sap.com/viewer/product/VISIBILITY_SERVICE/Cloud/en-US.*

6.3.2   Integrating SAP Cloud Platform Process Visibility with SAP Cloud Platform Workflow Management

SAP Cloud Platform Process Visibility works in close integration with SAP Cloud Platform Workflow Management and SAP Cloud Platform Business Rules, which help build, run, and manage simple and complex workflows spanning various organizations and applications. With standard applications and forms and custom-built interfaces, users can participate in decision-making processes and data entry.

SAP Cloud Platform as an enterprise business platform allows you to deploy and get visibility into all business processes without any development effort. Some important features available as part of SAP Cloud Platform Process Visibility are as follows:

- Streamlines the entire end-to-end processes and avoiding optimizing only parts of the process

- Enables identification and resolution of issues in real time to have an immediate and positive impact

- Clarifies reasons for process inefficiency and transforms processes based on facts and with confidence

- Creates and organizes process performance indicators to understand current and past process performance

- Provides out-of-the-box visibility into workflows from SAP Cloud Platform Workflow Management for immediate time to value

End users and admins of all business process management scenarios built on SAP Cloud Platform can see various stages of a process and look for improvements in processes based on various KPIs.

You can configure business scenarios to manage visibility, via the Configure Business Scenarios app, accessed via the tile shown in Figure 6.19.

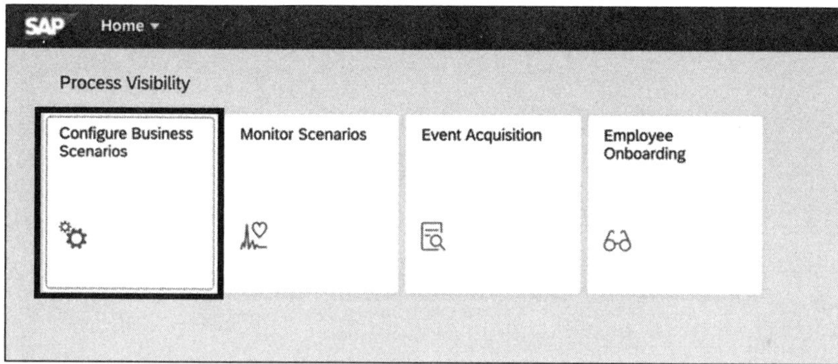

Figure 6.19 Configure Business Scenarios

In this app, an existing workflow definition can be added as shown in Figure 6.20 by clicking the **+** button ❶ and then selecting **Add SAP Cloud Platform Workflow** ❷. The workflow's various steps and milestones can be tracked in the visibility dashboard.

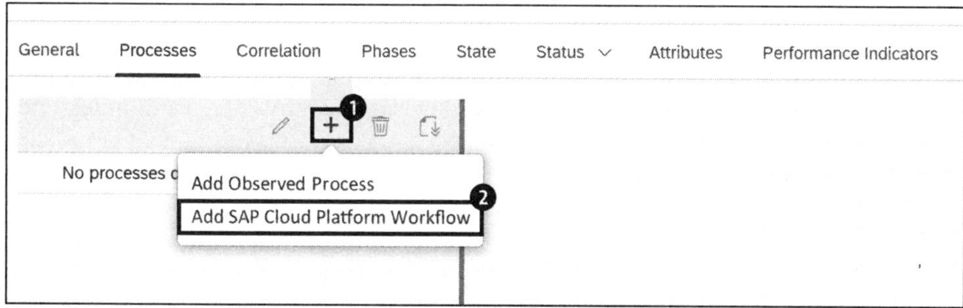

Figure 6.20 Add Workflow Process to Manage Visibility

6.3.3   SAP Cloud Platform Process Visibility Product Attributes

SAP Cloud Platform Process Visibility brings real-time insights into stuck processes on a regular or ad hoc basis to help process admins manage additional workload and ensure stuck processes can be completed on time.

In the long run, insights from these process logs can be used for process improvements and find potential areas of focus. Figure 6.21 shows an overview flow of combining process data from both cloud and on-premise solutions such as SAP S/4HANA, SAP Ariba, and non-SAP solutions with data from the cloud-based Qualtrics experience management solution. Before we walk through the steps involved in combining process and experience data within SAP Cloud Platform Process Visibility, let's take a look at an example use case that demonstrates why this approach makes good business sense.

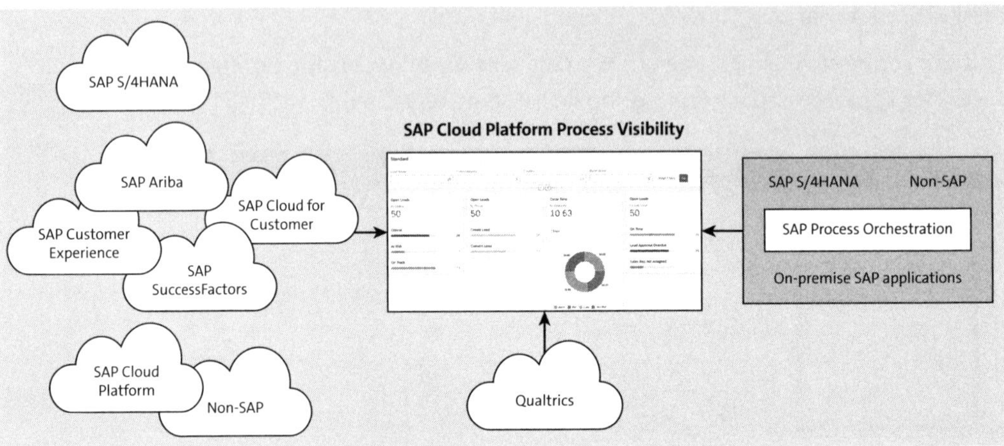

Figure 6.21 Combining Data from Various Sources

Using Qualtrics and SAP Cloud Platform Process Visibility, admins can sense the nerves of their business process by combining process and experience data relevant to their business processes (common intelligent processes like lead to cash, make to order, and source to pay) and enabling insights into process inefficiencies that were not previously identified.

In the context of our sales inquiry example scenario, process visibility would help lead creators to understand the statuses of various inquiries from various systems and whether all critical inquiries have met service-level agreements (SLAs) and milestones in the planned time or not. End-to-end visibility of process of lead time—that is, from receipt of a sales inquiry until making a connection with a customer and sharing a quote—enables an organization to cater to its customer's needs.

Visibility and monitoring of such critical business processes is key to staying competitive and offering better customer experiences. If any steps are managed by third-party workers in this flow, then SAP Cloud Platform Process Visibility's open APIs can offer visibility from external systems and give a good overview of process stages and whether all inquiries have been addressed.

An example dashboard for the lead-to-cash scenario is shown in Figure 6.22.

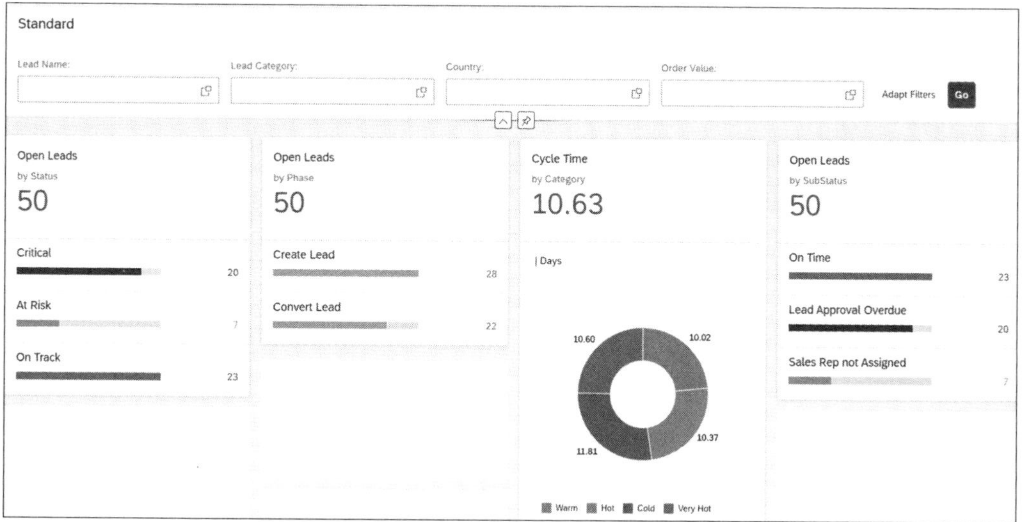

Figure 6.22 Dashboard for Lead to Cash Scenario

6.4   Automation Use Cases and Decision Matrix

This section focuses on various considerations for selecting business processes that make good cases for automation and business transformation. It also discusses scenarios delivered by SAP for automation and provides some specific examples of transformation. This section ends by discussing criteria for success in an automation project.

6.4.1   Selection

Selection and prioritization of processes to automate that can justify the value derived from automation is very subjective. Often, business process owners are most well-suited to understand the intricacies and advice on which processes are critical. Some key attributes to validate the scenarios are as follows:

- Repeatability of process
- Chances of error
- Feasibility of native integration
- Importance of scalability
- Time zone-based needs
- Frequency with which the process changes

- Reason for human errors
- Systems involved

Some key attributes and how they influence the selection of processes are mentioned ahead. A common topic during such discussions is the perspective that automation is focused on replacing humans. It's critical to understand that the intent of these tools is to remove repetitive and mundane tasks from the scope of employees so that they can focus on value-added work, improving the experiences of their customers and stakeholders.

Shortlisting processes that are good candidates for automation can be attributed to the following factors:

- Manual and repetitive
- High volume
- Multiple systems
- Workaround for native integration

6.4.2   SAP-Delivered Scenarios

As of September 2020, 125+ scenarios (scope items) are available to deploy and configure directly from SAP product teams. SAP provides SAP Intelligent RPA content for the following functional areas:

- **Finance**
  Scenarios like profit center maintenance, manage payment advice, mass asset operations, and automation of dispute management can be done using predelivered content.

- **Sourcing and procurement**
  Automation of processes like purchase requisition creation, purchase order confirmation, and uploading supplier invoices based on email attachments can be done using predelivered content.

- **Sales**
  Automation of creation of sales inquiries, sales orders based on Excel, delivery schedules, and returns can be done using predelivered content.

- **Supply chain**
  Supply chain operation processes like uploading inventory count list, posting goods movements, and post goods issue (PGI) for outbound deliveries can be automated.

- **Manufacturing**
  Purchase order operation confirmation, maintaining planned independent requirements, purchase order confirmation and completion, picking, and proof of delivery are some operations that can be automated in the manufacturing domain.

- **Professional services**

  In professional services, the following automation scenarios are available from the library: mass reassignment of profit centers, uploading timesheet entries, automation of lists of terminated employees, and more.

- **Data management**

  Automation of business partner master data checks, the process of product master data creation with references, creation of bill of materials (BOM) master data references, and more are available in the data management scope.

- **Cross-application**

  In context of cross-application scenarios, the scenario currently supported is the automation of the activation of a communication arrangement for SAP APIs in SAP S/4HANA.

There are many offerings from SAP that can run standalone, and SAP Intelligent RPA can integrate with various other intelligent offerings, including the following:

- SAP Conversational AI

- SAP Intelligent BPM

- SAP Process Mining by Celonis

- SAP Cloud Platform

- Spotlight by SAP

- Ruum by SAP

6.4.3 Transformation Cases

There are some easy ways for an enterprise to automate their processes and impact common scenarios to enable a future-ready enterprise. The following are some initiatives that can be implemented quickly:

- **Financial transformation**

  There are various challenges in financial receivables processing, like the following:

  - Highly manual accounts receivable processes

  - Incomplete data when automation uses static business rules, which exclude information from unstructured documents

  SAP Intelligent RPA uses bots to extract attachments from email. Machine learning can handle unstructured data from email or PDF and automate dispute creation, claim extraction, and dispute matching. Figure 6.23 demonstrates the automated accounts receivable process with deductions.

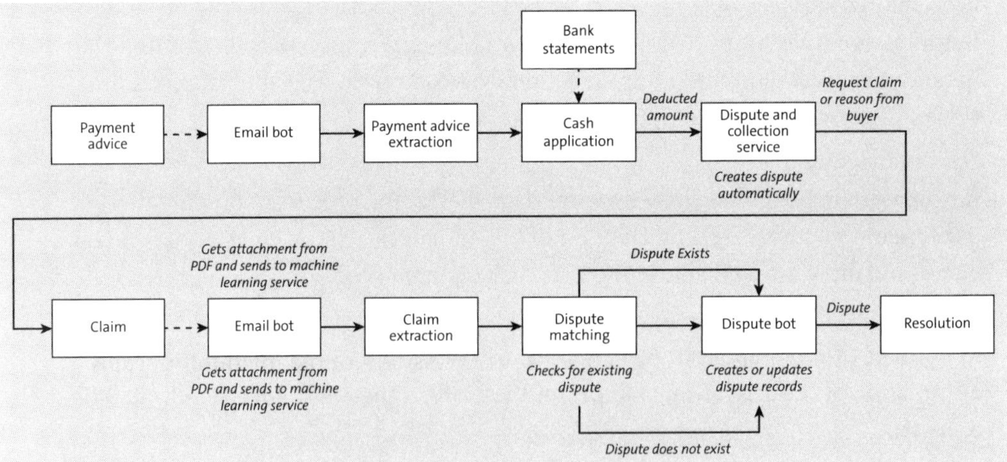

Figure 6.23 Automated Accounts Receivable Process with Deductions

- **Shared service automation**

  Shared service centers are established by many enterprises to cater to centralized activities like HR processes, offering a consistent experience to employees globally.

  This is a good case in which numerous administrative activities can be automated and administrators can focus on improving current processes and on outcomes rather than on performing mundane tasks.

- **Customer service and onboarding automation**

  Despite the rise of self-registration features, online banks and credit card companies require significant effort to onboard new customers. Typically, banks use external services for prechecks to validate credit risks and comply with regulatory requirements, and any resulting data is stored in separate systems.

  To solve such issues, SAP Intelligent RPA automates the collection of information from external systems and maintains backend systems during the provisioning process (see Figure 6.24).

  Deployment of bots in an attended scenario means they can be combined with manual tasks, creating a smooth transition between bots and employees. Figure 6.24 demonstrates the end-to-end automated customer service process.

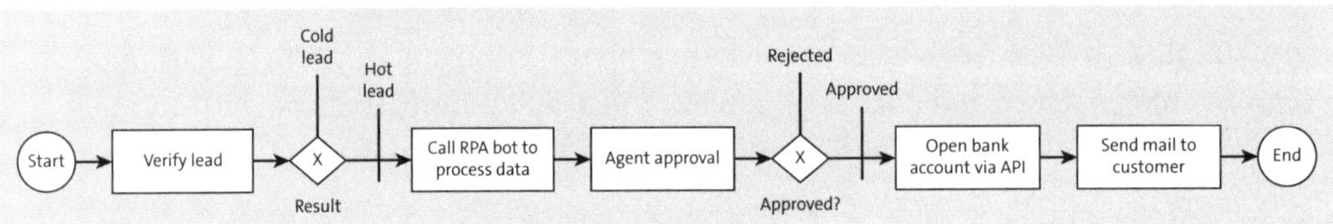

Figure 6.24 End-to-End Automated Customer Service Process

- **Operations automation**

  Service operation ensures timely and resourceful delivery of services without unanticipated interruptions. These are very high-volume scenarios, and it's difficult to plan for resources for unanticipated scenarios. Automation can help to solve this issue by deploying a scalable RPA solution. It also can help address key SLAs and service KPIs by improving service standards.

- **Data maintenance management**

  Many data maintenance activities can be automated—for example:

  - Business partner master data checks
  - Inconsistent new general ledger open item exclusions for the migration cockpit
  - Product master data creation with references
  - Creation of BOM master data references

6.4.4 Decisions and Criteria for Success

The first wave of adoption of robotic process automation processes was driven by the need to reduce the costs and errors associated with the involvement of humans in mundane, repetitive tasks in service centers or back-office functions. Robotic process automation enables the execution of a large number of tasks in a highly predictable way, and in the case of an increase in the number of tasks, enterprises have the option of deploying additional robots. From a process perspective, robotic process automation is suitable for structured and semistructured processes involving the following:

- High-volume transactions
- Workflow enablement
- Multiple systems or dual data entry
- Searching, collation, or information updates
- Data matching or comparison
- Simple or less-complex decisions that can be handled via a rules engine or algorithm (no judgement calls necessary)
- Repeatable and rules-based processes

Other factors impacting process suitability for robotic process automation include the following:

- Highly regulated activities
- Compliance and audit-related activities
- Data sensitivity
- Fluctuating volumes

Some key aspects to be successful in this initiative for a future-ready enterprise are to ensure that the following points are true:

- Understand the business impact of process before planning the automation steps.
- Decide which are critical processes based on impact, repeatability, scalability, and number of users impacted and then start with the right scenarios.
- If a process can be optimized before it's automated by simplification, then this initiative should be taken up by the business owners.
- Use machine learning cautiously based on performance of algorithms.
- Have a long-term enterprise automation strategy in place and review the same based on the changing needs of an organization.

Combining automation and artificial intelligence can deliver a step change in productivity and efficiency. This automation offers great potential to improve your customer experience and help your employees be more productive with higher-value tasks.

6.5 Summary

It's critical to have an enterprise-wide automation strategy based on current business processes and an automation roadmap. Identification of correct processes is key for initial rollouts, setting a path for establishing a platform that can be centrally utilized for all automation scenarios. How this platform manages the logging, execution, and security of the processes being automated is key to successful outcomes for an enterprise. It's also critical to have intelligent capabilities available as part of this platform to ensure that unstructured scenarios can be catered to and complex automation use cases can also be handled.

Chapter 7

User Experience and Mobile Consumption

This chapter further builds on the UX principles mentioned in Chapter 3. The best end user experience ensures business efficiency for your business and adoption of your enterprise system by end users. Thus, it's essential to consider the best UX strategy: one that's the least disruptive to end users, is intuitive, provides a consumer-grade experience, and can be accessed from anywhere and on any device for a modern and agile landscape. In this chapter, we'll cover various SAP Cloud Platform services that help create the best end user experience.

In this chapter, we'll look at an important aspect in your future-ready transformation journey: defining the experience for your end users using and interacting with your system. End users could be employees in your organization, your vendors or suppliers, or storefront users.

The user experience you provide determines how engaged your end users are and defines their level of satisfaction with your systems, which could be vital to your project's success. The challenges in this area are very demanding, and transformation may be ongoing because the user experience requirements and domain remains disruptive. For example, as mentioned in Chapter 1, the global COVID-19 pandemic in 2020 forced many organizations to extend remote-working possibilities and still help their employees remain productive in the best possible way. This demands that data and applications are available for employees to support their daily activities in a seamless and mobile-friendly way, which in turn demands that IT teams provide such features quickly and agilely while still meeting enterprise security requirements and needs.

In this chapter, we'll discuss how various SAP Cloud Platform services can help you and your organization create and define the best user experience for your end users in an agile and nimble way and still be future ready to adapt to changes quickly. Specifically, we'll cover how you can provide a consistent look and feel across your enterprise using the SAP Cloud Platform Launchpad, implementing adaptive content, and use SAP Cloud Platform mobile services to design and create a mobile user experience for end users in a quick, agile, and flexible way.

7.1   SAP Cloud Platform Launchpad

Let's consider the following use cases and scenarios to plan for in your organization:

- Your organization has multiple applications in its landscape, which includes both cloud and on-premise systems.

- Your organization is moving to SAP S/4HANA. You decide to move over by each application area. In other words depending on your project plan and schedule, there could be times when you have both SAP S/4HANA and SAP Business Suite running together.

- Your organization could have multiple installations of SAP S/4HANA and other on-premise systems segregated based on usage in different geographies. To further add to the challenge, different SAP S/4HANA systems might have different upgrade schedules.

- Your organization operates a core SAP S/4HANA system. Your organization has a subsidiary that operates another installation of SAP S/4HANA Cloud. There is data exchange and user interaction between these systems.

All these use cases and scenarios pose challenges for how you can provide a harmonized, seamless, and easy user experience to your end users, spanning different systems in your landscape and abstracting the underlying system transformation complexities from them.

Adding to these landscape and system challenges, frequently changing business requirements and needs in the world mean that IT and business must find ways to make these systems accessible from anywhere over the internet, while at same time ensuring their proper security, including authentication and authorization. All this must happen while making sure that these changes take place quickly and can adapt to the needs of business end users.

The *SAP Fiori Deployment Options and System Landscape Recommendations* guide (*http://s-prs.co/v515749*) outlines the solution to this challenge, using SAP Cloud Platform Launchpad as explained in Figure 7.1.

These requirements and needs to adapt exist not only for employees, but also for suppliers, vendors, and end customers who interact with your organization.

The SAP Cloud Platform Launchpad in the Cloud Foundry environment acts as a central entry point for your different SAP S/4HANA systems and custom applications. At the same time, SAP Cloud Platform services provide features to expose your on-premise systems and landscape in a secure and fast way without having to go through the process of changing ports and firewalls in your network, thus helping to adapt to changing business requirements at a quick pace. At the time of writing this book, the roadmap for this service is evolving fast, and more system and service integrations as out-of-the-box functionalities are planned.

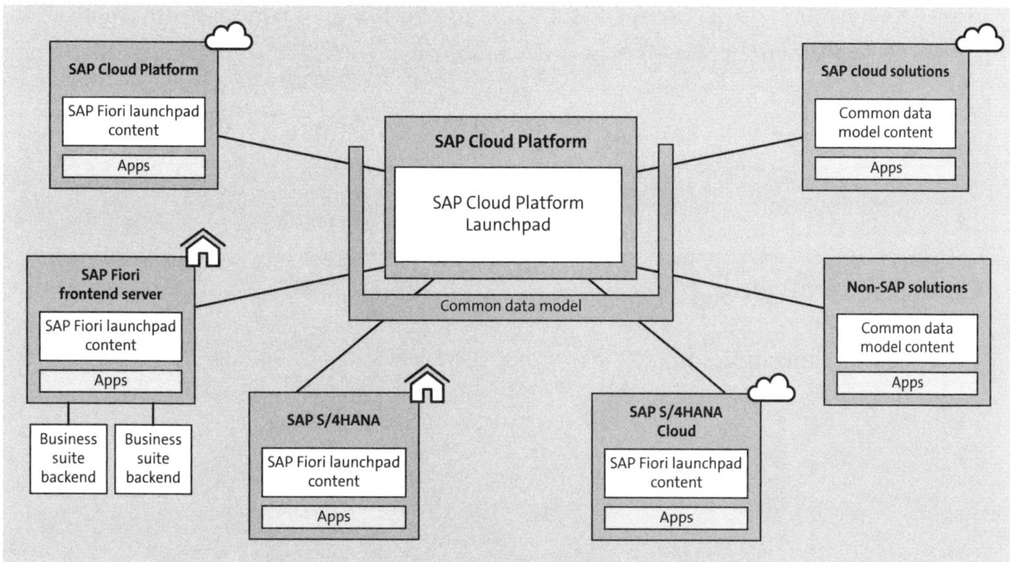

Figure 7.1 SAP Cloud Platform Launchpad Future Direction

Before you begin to set up the SAP Cloud Platform Launchpad, you must have a Cloud Foundry environment subaccount. You must also have set up entitlement for the SAP Cloud Platform Launchpad for this subaccount from the **Entitlements** area.

To begin your setup, follow these steps:

1. Navigate inside your subaccount to **Subscriptions** and choose the **Portal** service tile.

2. Click **Subscribe**.

3. Navigate to **Security • Role Collections**. The SAP Cloud Platform Launchpad has two roles, listed in Table 7.1.

Role Collection	Description
Portal_Admin	Administrator role that provides authorization to perform all administrator roles for the portal
Portal_External_User	Authorizations for external users for portals to enable access to vendors, suppliers, and so on

Table 7.1 SAP Cloud Platform Launchpad Role Collections

4. Now assign these role collections to users. For this, navigate to **Security • Trust Configuration**.

5. Select the identity provider, enter a user name, and choose **Show Assignments** (for the SAP ID service, the user name is your email address).

6. Click **Assign Role Collection** and assign the role collection to a user.

7. Navigate back to **Subscriptions** and choose **Portal**, then choose **Go to Application**. This will open the **Site Manager** view as shown in Figure 7.2, from where you can create sites.

8. Within a site, you can create apps for SAPUI5, SAP GUI for HTML, and Web Dynpro for ABAP, and integrate with web content to provide a single user experience for your end users.

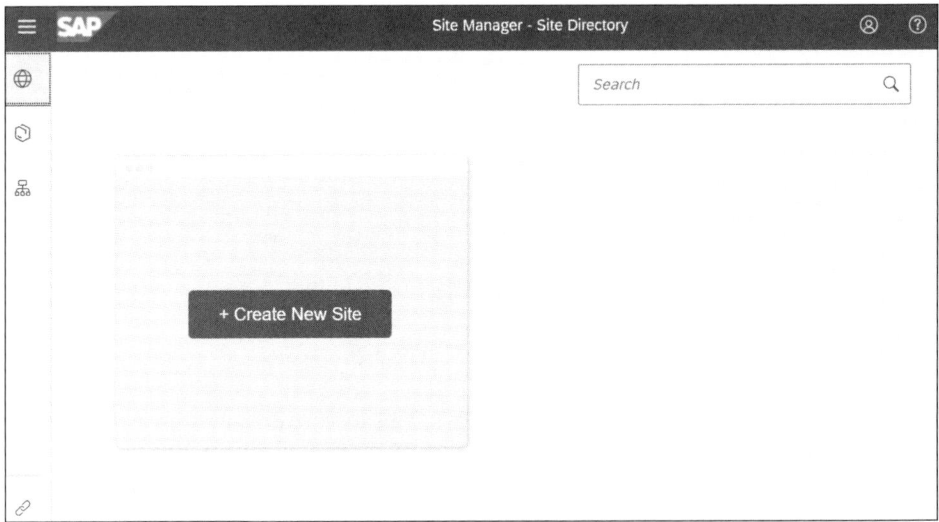

Figure 7.2 SAP Cloud Platform Launchpad Site Manager

7.2   SAP Work Zone

The need for a harmonized user experience is being redefined quite rapidly. Customers and employees of organizations are expecting a consumer-like experience with business systems as well. Users not only expect to have relevant business information and data from the systems presented in one place, but also expect the ability to further engage with that information through contextual recommendations and collaborate with other users through groups and chats around that information. This results in driving more end user productivity, efficiency, and satisfaction.

SAP Work Zone is a set of tools to provide such a user experience to your end users. SAP Work Zone aims to provide a simple, modern, and consistent user experience to your end users with data from enterprise systems.

From a technical standpoint, SAP Work Zone integrates a number of different business services available on SAP Cloud Platform, as follows:

- SAP Cloud Platform Portal
- SAP Cloud Platform Mobile Services
- SAP Cloud Platform Workflow Management

- SAP One Inbox
- SAP Conversational AI

It provides a harmonized user experience and design, all while making sure that your data is secure and follows your enterprise's security practices, as shown in Figure 7.3.

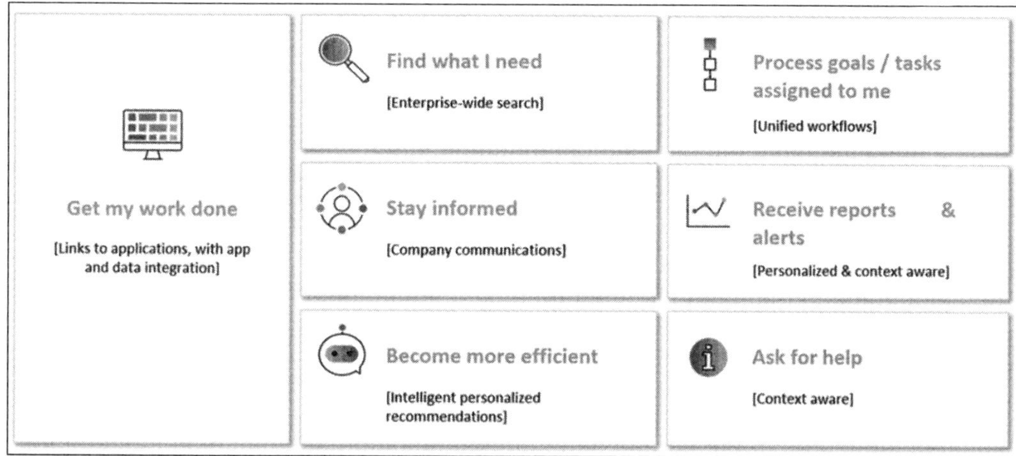

Figure 7.3 SAP Work Zone Overview

Broadly speaking, there are three different user personas who will use SAP Work Zone:

- **End users**

  End users are those to whom data is presented. These could be your employees, who are presented with modern user interface and collaboration capabilities, or your external users, like vendors, partners, and suppliers, who are provided with a modern and efficient way to integrate with your business system. End users can also be presented with mobile apps for a more engaging experience.

  End users have the ability to personalize and customize the experience to their unique needs and are presented data based on role-based authorizations.

- **Administrators**

  Administrators in an organization are responsible for design and manage SAP Work Zone for their organizations to make the user experience more seamless and harmonized for end users.

  An area administrator will have access to only those pages and workspaces for their particular area. A company administrator will have access to individual area administrator profiles. As an administrator, you can see information on user activities within SAP Work Zone.

- **Developers**

  Developers are responsible for creating content and deploying the content for use in SAP Work Zone's pages and workspaces. Developers use SAP Cloud Platform Business Application Studio to develop their contents.

At the time of writing this book, SAP Work Zone for HR is available and is a more employee-centered product. It brings different HR-related apps and services together in one place and provides the ability to collaborate on top of them, mobilize the data, and drive workflows, all in the same place.

> **UI5 Flexibility for Key Users**
>
> Another UX tool worth mentioning is UI5 flexibility for key users, which allows business users to adapt the UI to, for example, only show the fields most used by them. Using this tool, end users can change the UI without having to modify the underlying code. For more information, see *http://s-prs.co/v515755*.

7.3   SAP Cloud Platform Mobile Services

In this section, we'll dive into SAP Cloud Platform Mobile Services, which can help you create enterprise-grade business applications quickly. We'll discuss how various capabilities within this service can be used to quickly develop and deploy mobile applications for your end users, providing rich user experience. We'll first provide an overview of SAP Cloud Platform Mobile Services as a whole and its initial configuration. We'll then dive into some of the individual services: the mobile card kit (and the corresponding app, SAP Mobile Cards), the mobile development kit, SAP Cloud Platform SDK for iOS, and SAP Cloud Platform SDK for Android.

7.3.1   Product Overview and Configuration

SAP Cloud Platform Mobile Services provides various features and capabilities to empower your organization to quickly develop and deploy enterprise-grade secure mobile applications, thus enabling your business to offer the latest digital experience to customers and employees faster. SAP Cloud Platform Mobile Services supports development of different app types: native apps, hybrid apps, and web apps. Let's take a quick look at the various features that are available with SAP Cloud Platform Mobile Services:

- **Mobile cards**
  This is a capability within SAP Cloud Platform Mobile Services that enables developers to deploy a consumer-grade wallet or passbook-style app with little to no coding. Mobile cards provide a lot of easy-to-access and predefined templates that can be used to create simple apps.

  Mobile cards also enable end users to mobilize SAP Fiori app web content, which allows them to view business content in the form of cards with the most up-to-date business content and view the same while offline. End users can perform actions on this content, such as approve or reject. End users can subscribe and unsubscribe from business cards to personalize their list.

The SAP Mobile Cards app needs to be downloaded from the Apple App Store or Google Play Store. Using this feature, you can quickly enable your end users to perform tasks like approve or view a sales order or purchase order, view product information, view pay slips for your employees, approve leave requests, and many more.

- **Mobile development kit**

  This capability within SAP Cloud Platform Mobile Services provides a metadata-based application development platform. This is a rapid mobile app development tool that can help your organization quickly create native mobile apps by minimizing the coding needed.

  Applications can be developed rapidly with offline capabilities and robust business logic with native mobile support, and they provide an SAP Fiori user experience. Application development is done using web editor tools available in SAP Cloud Platform via the Mobile Development Kit editor plug-in for the SAP Web IDE full-stack service or SAP Cloud Platform Business Application Studio. The plug-in/template provides app templates, drag-and-drop features, additional wizards, and building blocks to create applications.

  This feature will enable your organization to simplify app lifecycle management with a lower learning curve and higher developer productivity, letting you create native apps across platforms (iOS and Android) cost effectively.

- **SAP Cloud Platform SDK for iOS**

  This capability in SAP Cloud Platform Mobile Services lets organizations develop feature-rich native iOS enterprise mobile applications using iOS development features.

  The SDK is based on Apple's Swift programming language and supplements the Swift SDK. Apps are developed using the Xcode IDE. This simplifies developing iOS-native enterprise mobile apps that can take full advantage of iPhone and iPad features with easy access to core SAP S/4HANA data and business processes with mobile services.

- **SAP Cloud Platform SDK for Android**

  This capability in SAP Cloud Platform Mobile Services provides tools and capabilities to create native Android applications. This SDK is built on top of Google's Android SDK and provides an easy way to develop native Android enterprise applications. The SDK comes with a plug-in for Android Studio (a native Android project development IDE) that simplifies the development of Android projects by generating source code templates.

- **Mobile transaction bridge**

  The mobile transaction bridge in SAP Cloud Platform Mobile Services provides the capability to expose parts of SAP transactions by recording user interaction flows running in the web GUI as OData services. This can enable you to quickly build, develop, and mobilize your SAP transactions in the web GUI, bringing immense value to your end users, both employees and customers.

  Thus, an exposed OData service can be consumed for building any applications, web or mobile, using the capabilities noted previously. This service is available only in the Cloud Foundry runtime.

Table 7.2 outlines additional features provided by SAP Cloud Platform Mobile Services.

Feature	Description
Mobile app catalog	This feature helps to manage apps during development, test apps for purposes like early beta release, and push apps to enterprise mobility management solutions to bring them to end users.
Mobile app update	This feature helps to manage and roll out updates for each application through its lifecycle.
Mobile client log update	This feature allows applications to upload application log files for analysis on the server.
Mobile client resources	This feature allows you to manage resources that can be accessed from mobile applications.
Mobile client usage and user feedback	This feature allows mobile applications to gather feedback from client usage to analyze further on the server.
Mobile cloud build	This feature allows you to easily build clients using the latest SAP SDKs in the cloud. This also allows you to create customized versions of standard SAP mobile applications like SAP Mobile Cards and implement your enterprise brand.
Mobile connectivity	This feature provides the functionality to securely access backend data and systems.
Mobile network trace	This feature enables applications to collect network trace information for debugging based on user name or content type.
Mobile offline access	This feature enables applications to run in offline mode by switching to a local data source on a device.
Mobile push notification	This feature enables you to notify your end users proactively of important events by registering devices to receive native push notifications. Mobile application developers don't need to implement specific code for the Apple Push Notification Service (APNS) or Firebase Cloud Messaging (FCM).
Mobile sample OData ESPM	This feature provides a sample OData service, which simplifies the application development.
Mobile settings exchange	This feature handles device registrations and provides an exchange of general settings between the mobile client and server.
Mobile transaction bridge OData	As explained earlier, this feature provides the capability to expose parts of SAP transactions by recording user interaction flows running in the web GUI as OData services.

Table 7.2 SAP Cloud Platform Mobile Services Features

Let's now look into how you can use and set up these various capabilities from SAP Cloud Platform Mobile Services to mobilize your business applications for your employees and customers, thus bridging the gap between IT and business.

To set up SAP Cloud Platform Mobile Services, you must meet the following prerequisites:

- You have a Cloud Foundry environment subaccount.
- You have assigned the SAP Cloud Platform Mobile Services plan to your subaccount from the **Entitlements** area.
- You have enough SAP Cloud Platform Application Runtime assigned to your space.
- You have an SAP Cloud Platform Destination service created for your SAP S/4HANA system already.

Once these prerequisites are met, you can begin your setup, as follows:

1. Inside your Cloud Foundry subaccount, navigate inside your space.
2. Navigate to **Services • Service Marketplace** and select the **Mobile Services** tile.
3. Click the **Support** link, enter **Organization** and **Space** information, and choose **Open**. This will open the SAP Cloud Platform Mobile Services cockpit page, as shown in Figure 7.4. During this process, SAP Cloud Platform Mobile Services performs a health check of your organization and will inform you of any issues it finds.

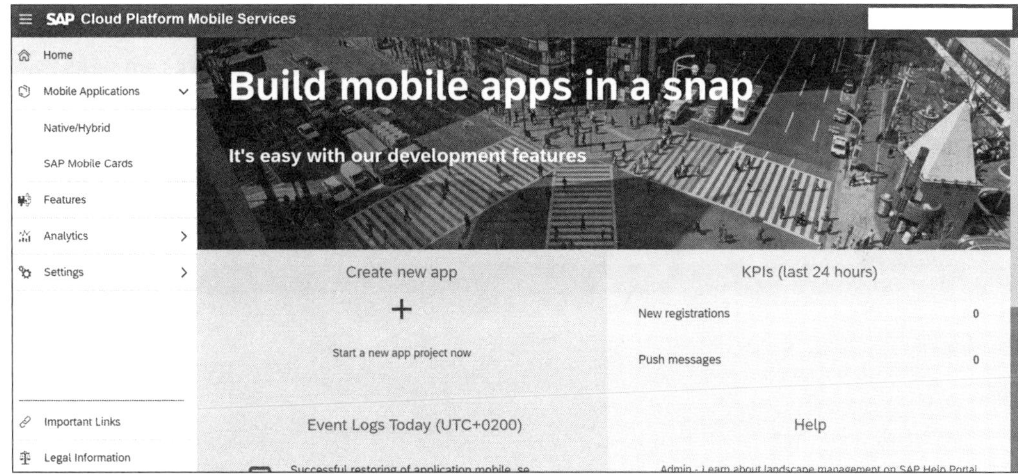

Figure 7.4 SAP Cloud Platform Mobile Services Home Screen

Note

Alternatively, you can also click the **Mobile Services** tile, navigate to **Instances**, and choose **New Instance**. This creates a new mobile application; every new mobile application created from within the mobile services cockpit will appear as instances in the future.

> Navigate through the steps in the pop-up window, enter an **Instance Name**, and choose **Finish**. A new service instance of the mobile service is created in the service instance. Click **Actions • Open Dashboard**. This opens the SAP Cloud Platform Mobile Services cockpit as shown in Figure 7.4.

7.3.2   Mobile Card Kit

In this section, we'll look at the details of how you can create a mobile card kit application. Before you begin, you must have downloaded the SAP Mobile Cards app from the App Store or Google Play on your device. You can now follow these steps:

1. In the SAP Cloud Platform Mobile Services cockpit, navigate to **Mobile Applications • SAP Mobile Cards**.

2. This will open the home page for the mobile card kit, where you'll find details about various **Features** assigned; **Security**, **Configuration**, and **Card Templates** already created; a **Template Manager** in which predefined templates from SAP are provided; and other settings for SAP Mobile Cards.

3. Under the **Features** tab, update the SAP S/4HANA service destination under **Mobile Connectivity**.

4. Choose the **APIs** tab and scan the QR code for iOS or Android devices to onboard the SAP Mobile Cards client application. Go through the steps in your mobile device to onboard the client.

5. Navigate to the **Template Manager** tab, select **Sample Sales Orders Template**, and choose the **Create Card Template** action from the icon on the right side.

6. This opens a screen to create new mobile cards, as shown in Figure 7.5. Here you should provide a **Name** and **Version** for the application. Make sure the **Destination**, **Template Source**, and **HTML Template** are correctly set.

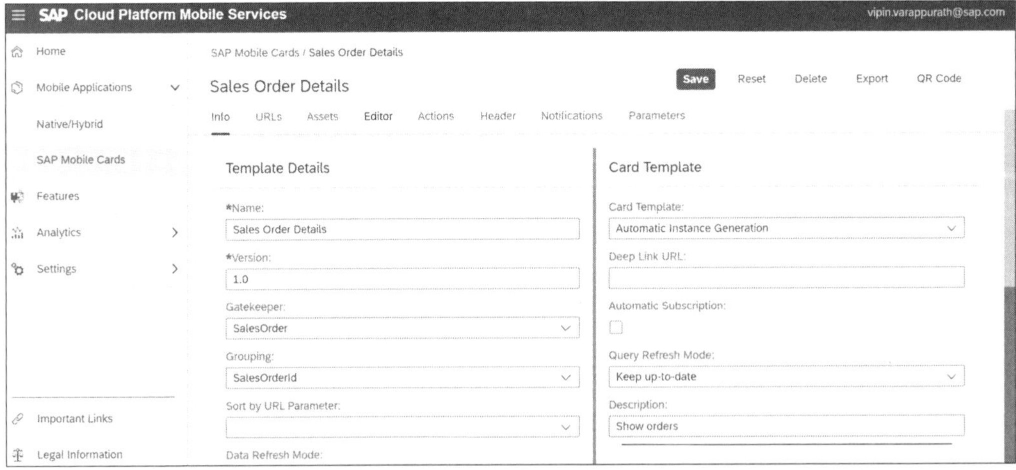

Figure 7.5 Mobile Card Kit Creation

7. You can choose to group your mobile cards in **Grouping**.

8. In **Card Template**, choose **Automatic Instance Generation**. This field provides the following options:

    - **Default**
      These are card templates created via an explicit call made to create them.

    - **Welcome Card**
      These are custom cards that are automatically downloaded after registration. Welcome cards are autosubscribed.

    - **Server Managed Card**
      These are similar to welcome cards, but end users need to explicitly subscribe to these cards.

    - **Web Page Matching**
      Allows the mobile cards kit plug-in installed in the SAP Fiori launchpad to automatically detect applications for which cards can be created.

    - **Automatic Instance Generation**
      These are card sets that are automatically created based on a query entity set. This option helps the user to download a few specified instances of the object automatically.

9. In the **URLs** tab, you'll find the query URL, which points to the data endpoint definitions for the element you want to display in a card inside your connectivity service destination.

Note

If the user doesn't have access to this query URL, they can't subscribe to the card. For this, you can define gatekeeper settings under **SAP Mobile Cards • Configurations**, as follows:

1. Click **+**, enter a **Name**, select the **Destination**, and provide the **URL** data endpoint definition of the card template.

2. Select this defined gatekeeper setting for your card template under the **Info** tab in the **Gatekeeper** field.

This way, you can create gatekeeper settings to limit card subscriptions to specific roles associated with users.

---

10. The following are other functionalities available on the **Mobile Card App Creation** screen:

    - **Assets**
      Add images for the card and define a separate resource Javascript file

    - **Editor**
      Make changes to HTML5 or CSS code for the card, or create custom handlebar function definitions

- **Data Mapping**
  Define handlebar bindings to an actual JSON path reference

- **Actions**
  Define an action to be performed from a mobile card

- **Headers**
  Create custom headers to be sent with the card

- **Notifications**
  Enable and define push notification rules for the card

- **Parameters**
  Define global parameters, URL parameters, or subscription parameters for the card

11. Click **Save**.

12. You can set the card into production from **Info • Versions**; select **Actions** to publish the card template to productive status. The new **State** is **Productive**.

You can test the card from the **Emulator** tab inside the mobile card kit application by choosing **Launch Emulator,** or subscribe to the card directly from your SAP Mobile Cards application on your mobile device as shown in Figure 7.6.

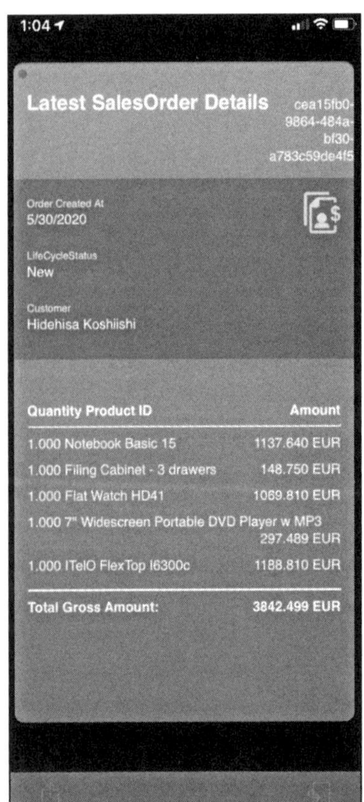

Figure 7.6 SAP Mobile Cards

7.3.3   Mobile Development Kit

As mentioned, the mobile development kit is a rapid mobile application development platform for creating native mobile applications. In this section, we'll look at how to create your first native application using the mobile development kit to help you bridge the gap between your business expectations and IT backlog. The mobile development kit has following logical components:

- Development plug-in for SAP Web IDE full-stack service and SAP Cloud Platform Business Application Studio
- Mobile development kit client build
- SAP Cloud Platform application metadata definitions and runtime

Before you begin, you must have downloaded the SAP Cloud Platform Mobile Services client from your chosen app store and have SAP Cloud Platform Business Application Studio assigned to your subaccount. To create your first application, follow these steps:

1. Inside the SAP Cloud Platform Mobile Services cockpit, navigate to **Mobile Applications • Native/Hybrid**.
2. Click **New** and provide the details indicated in Table 7.3.

Field	Description
ID	This is the unique identifier for the application, in reverse domain notation. The administrator uses this ID to register the application with SAP Cloud Platform Mobile Services, and client applications use the application ID when sending requests to the server. The recommendation is to provide an ID that contains a minimum of two periods—for example, *com.<org_name>.mobile.<app_name>*. If it's an Android application, the ID must follow Google's defined rules: it must have one or more periods, each segment should start with a letter, and all characters must be alphanumeric or underscores.
Name	An application name that can contain only alphanumeric characters, spaces, underscores, and periods and is up to 80 characters long.
Description (optional)	This is a description of the app.

Table 7.3 SAP Cloud Platform Native/Hybrid Application Definition

Field	Description
Vendor (optional)	The vendor that developed the application.
XSUAA Service	The XSUAA authentication and authorization service you want to use for the application. The default instance value would create a new instance, or you can select an existing service from the list.

Table 7.3 SAP Cloud Platform Native/Hybrid Application Definition (Cont.)

3. Click **Next**.

4. On the next screen, in the **Assign Features For** dropdown, select **Mobile Development Kit Application**, as shown in Figure 7.7.

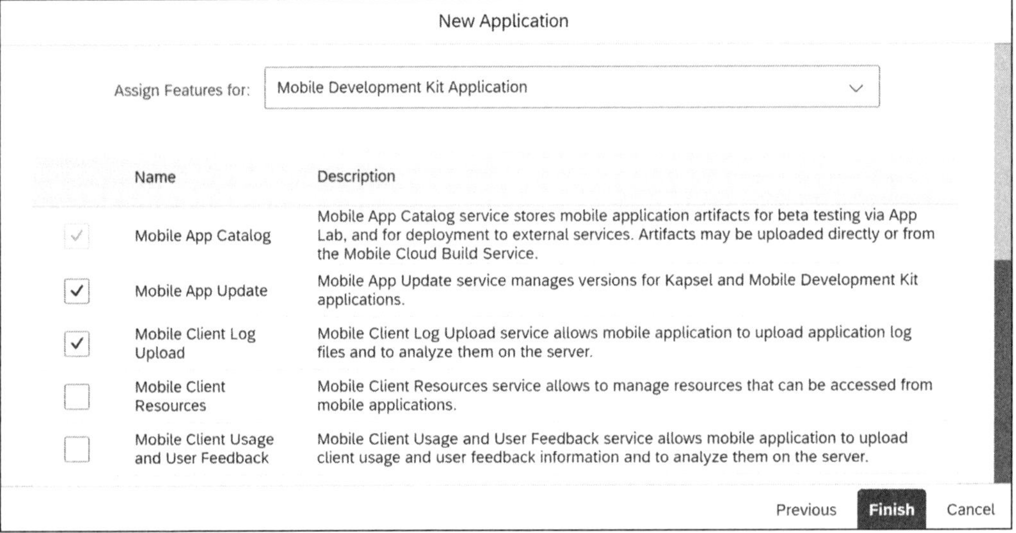

Figure 7.7 Mobile Development Kit Application Feature Selection

5. Select the features for the application.

6. Click **Finish** to create the application.

7. In the created application screen, update the **Mobile Connectivity** feature with the SAP S/4HANA destination.

Now you can further develop and configure this application using SAP Cloud Platform Business Application Studio, as follows:

1. Navigate inside your Cloud Foundry environment subaccount to **Subscriptions** and select **SAP Business Application Studio**.

2. Click the **Subscribe** button (if the service is not already subscribed).

3. Click **Manage Roles** and assign your user to the rights required to access SAP Cloud Platform Business Application Studio.

4. Click **Go to Application** to launch SAP Cloud Platform Business Application Studio.

5. Click **Create Dev Space**.

6. Enter a **Dev Space Name** and select the **SAP Cloud Platform Mobile Services** radio button as shown in Figure 7.8. Select other development tools you need in the space.

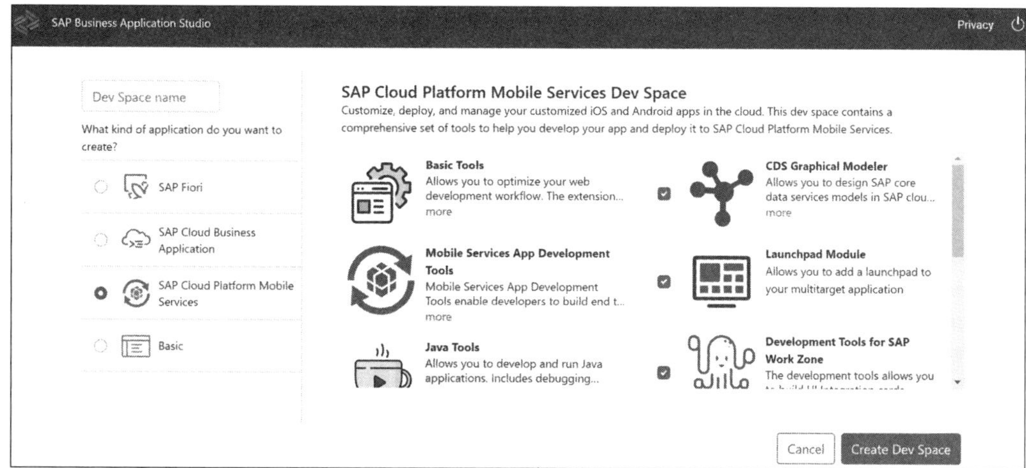

Figure 7.8 SAP Cloud Platform Mobile Services Dev Space Creation in SAP Cloud Platform Business Application Studio

7. Click **Create Dev Space** and a new dev space is created in SAP Cloud Platform Business Application Studio.

8. Select the space created. On the **Welcome** screen, select **New Project from Template**.

9. Specify a target path, select **MDK Project** for the template, and choose **Next**, as shown in Figure 7.9.

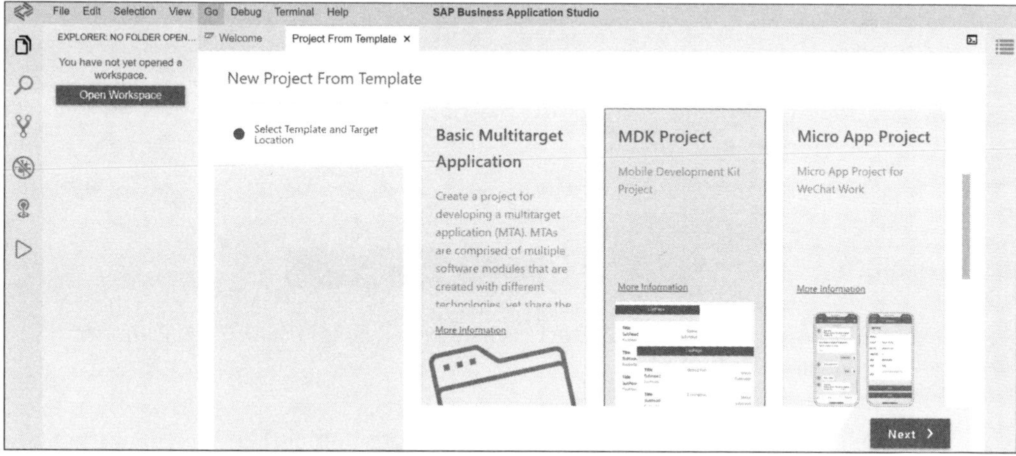

Figure 7.9 Create New Project from Template for Mobile Development Kit

10. Enter the **MDK Template Type**, **Project Name**, and **Application Name** and select **Next**.

11. Select your Cloud Foundry **Organization** and **Space** and choose **Next**.

12. Provide a **Service Name** and select the **Application ID** you provided while creating the application from the SAP Cloud Platform Mobile Services cockpit. Choose **Destination** and select **Next**.

13. Select the OData entity sets and choose **Next**. The project is generated with all necessary files.

14. Once you open the project in the workspace, you'll see the project configuration in an editor for mobile applications, as shown in Figure 7.10. The editor has following functionalities:

    – Edit pages with drag-and-drop controls, map them to OData service entity sets, and add custom controls or business logic.

    – Use the object browser to quickly locate and map elements to the screen, including actions, rules, services, styles, and UI elements.

    – Use the action editor to define custom actions like offline access, messages, and business logics.

    Here you define the application properties and application lifecycle events—for example, OnLaunch and OnExit in the *Application.app* file.

15. Right-click **Application.app** and choose **MDK:Deploy** to deploy the app to SAP Cloud Platform Mobile Services.

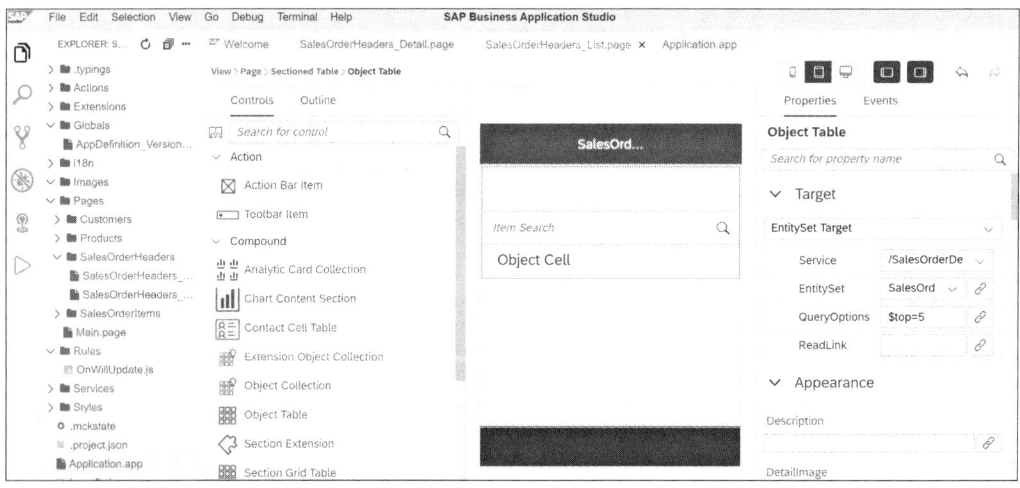

Figure 7.10 SAP Cloud Platform Mobile Development Kit Editor in SAP Cloud Platform Business Application Studio

You can now view this app in the SAP Cloud Platform Mobile Services client from any device.

> **Note**
>
> The mobile development kit provides the functionality to define and create your own mobile client to connect to your SAP Cloud Platform mobile application.
>
> Mobile apps generated by the cloud-build service can only be distributed using enterprise distribution mechanisms.
>
> To support this scenario, you can download the SDK for the mobile development kit from software downloads and build the client as explained at *http://s-prs.co/v515748*.

7.3.4   SAP Cloud Platform SDK for iOS

SAP Cloud Platform SDK for iOS is provided by SAP Cloud Platform Mobile Services. It's based on Apple's Swift programming language and helps in development of apps using the Xcode IDE. The SDK includes the SAP Cloud Platform SDK for iOS Assistant and comprehensive API documentation to support developers in quickly getting started and enabled to develop apps by generating code templates familiar to iOS developers.

The SDK Assistant builds the code template based on the components, and developers can further enhance their application code based on their requirements. SAP Cloud Platform SDK for iOS has following components for end-to-end application development:

- **SAPFiori framework**
  This provides and implements the user experience for iOS apps. It implements SAP Fiori for the iOS design language and helps customize and brand applications to meet organizational standards.

- **SAPFioriFlows framework**
  This framework provides the capability to complete user onboarding scenarios in the app.

- **SAPOData frameworks**
  This provides a data integration layer for both online and offline data access for your app. This framework parses OData payloads, produces OData requests, and handles responses.

- **SAPCommon and SAPFoundation frameworks**
  These provide components to integrate your app with the SAP Cloud Platform environment. This includes functionalities like backend connectivity, authentication and authorization, logging functionality, and so on.

- **SAPML framework**
  This framework provides general-purpose text-recognition capabilities, which helps in distributing new versions of core machine learning models to devices and handles local compilation and persistence. The models can be distributed remotely using SAP

Cloud Platform Mobile Services, thus allowing the app to use an updated version without redistribution of the app.

Developing in Xcode also allows your Apple developers to integrate the app with Apple UIKit or other third-party Swift components.

Before you begin to work with SAP Cloud Platform SDK for iOS, you must have met the following prerequisites:

- You have the Xcode editor installed (which only works on Macbooks).

- You have downloaded SAP Cloud Platform SDK for iOS from the download center at *https://developers.sap.com/trials-downloads.html*.

To set up and configure app development using SAP Cloud Platform SDK for iOS, follow these steps:

1. Extract the downloaded SAP Cloud Platform SDK for iOS on your device.

2. Open the DMG file to find the SAP Cloud Platform SDK for iOS Assistant.

3. Drag and drop **SAP Cloud Platform SDK for iOS Assistant** to the **Applications** folder on your Apple Macbook.

4. Click **SAP Cloud Platform SDK for iOS Assistant** in the **Applications** folder to launch the assistant.

5. In the SAP Cloud Platform Mobile Services cockpit, navigate to **Mobile Applications • Native/Hybrid**.

6. Choose **New**, enter application fields as shown in Table 7.3, and click **Next**.

7. Select **Native Application** under **Assign Features For**. Select the features required for the application and choose **Finish**. The app is created.

8. In the **Mobile Connectivity** feature, update your destination to the OData service you want to use for the application.

9. In the SAP Cloud Platform Mobile services cockpit, navigate to **Important Links** and click the **Import URLs Directly (Only Supported for iOS Assistant)** link. This will open your SAP Cloud Platform SDK for iOS Assistant with populated links for SAP Cloud Platform Mobile Services, as shown in Figure 7.11.

10. Enter a **Name** for your SAP Cloud Platform mobile account and a **Username** and **Password** to login to the account, and click **Save**.

11. In the SAP Cloud Platform SDK for iOS Assistant, click **Manage Accounts • Add new...**. For **Type**, select from the following:

    - Select **SAP Cloud Platform Developer Portal** to configure your SAP Cloud Platform API Management developer portal to use APIs exposed using this service while creating an iOS application.

    - Select **SAP Cloud Platform Translation Hub** to configure SAP Translation Hub, covered in the next section, to localize the application for various languages.

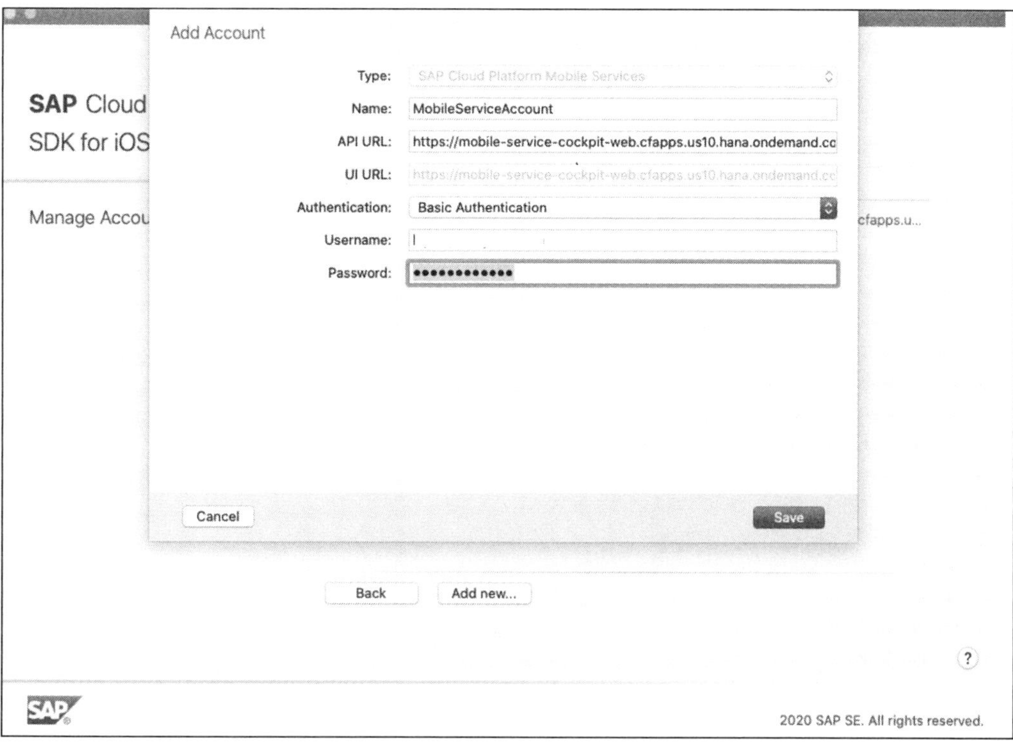

Figure 7.11 SAP Cloud Platform SDK for iOS Assistant: Manage Accounts

12. In SAP Cloud Platform SDK for iOS Assistant, click **Create New**. On the next screen, you can select a project template to create your application. For this example, select the **Reuse Existing Application** template.

13. In the next screen, **Account**, select the SAP Cloud Platform Mobile Services account you added in the previous step and click **Next**.

14. On the next screen, **Cloud Application**, select the cloud application you had created in the SAP Cloud Platform Mobile Services cockpit in previous step and choose **Next**.

15. On the next screen, **Xcode Project**, provide a **Product Name**, **Organization Name**, **Organization Identifier** (these names need not match the names that you have set in the SAP Cloud Platform Mobile Services cockpit), and **Path** to save the project. Click **Next**.

16. On the next screen, **Proxy Classes**, choose the destination to create the app and click **Next**.

17. On the next screen, **UI Configuration**, select the templates for UI generation and choose **Finish**.

18. The generated project will open in your Xcode IDE, where you can make required changes to the project and run the app in simulator mode, as shown in Figure 7.12.

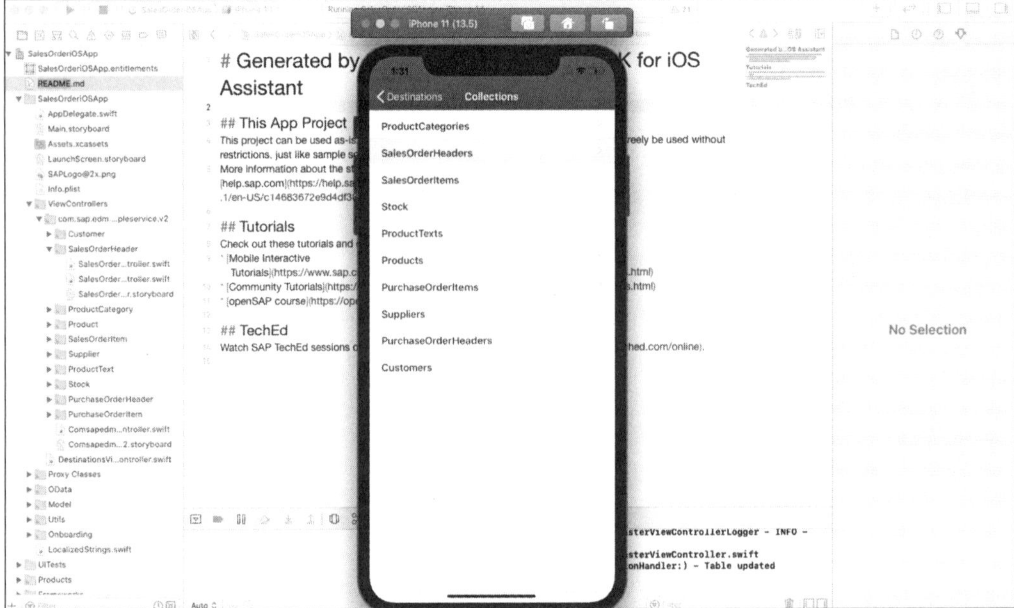

Figure 7.12 SAP Cloud Platform SDK for iOS Project in Xcode

Your first SAP Cloud Platform SDK for iOS-based app is now ready.

7.3.5 SAP Cloud Platform SDK for Android

SAP Cloud Platform SDK for Android provides tools and capabilities to create enterprise-grade native Android applications using SAP Cloud Platform Mobile Services. The SDK is based on Google's Android SDK and has an Android Studio project plug-in that simplifies project development.

The SAP Cloud Platform SDK for Android has the following components, similar to SAP Cloud Platform SDK for iOS:

- **SAP Fiori component**
  Provides a library of UI modules for creating mobile applications that follow the SAP Fiori design language and Android Material Design

- **Flows component**
  Provides implementation for an app's onboarding scenarios

- **OData component**
  Provides capabilities that allow your app to access OData for both online and offline mode

- **Foundation component**
  Provides capabilities that allow your app to use features and capabilities of SAP Cloud Platform like authentication, push notification, logging, and more

Before you begin to work with SAP Cloud Platform SDK for Android, you must have met the following prerequisites:

- You have the Android Studio IDE installed.
- You have downloaded SAP Cloud Platform SDK for Android from the Download Center at *https://developers.sap.com/trials-downloads.html*.

To set up and configure app development using SAP Cloud Platform SDK for Android, follow these steps:

1. Extract the contents of the downloaded Android SDK and run the installation script.
2. Set the SAP_ANDROID_HOME environment variable to your installation folder (optional).
3. From the SAP Cloud Platform Mobile Services cockpit, navigate to **Mobile Applications • Native/Hybrid**.
4. Choose **New**, enter application fields as in Table 7.3, and click **Next**.
5. On the next screen, select **Native Application** in the **Assign Features For** dropdown. Select the features required for the application and choose **Finish**. The app is created.
6. In the **Mobile Connectivity** feature, update your destination to the OData service you want to use for the application.
7. From the SAP Cloud Platform Mobile Services cockpit, navigate to **Important Links** and select **Copy Admin API** and **Copy Admin UI**.
8. Launch Android Studio. You'll see the wizard for SAP Cloud Platform Mobile Services-based Android projects, as shown in Figure 7.13.

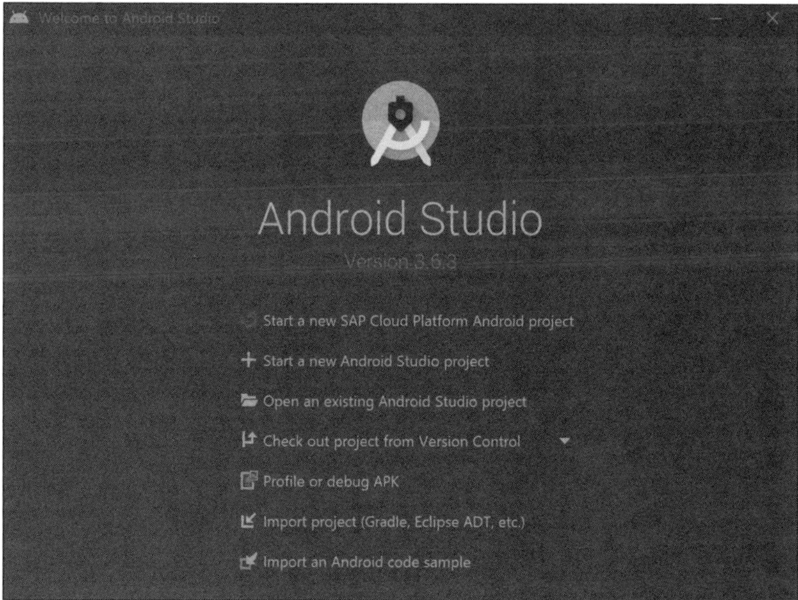

Figure 7.13 Android Studio with SAP Cloud Platform SDK for Android Installed

9. Select **Start a New SAP Cloud Platform Android Project** and provide the following details on the screens that follow:

   - **Server Configuration**
     Provide the **Account name**, the **Admin API URL** and **Admin UI URL** copied from the previous step, and authentication details to log in to your SAP Cloud Platform Mobile Services account, then click **Next**.

   - **Cloud Configuration**
     Choose to use an existing app, create from scratch, or create a sample app from here. Because you created an existing app from the SAP Cloud Platform Mobile Services cockpit earlier, choose **Use Existing App**, provide the **Application ID**, and click **Next**.

   - **OData Services**
     Select the OData service for the application and click **Next**.

   - **Android Studio Project**
     Provide the **Project Name**, **Project Namespace**, and **Project location** and choose the target language (Java or Kotlin) for your project to be created. Click **Next**.

   - **Project Features**
     Select the relevant project features, like enabling logging, usage reporting, push notification configuration, OData access for online or offline use, and so on.

10. Click **Finish**.

This will generate a project template with lines of code that supports CRUD access to the OData service selected. The project will include SAP Fiori screens and OData and foundation libraries used for the project's specific activities. You can open the project in Android Studio and run it in the emulator to test it.

7.4   SAP Translation Hub

In previous sections, we've explained why an optimum user experience is important in this digital transformation era and how you can use certain tools to create these experiences for your customers and employees quickly. What also contributes to the end user experience and further enhances it is the ability to provide a UX in the user's native language. A portal application or mobile application presented in a user's native language not only enhance the user experience, but also helps drive user adoption and end user satisfaction.

For enterprise applications, it's quite important to understand the business context for a translation. This is defined as a *domain* in SAP Translation Hub. SAP Translation Hub helps you translate your UI texts via SAP-approved translations and terminology that take the business context (the domain) into consideration during machine translation.

The translation can be done using this service at a quick pace, thus enabling you to roll out translated screens and user experiences to end users quickly. You can access the translation resources by using integrated workflow scenarios in a UI or by consuming a range of API methods. SAP Translation Hub integrates natively with the following SAP Cloud Platform user experience categories to help quickly translate applications:

- SAP Cloud Platform Mobile Services
    - Mobile development kit
    - SAP Cloud Platform SDK for iOS
    - SAP Cloud Platform SDK for Android
- Git repository
    - On-premise Git repository
    - Web-based Git repository
    - SAP Cloud Platform Git repository

SAP Translation Hub comprises the following translation providers:

- A repository of multilingual texts from SAP applications referred as the SAP-provided MLTR (multilingual text repository). You can optionally create your own multilingual repository by uploading your own language data to SAP Translation Hub to use this as the first-choice translation provider. This is referred to as *company MLTR.*

- Machine translation, which is verified by SAP-certified language experts.

SAP Translation Hub runs in the Neo environment. Before you begin work, you must have a subaccount based on the Neo environment. To start using SAP Translation Hub, follow these steps:

1. Navigate inside your subaccount to **Services** and select the **SAP Translation Hub** tile.
2. Click **Enable**.
3. Click **Configure Service** and navigate to **Roles**. Make sure your user is assigned to the predefined role called *user.*
4. Click **Go to UI for Translation Workflow** to open the translation project home screen, as shown in Figure 7.14.

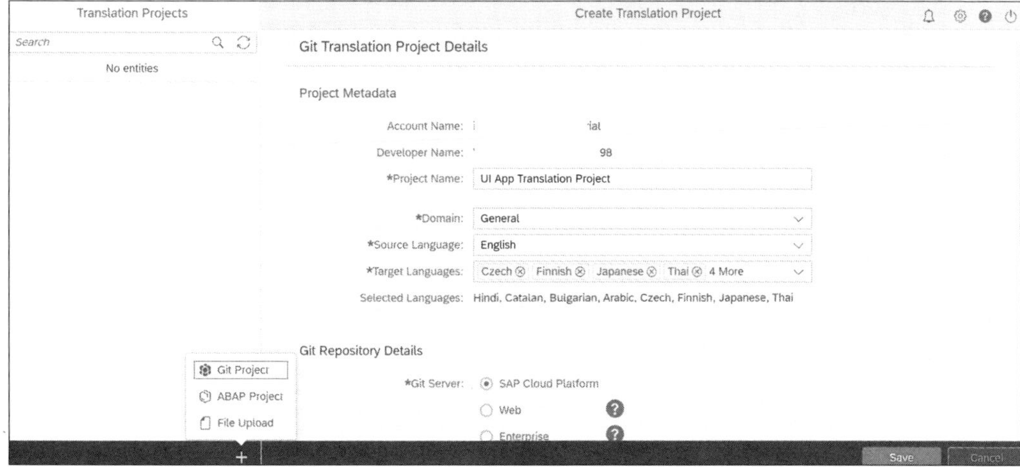

Figure 7.14 SAP Translation Hub

5. Click **+** to create a translation project from a **Git Project**, from an **ABAP Project**, or via **File Upload**. For this example, choose **Git Project**.

6. On the next screen, enter a **Project Name, Domain, Source Language, Target Language**, and **Git Repository Details**. You can select multiple target languages as required.

7. You can choose additional settings under **Project Configuration** for **Keep Translations, Post-Editing Required, Review Required**, or **Use Company MLTR**.

8. Select **Save**. An overview of the translation project is displayed.

9. To start the translation project, choose **Translate and Push**.

10. You'll be asked for your Git repository password. Enter it and click **Submit**.

11. On the subsequent screen, choose the **Translations** tab to see the translated text. This can be updated based on the *quality index*, a score of 1–100 indicating the quality of the translation.

12. Click **Push** to push the translation to the Git repository.

13. The translation is made available to the project by executing a pull request from Git. You'll find the translations in the following places:

    – For SAP Fiori apps, the i18n folder in your project now contains a new properties file for translated languages.

    – For SAP Cloud Platform SDK for iOS apps, the translations are defined in `Localizable.strings`.

    – For SAP Cloud Platform SDK for Android apps, the translations are defined in *strings_localized.xml*.

> **Note**
>
> As mentioned, SAP Translation Hub integrates with other SAP Cloud Platform services like SAP Cloud Platform SDK for iOS. Hence the translation project can be triggered from those services as well from a developer/user flow perspective.

7.5   Summary

Defining a consistent and seamless user experience for end users of your system is very critical in your transformation journey. This ensures end user adoption and satisfaction, adding value to your business. Furthermore, the user experience and expectations are fast changing in the current world and you need tools and technologies to catch up with these ever-changing requirements.

In this chapter, we looked at the SAP Cloud Platform Launchpad service, which can bring applications from your multiple SAP systems and custom applications together in a single place, providing your employees with a consistent user experience. We also looked into SAP Work Zone, which brings together different SAP Cloud Platform capabilities in a coherent way, providing a seamless and delightful end user experience.

Because mobile devices are changing UX expectations and as organizations are considering *work from anywhere* approaches, it becomes increasingly important for organizations to provide enterprise application access on mobile devices. We looked at various SAP Cloud Platform Mobile Services capabilities for different use cases and developer experiences, which can quickly and rapidly create mobile apps in both online and offline mode and provide a delightful user experience.

Finally, we saw how you can enable business translations for your user experience applications in an easy and consistent way across the enterprise, thus further helping you to enhance your business user experience and drive adoption and satisfaction.

Chapter 8
Security

Security is of prime importance for an enterprise during the digital transformation to become a future-ready enterprise. In this chapter, we'll talk about how you can easily set up different security settings in your organization for secure application development and platform management.

Enterprise organizations have and need very well-defined security policies and best practices, especially in cloud environments. The security policy is relevant not just for the applications that you develop in the cloud, but also for the overall platform and development environment.

In this chapter, we'll talk about how you can secure your applications that you develop in SAP Cloud Platform, define security measures on SAP Cloud Platform itself, and set up an overall security governance model to comply with security regulations in your organization.

We'll look in detail at the different services available within SAP Cloud Platform to achieve this and investigate various configurations. We covered in earlier chapters how you can develop applications in the cloud. In this chapter, we'll go into depth on best practices to secure these applications for your end users. In particular, Section 8.1 through Section 8.4 cover application-level security, whereas Section 8.5 covers platform-level security.

> **Note**
>
> There are two types of users in SAP Cloud Platform: application/business users and platform users:
>
> - Application or business users are the end users of the applications deployed to SAP Cloud Platform.
> - Platform users are developers, administrators, security administrators who set up security policies, and operators who deploy, administer, and troubleshoot applications and services in SAP Cloud Platform.

8.1   Authentication Management

In this section, we'll cover how authentication management can be handled in the applications on SAP Cloud Platform. SAP Cloud Platform supports SAML 2.0 identity providers, so you can use any SAML 2.0-compliant identity provider in your landscape.

In the following sections, we'll explain how you can set up authentication management in SAP Cloud Platform in the Cloud Foundry environment and work through setting up trust settings with SAP Cloud Identity Services, Identity Authentication for authentication of your application and services.

8.1.1   Cloud Foundry Environment

In SAP Cloud Platform in the Cloud Foundry environment, the User Authentication and Authorization (UAA) component of the runtime platform manages the business user authentication and authorization management. The users are stored in SAML 2.0 identity providers, whereas authorizations are stored in the UAA component.

The Cloud Foundry environment in SAP Cloud Platform distinguishes between platform users and business users. Business users are always bound to a specific tenant, which holds the information about users' authorizations. Tenants, business users, and their authorizations are managed by separate UAA instances using the extended service for UAA (XSUAA). This component provides a programming model for authentication and authorization in business applications. Handling of platform users is covered in the next section of this chapter.

To secure the application, two components are involved in a Cloud Foundry application: the XSUAA service and the application router. The application router works as a central entry point for the application and is used in combination with the XSUAA service to authenticate the business user and route the user to the secured application. To integrate authentication into an application, you must create a service instance of the XSUAA service and bind it to the application router and the application containers. The application router helps you handle the OAuth internal authentication in the Cloud Foundry runtime with the authentication component and application container. The authentication process itself is triggered by the application router component, which is defined in the xs-app.json design-time artifact if required.

> **Tip**
>
> The xs-app.json routing configuration file syntax is defined in the help document found at *http://s-prs.co/v515736*.

8.1.2   Default Authentication in SAP Cloud Platform

The default identity provider for SAP Cloud Platform is the SAP ID service. SAP Cloud Platform is delivered with default trust established with the SAP ID service. To see the default identity provider, you can navigate to **Security • Trust Configuration** in your Cloud Foundry subaccount.

Note

To see this setting, you must have either created this subaccount or have the security administrator role assigned to your users, which we'll cover in details in Section 8.2.

In SAP Cloud Platform, you can set up trust and authentication with SAML 2.0-compliant identity providers. This configuration and set up consists of two steps:

1. Establish trust with the SAML 2.0 identity provider with the SAP Cloud Platform subaccount.

2. Register the SAP Cloud Platform subaccount in the SAML 2.0 identity provider.

In the following section, we'll cover how this set up can be done with the Identity Authentication service.

8.1.3   Identity Authentication Service

Identity Authentication, part of SAP Cloud Identity Services, is a service in SAP Cloud Platform that helps set up authentication for your applications. It was previously known as SAP Cloud Platform Identity Authentication, and as of the time of writing, the old name still appears in the system in some places. It allows authentication decisions to be removed from the application and handled in a central service. Users can benefit from features such as a user base, social ID integration, and corporate branding, and it provides an easy way for applications to delegate authentication and identity management.

To set up Identity Authentication, you must meet the following prerequisites:

- You already have a Cloud Foundry subaccount created and either have administrator privileges or at least security administrator rights in the subaccount.

- You have SAML 2.0 metadata available from your Identity Authentication tenant as an XML file. You can download this using the following URL and save it in XML format: *https://<IdentityAuthentication_Tenant>.accounts.ondemand.com/saml2/ metadata*.

To begin your setup, follow these steps:

1. Inside your subaccount, navigate to **Security • Trust Configuration** and choose **New Trust Configuration**.

2. In the screen that appears, **Upload** your Identity Authentication SAML 2.0 metadata. Choose **Parse** in the screen to validate the metadata, which will fill the **Subject** and **Issue** fields with relevant data.

3. Provide a **Name** and **Description** and choose **Save**.

4. You can choose to set the status of the SAP ID service to **Inactive** (this is optional).

5. Now you'll get SAML metadata for your subaccount to set up trust for Identity Authentication. For this, choose the **SAML Metadata** button, as shown in Figure 8.1. This will download the metadata file for your subaccount in XML format with a name in the following format: *saml-<subdomain>-sp.xml*.

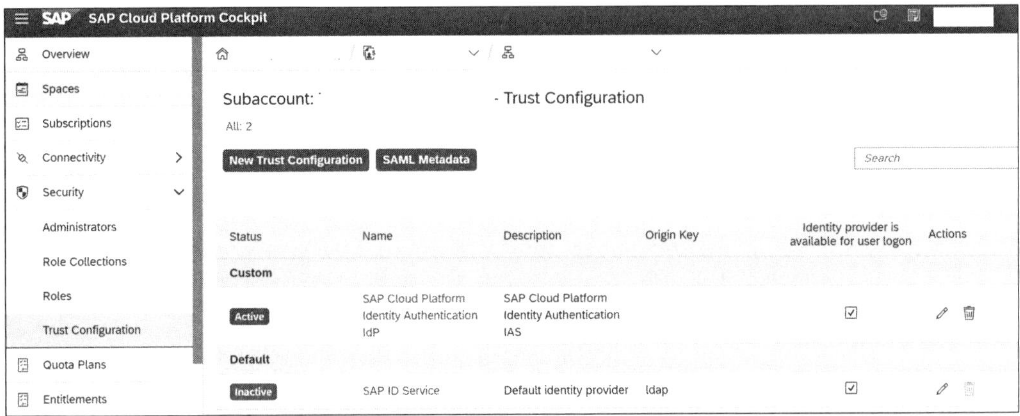

Figure 8.1 Trust Configuration and SAML Metadata in SAP Cloud Platform Subaccount

6. Open the administration console of Identity Authentication at *https://<Identity_ Authentication_Tenant>.accounts.demand.com/admin*.

7. Inside Identity Authentication, navigate to **Application & Resources • Applications**.

8. Click the **+Add** button to create a new application for a new SAML service provider. Enter a **Name** in the pop-up and click **Save**, and the application is added, as shown in Figure 8.2.

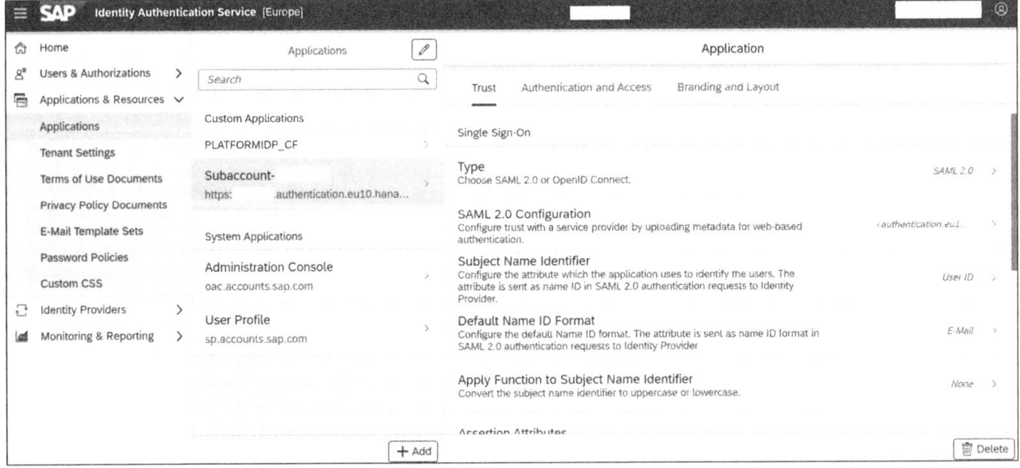

Figure 8.2 Application in Identity Authentication Service

9. The name is seen by the user on the logon screen when authentication is requested by the UAA service, by which users identify which application they will access after authentication.

10. Choose **Type** and select SAML 2.0.

> **Tip**
>
> You can use Identity Authentication to authenticate users in OpenID Connect–protected applications as well. You can set the **Type** to **OpenID Connect**.

11. Choose **SAML 2.0 Configuration** and import the subaccount metadata XML file downloaded in the previous step by clicking **Browse**.

12. Choose **Save**.

13. Choose **Default Name ID Format** and select **E-mail** as a unique attribute. Click **Save**.

14. Choose **Assertion Attributes**, use **+Add** to add a multivalue user attribute, and enter "Groups" as the assertion attribute name for the groups user attribute. Choose **Save**. You can also define custom attributes from the list.

15. Figure 8.3 shows the various configurations required and set in the Identity Authentication tenant.

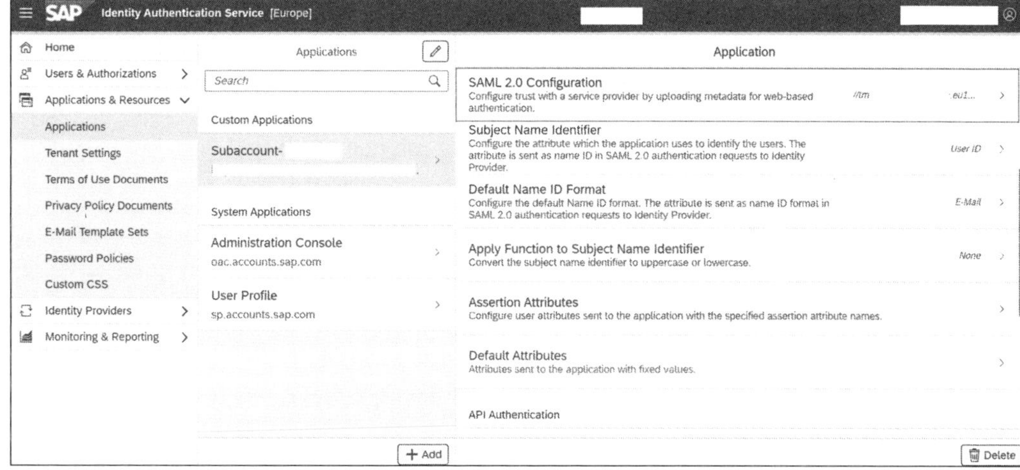

Figure 8.3 SAP Cloud Platform Identity Authentication Tenant SAML Configuration Settings

> **Tip**
>
> The approach explained so far details steps to establish trust between your SAP Cloud Platform account and the Identity Authentication tenant. These steps are applicable for setting up trust with any SAML-compliant identity provider in SAP Cloud Platform.
>
> However, the SAP Cloud Platform Cloud Foundry environment also provides an easy and simple way to establish trust with the Identity Authentication service using a

one-click approach. Follow these steps to configure trust with the Identity Authentication service through this approach:

1. Navigate inside your subaccount to **Security • Trust Configuration**.
2. Click **Establish Trust**. In the pop-up window, select the Identity Authentication tenant and select **Establish Trust**.

Note

The configuration of Identity Authentication in the Neo environment is very similar to the configuration for Cloud Foundry environment. Follow these steps to configure this trust between your SAP Cloud Platform Neo-based subaccount and the Identity Authentication tenant using a one-click approach:

1. To perform the configuration, navigate inside your Neo subaccount to **Security • Trust**.
2. Click the **Local Service Provider** tab, click **Edit**, and change the **Configuration Type** to **Custom**.
3. Click the **Application Identity Provider** tab and click **Add Identity Authentication Tenant**.

For more information on these settings, visit *http://s-prs.co/v515737*.

8.1.4   SAML 2.0 Identity Provider

You can use any SAML 2.0-compliant identity provider with SAP Cloud Platform for authentication, user store, and federation.

Before you begin, you must meet the following prerequisites:

- You have access to a Cloud Foundry subaccount as the subaccount administrator or security administrator.
- You have SAML 2.0 metadata for your identity provider.
- You have access to the administration console of your SAML 2.0 identity provider.

The process to set up the trust is identical to the one explained in Section 8.1.3 for the Identity Authentication service. The only difference is that except for uploading the SAP Cloud Platform metadata to the Identity Authentication tenant, you upload it to a third-party identity authentication tenant.

Recommendation

If your business users are stored in multiple corporate identity providers and you want to grant users access to SAP Cloud Platform, we recommend using Identity Authentication as a hub.

In this scenario, you would first connect Identity Authentication as single custom identity provider to SAP Cloud Platform. Then you connect Identity Authentication to integrate your corporate identity providers. In this scenario, the Identity Authentication service acts as proxy. You establish this setting in your Identity Authentication tenant admin console under **Identity Providers • Corporate Identity Providers**. You can have one of the following identity provider types supported:

- SAML 2.0 compliant
- SAP Single Sign-On
- Microsoft ADFS/Azure AD

For more information on setting this configuration, visit *http://s-prs.co/v515738*.

Note that you need the *manage corporate identity providers* role assigned to your user in Identity Authentication to do this configuration.

You can also configure social media authentication providers with the Identity Authentication service to enable users to log on to applications using their social media credentials. The following social media providers are supported:

- Twitter
- Facebook
- LinkedIn
- Google

These settings are especially beneficial for B2C scenarios. Thus, SAP Cloud Identity Services: Identity Authentication can be configured as single hub to manage B2E, B2B, and B2C scenarios together.

8.2   Authorization Management

Authorization is the process of defining what end users are or are not allowed to see in an application. In this section, we'll discuss various options in SAP Cloud Platform to manage authorizations, both in the Cloud Foundry environment and in the Neo environment.

8.2.1   Cloud Foundry Environment

In the Cloud Foundry environment, application developers create and deploy application-based authorization artifacts for business users. Administrators use this information to assign roles, build role collections, and assign these collections to business users or user groups.

Before you begin, you must already have an application that you want to deploy in the Cloud Foundry environment of SAP Cloud Platform.

In the following sections, we'll discuss the various concepts in Cloud Foundry authorization management, the relationships among them, and how to use these concepts while building applications.

Terminology

Application security is maintained in the *xs-security.json* file, which we call the application security descriptor. This file defines authentication methods and authorization types to use for access to your application. Before we dive into the details of the configuration, let's look at the key Cloud Foundry authorization artifacts:

- **Role collections**
  Role collections group together different roles that can be applied to business end users. You assign multiple roles to a role collection.

  Role collections can be used to group various roles that are semantically aligned to a job function. You can add roles from multiple applications to a role collection.

  You can also assign users provided by the SAP ID service or by your identity provider to a role collection, and you can assign SAML 2.0 groups from your identity provider (not with the SAP ID service) to a role collection.

- **Roles**
  Roles are used to define the type of access granted to an application. You build a role based on role templates—or, a role is an instance of role template. Figure 8.4 shows the relations among role collections, roles, and users.

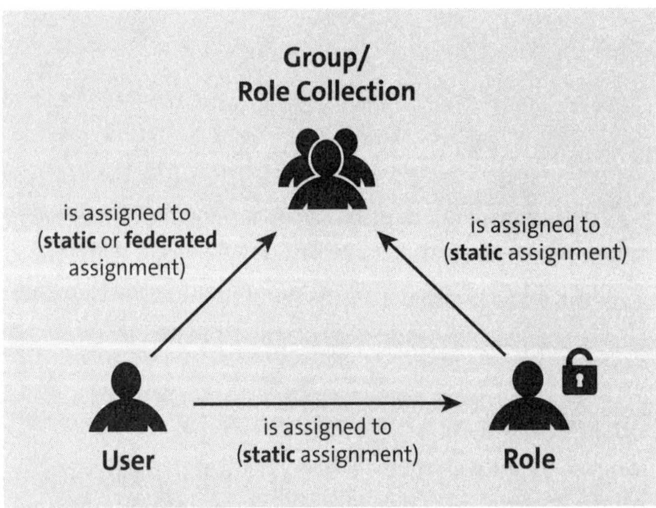

Figure 8.4 Role Collections, Roles, and Users

- **Role templates**
  Role templates define the type of access permitted for an application. Role templates

are defined in the application security descriptor (the *xs-security.json* file) of applications that have been registered as OAuth 2.0 clients in the User Account and Authentication service during application deployment.

> **Note**
>
> The User Account and Authentication service automatically creates default roles for role templates that do not include attribute references. They have the same name as the role templates and can't be deleted. Their description contains a default instance.

- **Scope**

  Scopes cover the business user's authorization in a specific application. Scope checks return Boolean results. Scopes are assigned to users by means of security roles, which are mapped to the user group(s) to which the user belongs.

  Scopes are used for authorization checks by the application router. URL prefix patterns can be defined and associate a scope with each pattern. If one or more matching patterns are found when a URL is requested, the application router checks whether the OAuth 2.0 access token included in the request contains at least the scope associated with the matching patterns. If not all scopes are found, access to the requested URL is denied.

  Authorization in the runtime container is also handled by scopes. The container security API includes the `hasScope` method, which allows you to programmatically check whether the OAuth 2.0 access token used for the current request has the appropriate scope.

- **Attributes**

  Attributes are defined in application security descriptor, the xs-security.json file, to check authorization based on certain attributes associated with end users. For each attribute, administrators have the following options to specify the value that restricts data access:

  - **Static attributes**

    These attributes are stored in the role. The attribute value is given when the administrator assigns the role collection with this role to the user.

  - **SAML attribute or attributes from a custom identity provider**

    You can dynamically reference all attributes that come with SAML 2.0 assertion of business users from the identity provider. You define the attribute and attribute values in the identity provider, and the SAML assertion attributes are used to further refine the roles.

  - **Unrestricted attribute**

    You can also define no value for the attribute, in which case the behavior is same as if the attribute does not exist for the role.

> **Note**
>
> For more about the syntax of the application security descriptor file, visit *http://s-prs.co/v515739*. Figure 8.5 shows example code output from the application security descriptor file.
>
> ```
> 📁
>   📁  ui
>        webapp
>        Gruntfile.js
>        neo-app.json
>        package.json
>        xs-app.json
>     mta.yaml
>     xs-security.json
> ```
>
> **Figure 8.5** Project Structure with Application Security Descriptor

Figure 8.6 shows the relationships among these artifacts in the Cloud Foundry environment.

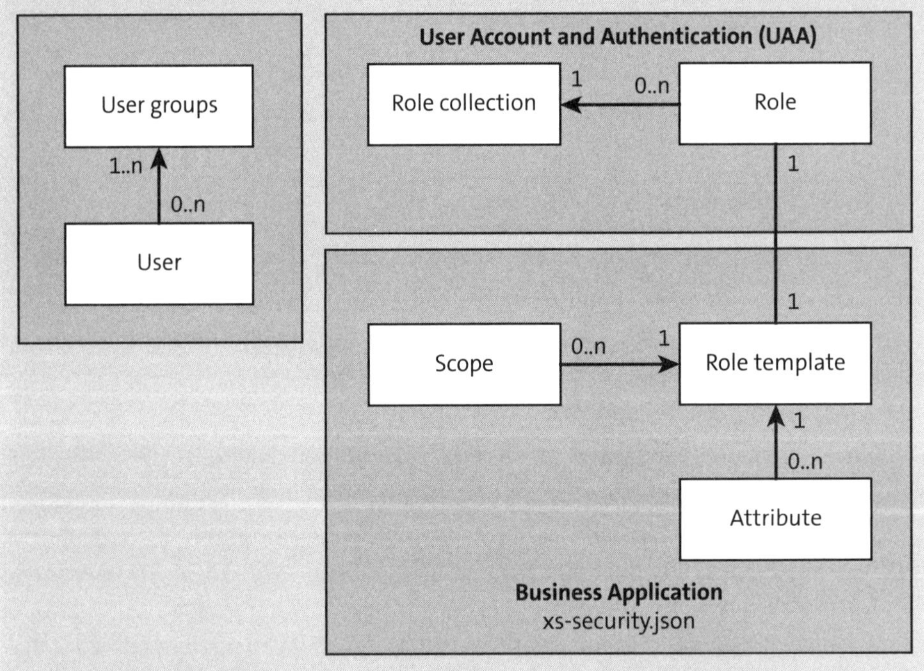

Figure 8.6 Relations among Different Cloud Foundry Authorization Artifacts

After the developer has created role templates with defined scopes and attributes and deployed them to SAP Cloud Platform, it is an administrator's task to use the role templates to further build roles, collect and aggregate roles into role collections, and assign role collections to end users of the application directly or through user groups. Let's now look at how these artifacts can be created in SAP Cloud Platform.

To start, you must have an application deployed in Cloud Foundry with role templates, scopes, and attributes defined in xs-security.json.

Create Role

To create a role using the role templates deployed as part of the application, follow these steps:

1. Inside your SAP Cloud Platform cockpit, navigate inside your global account to your Cloud Foundry subaccount/organization.

2. Navigate to the space where the application is deployed.

3. Navigate to **Applications** and select the application.

4. Navigate to **Security • Role** and click **+**.

5. In the **Create Role** pop-up that appears, provide a name for the role, an optional **Description**, and select the **Role Template** as shown in Figure 8.7. Click **Next**.

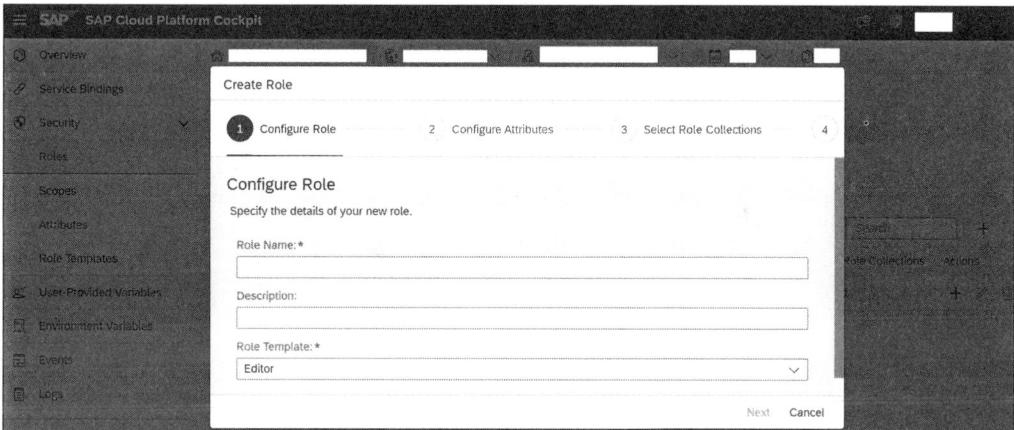

Figure 8.7 Create Role

6. In the next step, **Configure Attributes**, configure the attributes defined in the application security descriptor. You can provide static values or map to the values from an identity provider, as shown in Figure 8.8. Click **Next**.

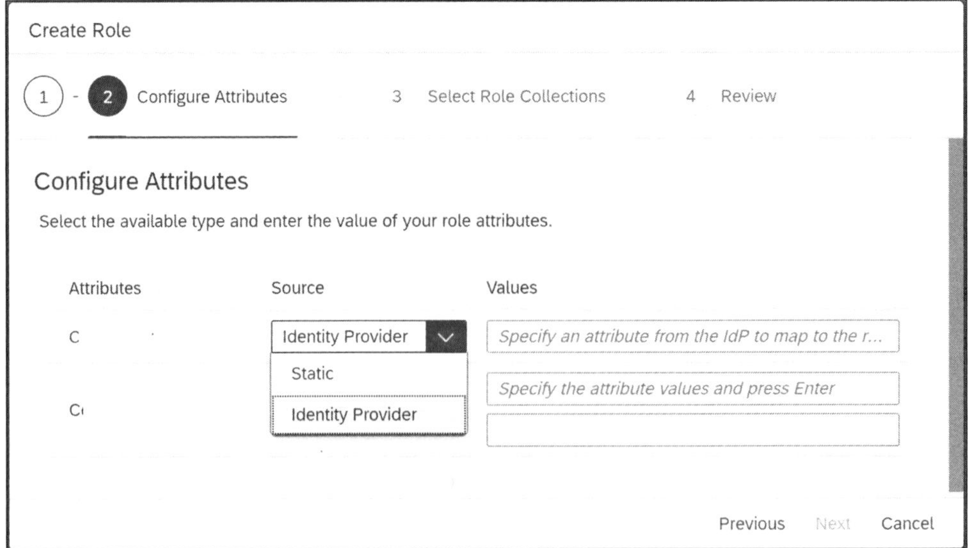

Figure 8.8 Configure Attributes

7. Now select a role collection if one already exists. If no role collections exist, we'll cover how to create role collections and assign roles in the next section. Click **Next**.

8. Review the information provided and click **Finish**. The role is created.

Tip

Alternatively, you can create roles from your subaccount with the following menu options:

1. Inside your subaccount, navigate to **Security • Roles**. You'll find all the role templates associated with all the applications inside the subaccount.

2. Click **+** next to the application and role template you want to create a role for and follow the process described previously mentioned.

Note

Other information that you may need to access can be found via the following menu paths:

- Navigate to **Security • Scopes** to find the scopes associated with the application.
- Navigate to **Security • Attributes** to find the attributes defined in the application.
- Navigate to **Security • Role Templates** to find the role templates defined in the application.

Create Role Collections

In this section, we'll discuss how to create role collections, as follows:

1. Inside your SAP Cloud Platform cockpit, navigate to your Cloud Foundry subaccount.
2. Navigate to **Security • Role Collections**.
3. Click **+**. In the pop-up, enter a **Name** and click **Create**, as shown in Figure 8.9. A new role collection is created.

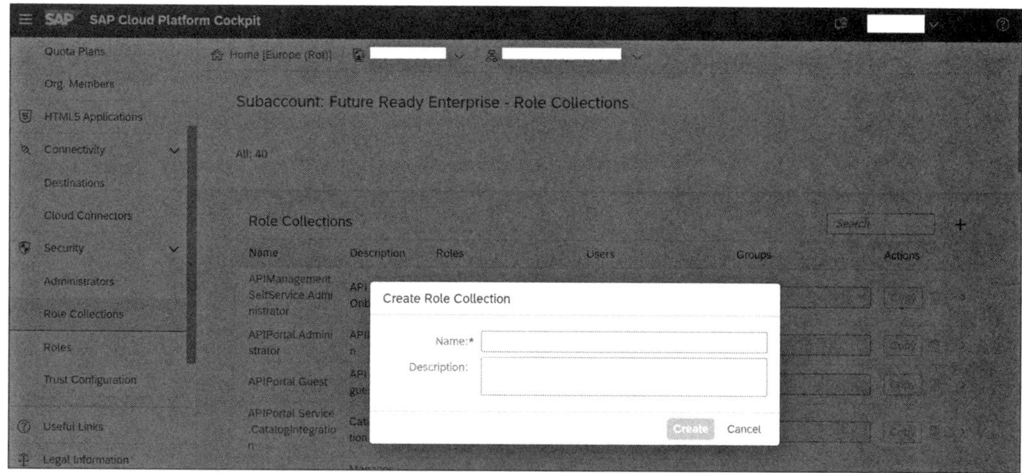

Figure 8.9 Create Role Collection

4. To assign a role to this new role collection, click on the role collection to open up its overview.
5. Choose **Edit**.
6. Choose **+** under **Roles**.
7. Select the required **Role Name**.
8. Click **Save**.

Map Role Collections to User Groups in the Identity Provider

In this section, we'll look into how to map role collections to user groups in the identity provider. First, you must meet the following perquisites:

- You have set up your custom SAML 2.0 identity provider and established trust with a subaccount, as explained in Section 8.1.
- You have configured groups as an assertion attribute in your identity provider.
- You have already created role collections and embedded roles within them as explained previously.

To set up the mapping, follow this process:

1. Navigate to your Cloud Foundry subaccount and choose **Security • Trust Configuration**.

2. Choose your custom trust settings and navigate to **Role Collection Mappings.**

3. Click **New Role Collection Mapping**. In the pop-up, enter the role collection you want to map, and enter the name of the user group in the **Value** field.

4. Click **Save.**

> **Note**
>
> You can do this step only if you have a customer identity provider established other than the default SAP ID service trust. The identity provider provides the user groups to which you want to assign role collections. You must know the name of the user group in the identity provider.
>
> In Identity Authentication, you can find the user groups in the administration console under **Users & Authorizations • User Groups.**

Directly Assign Role Collections to Users from an Identity Provider

You can directly assign role collections to business users provided by an SAML 2.0 identity provider. You need to have the prerequisites noted in the previous section met to configure this option.

Follow these steps to configure this option:

1. Navigate to your Cloud Foundry subaccount and choose **Security • Trust Configuration**.

2. Choose your custom trust settings and navigate to **Role Collection assignment**.

3. Enter the business user's name in the **User** field. Enter the name according to the name ID format configured for the identity provider. For the SAP ID service, use the email address.

4. To see the role collections that are currently assigned to a user, choose **Show Assignments**.

5. To assign a role collection, choose **Assign Role Collection**. Choose the name of the role collection you want to assign in the pop-up.

6. Click **Assign Role Collection** to save your changes.

Figure 8.10 outlines the relation between the authentication and authorization setups. You can see that the authentication and authorization artifacts exist together to provide end-to-end security for your application. The role templates defined in applications are assigned to your SAML identity provider users/user groups to manage authorizations, whereas the SAML identity provider is used for authentication to your application by establishing trust with your SAP Cloud Platform account.

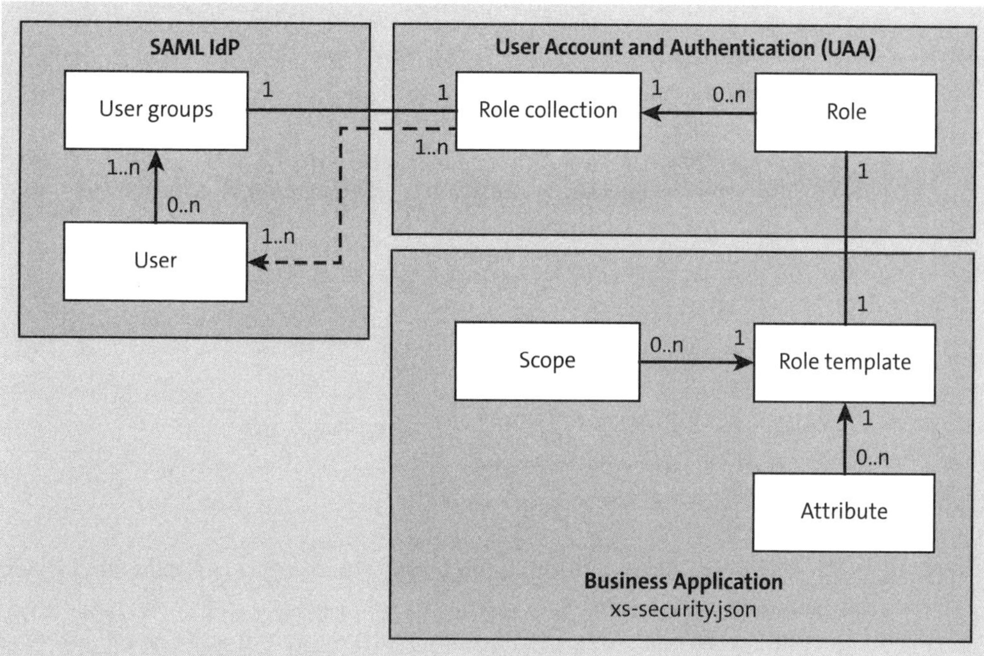

Figure 8.10 Mapping between Authentication and Authorization Artifacts

Assign Scopes to an Application

To assign scopes to an application, you need to perform the following steps:

1. Create an instance of the XSUAA service. You can do this in two different ways:

   – **Via mta.yaml**

   In your application's mta.yaml file, which we'll cover in detail in Chapter 11, you can specify a resource with the path to the *xs-security.json* file as discussed ahead.

   This step will create a service instance and bind the service instance to the application upon deployment of the application. You can find a sample resource definition in Listing 8.1.

```
resources:
  - name: <service_name>
    parameters:
        path: ./xs-security.json <path to xs-security.json file>
        service-plan: application
        service: xsuaa
    type: org.cloudfoundry.managed-service
```

Listing 8.1 Sample Resource Definition in mta.yaml

   – **Using Cloud Foundry command line interface (CLI)**

   Inside the CLI, go to the application folder where the *xs-security.json* file is stored and execute the command in Listing 8.2.

```
cf create-service <service_name> <service_plan> <service_instance_name>
-c xs-security.json
```

Listing 8.2 Cloud Foundry Command to Create Service Instance from xs-security.json

2. Bind the service instance to the application. If you have chosen to create an XSUAA service instance separately (rather than in *mta.yaml*), you need to bind the created service instance with an application.

   You can do this from the Cloud Foundry CLI by issuing the command in Listing 8.3.

```
cf bind-service <application_name> <service_instance_name>
```

Listing 8.3 CF Command to Bind Service Instance with Application

You can also perform this step from the SAP Cloud Platform cockpit.

8.2.2   Neo Environment

In this section, we'll cover an authorization management concept applied in the Neo environment and custom role creation. You can add a custom role to an application to configure additional access permissions for it without modifying the application's source code.

Custom roles are visible and applicable within the subaccount in which they're created. Creation of custom roles can be done as described in the following sections.

Define the Roles

In the following steps, you'll create various application-specific roles in the Neo environment:

1. Navigate inside your global account to a Neo environment subaccount.

2. Inside the subaccount, navigate to **Applications** and then to **Java Applications** or **HTML 5 Applications** (based on your application type).

3. For Java applications, navigate to **Security • Roles**. For HTML5 applications, navigate to **Roles**.

4. Click **New Role**, enter a role name, and click **Save.** A new role is created as shown in Figure 8.11.

⌂ New Role		
Name	Type	Actions
	Custom	🗑
	Custom	🗑
	Custom	🗑
	Custom	🗑
	Custom	🗑

Figure 8.11 Role Creation in Neo Environment

Create User Groups (Optional)

The next step is to assign users to the roles created in the previous step. You can directly assign users to the custom roles as explained ahead, but groups provide an easy way to manage role assignments to collections of users. To create groups, follow these steps:

1. Inside your subaccount, navigate to **Security • Authorizations**.
2. Open the **Groups** tab.
3. Select **New Group**. Enter a group name and click **Save**.

Assign Users or Roles to Group

To complete the configuration, you can now assign users and roles to a group. To assign roles, follow these steps:

1. Inside your subaccount, navigate to **Security • Authorizations**.
2. Open the **Groups** tab and select the group created previously.
3. In **Roles** section, click **Assign**.
4. In the pop-up, select the **Subaccount**, **Application**, and **Role** created previously and click **Save**.

To assign users directly to the group, follow these steps:

1. In the **Individual Users** section, click **Assign**.
2. In the pop-up, enter user information and click **Save**.

> **Tip**
>
> Alternatively, you can perform these steps inside the application view of the subaccount as well. Relationships among roles, users, and groups are as depicted in Figure 8.4.

> **Note**
>
> For securing SAP HANA applications in the Neo environment, refer to the guide at *http://s-prs.co/v515740*.

Map Group to Role when Using an Identity Provider

You can map the group created in SAP Cloud Platform to groups from an identity provider. Follow these steps to do so:

1. In SAP Cloud Platform, in the Neo environment subaccount, navigate to **Security • Authorizations • Groups** and choose **Add Default Group**. Default groups are the groups all users logged by this identity provider will have.
2. From the dropdown list, choose the required group.

3. Navigate to **Security • Authorizations • Groups** and choose **Add Assertion-Based Group**.

4. In the row that appears, enter the name of the group to which users will be mapped and define the rule for this mapping.

5. In the first field of **Mapping Rules**, enter the SAML 2.0 assertion attribute and choose the comparison operator. In the last field of **Mapping Rules**, enter the value with which you want to compare the specified SAML 2.0 assertion attribute.

> **Tip**
>
> Different services on SAP Cloud Platform are considered applications and have pre-defined roles for managing access and authorization.
>
> The authentication management and role management topics discussed in earlier sections also can be applied to the predefined roles from these services.

8.3   Credential Store Service and Keystore Service

SAP Cloud Platform provides a service to securely store your cryptographic keys, passwords, and certificates used for applications. In the following sections, we'll look at its setup in both the Cloud Foundry and Neo environments.

8.3.1   Cloud Foundry Environment

SAP Cloud Platform Credential Store provides a secure repository for passwords and keys for applications running on SAP Cloud Platform in the Cloud Foundry environment. It enables applications to retrieve credentials and use them for authentication to external services or to perform cryptographic operations and TLS communication.

You can create only one service instance per space for this service. The data in this service is logically isolated using namespaces. A namespace can correspond to a customer, a subaccount, or anything else specific to an application, and each credential operation is executed in the context of a namespace. If a service instance is shared across multiple applications, you can define different access permissions for each service binding based on the namespace.

This service is exposed to applications using a REST API. API calls are always executed in the context of a namespace.

Before you begin, you must meet the following prerequisites:

- You have a space in Cloud Foundry subaccount/global account.

- You have assigned the SAP Cloud Platform Credential Store as an entitlement to your Cloud Foundry subaccount.

To set up this service, follow these steps:

1. Inside your Cloud Foundry space, navigate to **Services** • **Service Marketplace**.

2. Choose the **Credential Store** service tile and navigate to **Instances**.

3. Click **New Instance** and follow through the steps in pop-up to create a service instance and assign it to an application.

4. Navigate to the service instance and choose **Credential Store**.

5. Choose **Create Namespace**, select the **Credential Type**, and click **Next**.

6. Enter a **Namespace** name, credential name, and values, and choose **Create**, as shown in Figure 8.12.

7. To further create credentials within a namespace, choose the namespace created, choose **Create Credential**, and choose a credential type: **Password** or **Key**.

8. Enter name of the credential and the value, then choose **Create**.

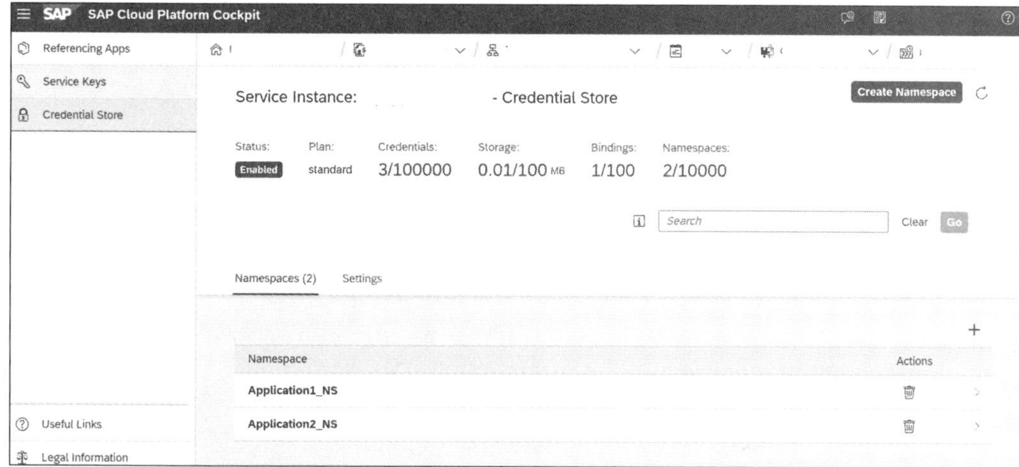

Figure 8.12 Credential Store Service in SAP Cloud Platform Cloud Foundry Environment

8.3.2   Neo Environment

Applications in SAP Cloud Platform in a Neo environment use the SAP Cloud Platform Keystore service as a repository for cryptographic keys and certificates. When using this service, applications can easily retrieve keystores and use them in various cryptographic operations and for performing SSL communication.

SAP Cloud Platform Keystore stores and provides keystores encoded in the following formats:

- Java Keystore (JKS)
- Extended Java Keystore (JCEKS)
- PKCS # 12 file (P12)
- Privacy-Enhanced Mail Certificate (PEM)

> **Note**
>
> To learn more about configuration and consumption of the Keystore service, visit *http://s-prs.co/v515741*.

8.4   OAuth 2.0 Application Protection

You can protect applications using the OAuth 2.0 protocol with the OAuth 2.0 service in the SAP Cloud Platform Neo environment.

OAuth 2.0 is a widely adopted security protocol for protection of resources over the internet. It allows applications to request authentication on behalf of users with third-party accounts, without the user having to grant credentials to the application.

SAP Cloud Platform supports two basic OAuth 2.0 flows:

1. **Authorization code grant**

   A human user authorizes a mobile application to access resources on a user's behalf.

2. **Client credentials grant**

   There is no human user but a device instead. In such a case, the access token is granted on the basis of client credentials only.

> **Note**
>
> You can read more about setting up OAuth 2.0–based protection for your application in the help document found at *http://s-prs.co/v515742*.

8.5   Secure Platform Operations

In this section, we'll look into various ways to secure your platform operations. As already mentioned, a platform has two types of users: business users and platform users. This section will cover platform users, who are responsible for the operation and management of the platform in the organization.

8.5.1   Predefined Roles and Functions

SAP Cloud Platform provides certain predefined roles with defined scopes. Based on the definition and scope of these roles, you can manage and segregate the access within your organization for different individuals.

In the following sections, we'll look at various predefined roles and respective functions provided by SAP Cloud Platform at the global account, subaccount, organization, and space levels.

Global Account

At the global account level, you have the role of administrator. As an administrator for the global account, you can perform following tasks:

- Create new subaccounts
- Assign entitlements to each subaccount
- Register systems
- Add other members
- In a consumption-based agreement model, monitor the overall credit consumption and manage costs

Subaccounts

Subaccounts in Neo and Cloud Foundry runtimes have different roles based on their purpose. In a Cloud Foundry-based subaccount, you have the predefined roles listed in Table 8.1 available.

Roles	Description
Organization manager	This role helps with the following: - Managing security - Trust configuration - Creating spaces - Cloud connector - SAP Cloud Platform Destination - Managing subscription to services - Creating space quota plans and assigning them to individual spaces You assign this role when adding a member at the subaccount level from the **Members** option in the sidebar in the subaccount.
Organization auditor	Auditor roles have a read-only view of the capabilities mentioned in the administrator role. You assign this role when adding a member at the subaccount level from the **Members** option in the sidebar in the subaccount.
Users and role administrator	This is an additional role available at to manage all settings related to security at the subaccount level. This role is relevant for security administrators to manage authentication and authorization in this subaccount. You assign this to users in a subaccount using by selecting **Security • Administrators** in the sidebar.

Table 8.1 Predefined roles in Cloud Foundry Environment Subaccount

In Neo-based subaccounts, you have the predefined roles listed in Table 8.2 available.

Roles	Description
Administrator	This role allows you to manage subaccount members. You can also manage subscriptions, trust, authorizations, and OAuth settings, and restart SAP HANA services on SAP HANA databases. Furthermore, you can view heap dumps and download a heap dump file. In addition, you have all permissions granted by the developer role, except the debug permission.
Developer	Supports typical development tasks, such as deploying, starting, stopping, and debugging applications. You can also change loggers and perform monitoring tasks, such as creating availability checks for your applications and executing MBean operations.
Support user	Designed for technical support engineers, this role enables you to read almost all data related to a subaccount, including its metadata, configuration settings, and log files. For you to read database content, a database administrator must assign the appropriate database permissions to you.
Application user admin	Assigned by the subaccount administrator to a subaccount member. Manage user permissions on the application level to access Java, HTML5 applications, and subscriptions. You can control permissions directly by assigning users to specific application roles or indirectly by assigning users to groups, which you then assign to application roles. You can also unassign users from roles or groups.
Cloud connector admin	Open secure tunnels via the cloud connector from on-premise networks to your subaccounts.

Table 8.2 Predefined Roles in Neo Environment at Subaccount Level

Spaces

Spaces (available in the Cloud Foundry subaccount) have the predefined roles listed in Table 8.3 available.

Roles	Description
Space manager	Administers a space within an organization
Space developer	Manages apps and services in a space
Space auditor	Read-only access to a space

Table 8.3 Predefined Roles at Space Level in Cloud Foundry Environment

8.5.2   Custom Platform Roles

An SAP Cloud Platform Neo environment allows further segregation of platform roles to manage the platform using custom role capabilities. You have an administrator role in a Neo environment subaccount.

> **Note**
>
> At the time of writing this book, this feature is available only in the Neo environment.

To begin your setup, follow these steps:

1. Inside the subaccount, click **Platform Roles** in the sidebar menu, as shown in Figure 8.13.

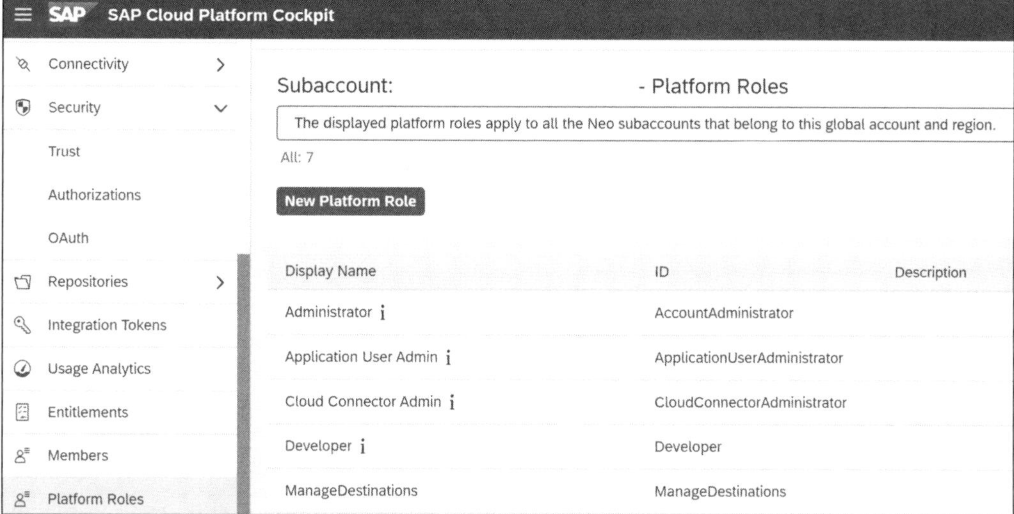

Figure 8.13 Platform Role Configuration

2. Click **New Platform Role**.
3. In the pop-up screen, enter an **ID** and (optionally) a **Display Name** and select the **Scopes** you want to provide to this role.
4. **Save** changes.
5. Optionally, you can make further changes (via the **Edit**, **Copy**, and **Delete** options) to this role.

> **Note**
>
> The platform role defined in one subaccount is applicable to all Neo environment subaccounts under this global account.

8.5.3   Platform Identity Provider

SAP Cloud Platform in Cloud Foundry- and Neo-based environments allows you to use a user store from your identity provider to log in to the SAP Cloud Platform cockpit instead of using the default login with an S-user or P-user.

To begin your setup, follow these steps:

1. Navigate inside your Neo-based subaccount to **Services** and search for "Platform Identity Provider".

2. Choose the **Platform Identity Provider** tile and **Enable** the service.

3. Navigate inside the same subaccount to **Security • Trust**.

4. Choose the **Platform Identity Provider** tab as shown in Figure 8.14.

Figure 8.14 Platform Identity Provider

5. Choose **Use Identity Authentication Tenant**. Select the required tenant and save.

You can now add users to this subaccount in the **Add Members** section by selecting **SAP Cloud Platform Identity Authentication Service** as the **User Base**.

Access the SAP Cloud Platform cockpit via the following URL: *https://account-<sub-account-name>.<SAP Cloud Platform host>*, where the subaccount name is your sub-account technical name and the SAP Cloud Platform host is the host address of SAP Cloud Platform (for example *us2.hana.ondemand.com* for US West (Chandler) DC).

> **Note**
>
> For setup and configuration of this feature for your Cloud Foundry-based global account, visit *http://s-prs.co/v515743*.

> **Note**
>
> This scenario setup needs the Identity Authentication service.
>
> If you want to use your own corporate identity provider as the user store, you need to set up your corporate identity provider with Identity Authentication. This is explained in Section 8.1.3.

8.6   Identity Provisioning Service

In the previous sections, we've explained how to set up authentication and authorization in SAP Cloud Platform for business users and platform rules. Now you need an automated way to provision the users from different systems and update permissions based on user role changes.

SAP Cloud Identity Services: Identity Provisioning automates identity lifecycle management processes. It helps you provision identities and their authorizations for various cloud and on-premise business applications. You can provision users and groups for various business applications that you can add as source and target systems by applying transformations and running them as scheduled or manual jobs. Your source system is usually your existing corporate user store, which can be in the cloud or on premise, and your target system is the cloud or on-premise system in which you want to populate the entities from the source system. You can provision entities from one SCIM-based system to another external non-SCIM one using a proxy connector without making a direct connection between them.

Before beginning your setup, you must have met the following prerequisites:

- You have administrator access to SAP Cloud Platform in a Neo-based subaccount.

- You have activated the Identity Provisioning service from **Services** in the Neo subaccount.

To begin your setup, follow these steps:

1. From the Neo subaccount, click the **Identity Provisioning Service** tile and choose **Go to Service**.

2. This opens the service's home page, as shown in Figure 8.15. The configuration of the service includes setting up various source Systems and target systems for which the user provisioning needs to happen.

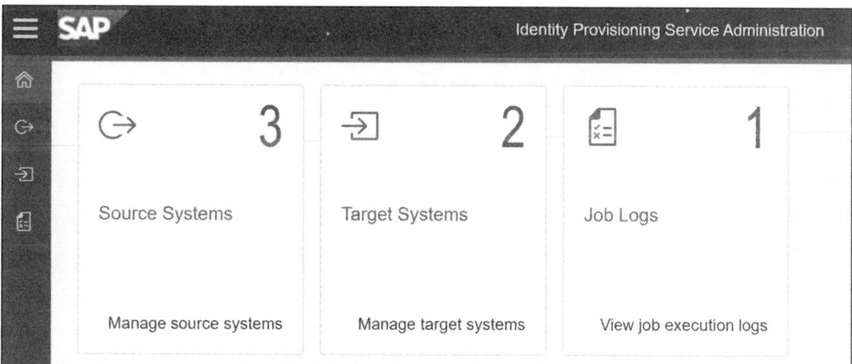

Figure 8.15 Identity Provisioning Service Home

3. To add a source system, choose the **Source System** tile from the home page or side navigation.

4. Choose **+Add**, select the **Type** for various systems supported, and provide a **System Name**.

5. To set up connectivity with the system, you can either create a destination in your SAP Cloud Platform's Neo subaccount and provide the **Destination Name** or set up connectivity parameters in the **Properties** tab.

> **Note**
>
> Destination setting is mandatory only for SAP NetWeaver Application Server ABAP systems.

6. Choose **Save**.

7. You can see the initial (default) transformation logic that converts the source system-specific JSON to a common JSON format. Adjust the transformations as needed, especially for initial setup cases, in the **Transformations** tab.

8. To add a target system, choose **Target System** from the home page or side navigation.

9. Choose **+Add**, select the **Type** of various systems supported, and provide a target **System Name**.

10. Provide a **Destination Name** (if already configured) to establish connectivity with a system, or enter connection details in the **Properties** tab.

11. Select **Source Systems**.

12. Choose **Save** to create the target system configuration, as shown in Figure 8.16.

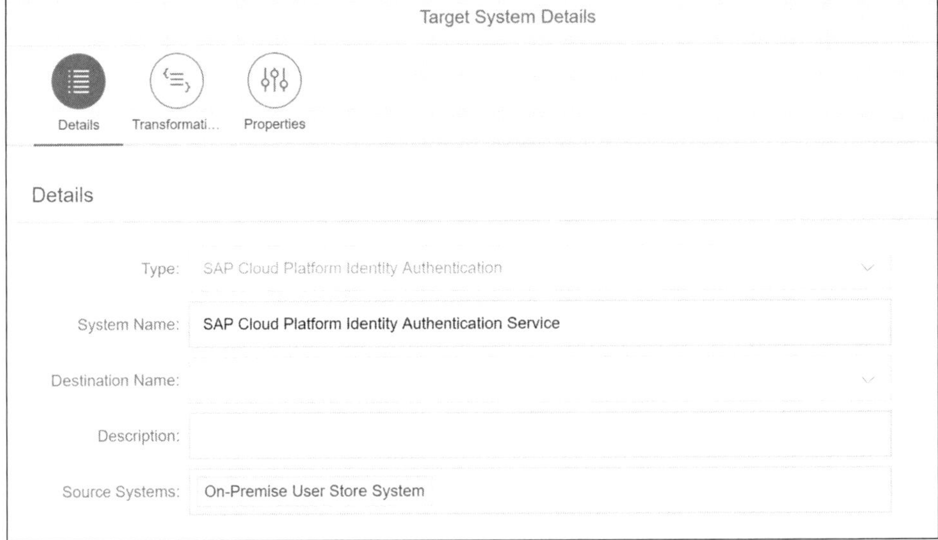

Figure 8.16 SAP Cloud Identity Services: Identity Provisioning, Sample Target System Configuration

13. Adjust the transformations as needed from the **Transformations** tab. This is where the common JSON format from the source system is passed to the target system.

14. To schedule provisioning of users, navigate to **Source Systems**, choose the source system from which the user provisioning needs to be scheduled, and go to the **Jobs** tab.

Note

You can read more information about Identity Provisioning at *http://s-prs.co/v515744*.

8.7   SAP Cloud Identity Access Governance

SAP Cloud Identity Access Governance enables you to use the following services to create access requests, analyze risks, and design roles. SAP Cloud Identity Access Governance service offers the following services:

- SAP Cloud Identity Access Governance, access analysis service
- SAP Cloud Identity Access Governance, access request service
- SAP Cloud Identity Access Governance, role design service
- SAP Cloud Identity Access Governance, access certification service

SAP Cloud Identity Access Governance integrates with other SAP Cloud Platform services and connects with cloud and on-premise target applications. SAP Cloud Identity Access Governance solutions use Identity Authentication for user authentication and to manage access to SAP Cloud Identity Access Governance apps. Security and permissions are maintained in groups and roles.

The access request service integrates with the following SAP Cloud Platform services to manage workflow management and business rules:

- SAP Cloud Platform Business Rules to provide decision-making and business logic
- SAP Cloud Platform Workflow Management to enable workflows for user access approvals and review
- The Identity Provisioning service to provision access requests to target systems

SAP Cloud Identity Access Governance supports integration with the Identity Authentication service and SAP Cloud Platform. This enables Identity Authentication users and SAP Cloud Platform users to initiate access requests that are then further provisioned to target applications. In addition, it also supports multiple other integration scenarios, including with SAP S/4HANA, which allows SAP S/4HANA users to use SAP Cloud Identity Access Governance services, such as the access request service and access analysis service, and features such as autoprovisioning and auditable workflows.

8.8 Audit Logging

Your organization will need, from a security and compliance perspective, to regularly audit SAP Cloud Platform's application operations. SAP Cloud Platform provides an Audit Log Retrieval API, which allows you to retrieve the audit log for your environment. It provides audit log results as a collection of JSON entities.

In the following sections, we'll look in detail at how to configure the audit logging service for Cloud Foundry and Neo environments in SAP Cloud Platform.

8.8.1 Cloud Foundry Environment

The audit log data stored for your account will be retained for 30 days, after which it is deleted. If you want to retain and store it for more than 30 days, you can retrieve it using the Audit Log Retrieval API as follows and store it in persistent storage:

1. Navigate inside your Cloud Foundry space to **Services • Service Marketplace**.

2. Choose **Auditlog Management** and navigate to **Instances**.

3. Choose **New Instance** and follow through the pop-up windows to create a service instance.

4. Choose the service instance created and navigate to **Service Keys**. Choose **Create Service Key**, enter a **Name**, and **Save** the changes.

5. Note the values for `uaa:url`, `uaa:clientid`, and `uaa:clientsecret` from the service key and create an OAuth access token. The access token is valid for 12 hours.

6. You can retrieve the audit logs by executing the following HTTP GET request: *<url_ from_service_key>/auditlog/v2/auditlogrecords*.

7. Provide the OAuth access token as an `"Authorization"` header (`"Authorization: Bearer <access_token>"`) and you will receive the audit logs in JSON format.

8. The Audit Log Retrieval API supports server-side chunking. Hence, if a given query produces large results, the results will be "chunked." That means you can provide following parameters while calling the API :

   - `time_from` and `time_to`
     The times are in UTC, and if no time is specified, it returns the value equal to the last 30 days.

   - `handle`
     The result is chunked if the result is too large. Then `handle=<value>` can be provided as a request parameter to retrieve the next set.

   The API documentation can be found in the SAP API Business Hub at *https:// api.sap.com/api/CFAuditLogRetrievalAPI/resource*.

SAP Cloud Platform's Cloud Foundry environment also provides a feature to display the audit logs for your Cloud Foundry account. These logs are produced by SAP applications and services you have subscribed to.

To activate this service, follow these steps:

1. Inside your subaccount, navigate to **Subscriptions** and select **Audit Log Viewer**, as in Figure 8.17, then click **Subscribe**.

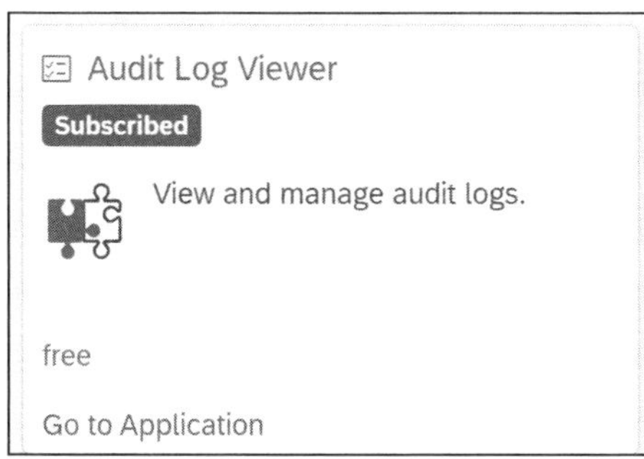

Figure 8.17 Audit Log Viewer in Cloud Foundry

2. Navigate to **Security Role Collections**, select a role collection, and add the roles shown in Figure 8.18.

Role Name	Role Template	Application Identifier
Auditlog_Auditor	Auditlog_Auditor	auditlog-management!
Auditlog_Auditor	Auditlog_Auditor	auditlog-viewer!t

Figure 8.18 Audit Log Viewer Roles

3. Navigate back to **Subscriptions**, choose **Audit Log Viewer**, and click **Go to Application**.

4. You can see the various logs written for your account for a given time frame, download the logs, and get more information on each selected audit log.

8.8.2   Neo Environment

The Audit Log Retrieval API for Neo environment supports the OData 4.0 standard and thus provides the result as OData with a collection of JSON entities. The Audit Log Retrieval API is protected with OAuth 2.0 client credentials. To call the API, create an OAuth client and obtain an access token. The retrieval API is documented in the SAP API Business Hub at *https://api.sap.com/api/AuditLogRetrievalAPI/resource*.

The Neo environment also provides an Audit Log Retention API which allows you to view your currently active retention period for all the audit log data stored in your account. Using this API, you can modify the default retention period to meet your business requirements. The retention API is documented in the SAP API Business Hub at *https://api.sap.com/api/AuditLogRetentionAPI/resource*.

8.8.3   Identity Authentication Logs

The Identity Authentication service provides logs. To retrieve them, follow these steps:

1. Inside the Identity Authentication admin console, navigate to **Monitoring & Reporting • Logs**.

2. Under **Generate Client Credentials for Audit Logs**, choose the **Generate** button. A dialog box with the generated **Client ID** and **Client Secret** appears.

3. Use the client ID and client secret to obtain an access token. Use the Audit Log Retrieval API for the Neo environment with these details to retrieve Identity Authentication-specific audit logs. Categories relevant for Identity Authentication are as follows:

   - `audit.authentication`
     Audit logs related to authentication

   - `audit.data-change`
     Audit logs related to changes in personal data

4. Under **Export Change Logs**, you can export logs as CSV files containing the history of operations for tenant administration, and configurations for identity providers and service providers.

> **Tip**
>
> To learn how to get an OAuth access token for the Neo API, visit *http://s-prs.co/v515745*.

8.9   Summary

In this chapter, we covered various SAP Cloud Platform security settings for business and platform users for managing your extensions, integrations, and the platform itself.

We covered authentication and authorization management in SAP Cloud Platform. We discussed how to set up authentication for your business and platform users for SAP Cloud Identity Services: Identity Authentication and how to set up authorizations using role collections/groups. We also covered on how you can secure your application passwords and keys using SAP Cloud Platform Credential Store or SAP Cloud Platform Keystore.

In addition, we covered how you can use provisioning and automate identity lifecycle management using SAP Cloud Identity Services: Identity Provisioning and covered how SAP Cloud Identity Access Governance can be used to analyze access information and process access requests, providing transparency and governance for an organization.

Finally, we covered how you can use audit log APIs in SAP Cloud Platform to retrieve various audit-logging events for security administration.

Future Creation

Chapter 9
Developing SAP S/4HANA Extensions

In previous chapters, you've learned about various design and architectural principles that need to be applied during your business transformation journey to build a future-ready enterprise. In this chapter, we'll look deeper into how you can create extensions in your landscape to support and build additional business functionalities in order to differentiate your company in the market in an agile and flexible way. We'll look at various functionalities available in SAP Cloud Platform to support you in this journey based on different business requirements.

In the previous chapter, we investigated how organizations need to start defining and realize various design principles in the future-aware stage to become future-ready enterprises. In this chapter, we focus on the future-focused stage, in which organizations build on the foundations from previous stages to enable for future digital transformation needs.

We'll look into how organizations can develop and create specific extensions for their standard-delivered software, like SAP S/4HANA, to meet their business demands and adapt to changing market needs. Organizations look to use standard processes and standard software as much as possible and whenever possible. However, organizations often run into requirements to adapt standard business software to their specific business needs. The adaptation requirements are triggered by one or more of the following needs:

- Business processes specific to your industry
- Differentiation from competitors
- Competitive advantage
- Simplifying the experience for end users
- Creating business processes specific to a region
- Adaptation to legal regulations in an operating market unit
- Developing new business models to meet changing market needs

In this chapter, we'll look at various methods by which organizations can adapt their standard SAP S/4HANA software to suit their business needs.

We cover in depth how various SAP Cloud Platform services can help your organization adapt its business software to suit your needs and requirements with applications and extensions that are loosely coupled with your core and support lifecycle stability with released APIs from SAP S/4HANA. Thus, you can build extensions, that are independent and abstracted from your SAP software updates and require less maintenance after upgrades. In addition, we'll cover how SAP Cloud Platform can help you build these extensions to tie in best way to standard business objects and processes in SAP S/4HANA.

We'll also look into how applications and extensions that you build using SAP Cloud Platform can not only be adapted to your changing requirements but also be scalable and resilient in the cloud. In the following sections, we'll look into SAP Cloud Platform as a place to simplify extension development, plus various runtime services available within SAP Cloud Platform to support your application runtime.

9.1   Building Extensions to Differentiate

As previously mentioned, while organizations try to use standard processes and standard business software, they often need to customize the software for their business needs.

Gartner approaches businesses differentiating themselves from their competition through its PACE layering strategy with three layered systems: systems of record, systems of differentiation, and systems of innovation. Per Gartner, the PACE layering application strategy is a methodology for categorizing, selecting, managing, and governing applications to support business change, differentiation, and innovation. Gartner claims that adopting this strategy can help in building a business application strategy that accelerates innovation with faster response and better ROI, without sacrificing integration, integrity, or governance. Thus, organizations often must customize their software by creating extension applications.

These customizations can be of different types based on the requirements and at a broader level can be classified into the following groups:

- In-app extensibility, done within your core business system, SAP S/4HANA
- Side-by-side extensibility, which cannot be done in the core system and is performed using SAP Cloud Platform extension capabilities

The concepts of in-app and side-by-side extensions are applicable for both SAP S/4HANA Cloud and SAP S/4HANA systems. An important aspect to consider during extensions is lifecycle stability, which they provide through decoupling. Decoupling means that only extensions in the SAP S/4HANA core that do not affect upgrades of the core system are permitted.

Table 9.1 outlines the different use cases in each category while your organization is developing applications and extensions.

In-App Extensions	Side-By-Side Extensions
Create custom fields/adapt user interface for end users	Create new experience to differentiate for B2E, B2B, and B2C
Create custom business objects	Create applications and extensions that utilize data from multiple systems (cloud or on-premise)
Create forms/email templates	Create applications that require preprocessing of data before entering the SAP S/4HANA core—for example, for processing IoT data or data from other event streams or checking for correctness of data before entering it in the core
Create custom CDS views	Create applications that require postprocessing of data—for example, for triggering specific rules and workflows on creation of a sales order or business partner in SAP S/4HANA
Create custom analytics using existing CDS views	Create data lake and analytical scenarios that consume data from different systems or provide an analytical dashboard with only relevant data in databases physically closer to third-party end user like vendors/suppliers
Add application logic to enhance existing business processes	Create a completely new application that can create a new business model/process or substitute an existing business process in SAP S/4HANA
N/A	Create applications for external vendors and suppliers to manage business processes for your organization directly
Possible for scenarios that don't impact the SAP S/4HANA upgrade process	Applications and extensions are decoupled with core systems through stable and released APIs and events
Target audience: key users with deep knowledge of business systems	Target audience: specialized ABAP developers or cloud-native developers

Table 9.1 In-App and Side-By-Side Extensions Use Cases

Figure 9.1 outlines the application patterns based on architectural best practices for building these applications and extensions.

Side-by-side extensions are developed in SAP Cloud Platform, and in subsequent sections, we will look at various options to develop these extensions for your SAP S/4HANA system with SAP Cloud Platform's extension capabilities.

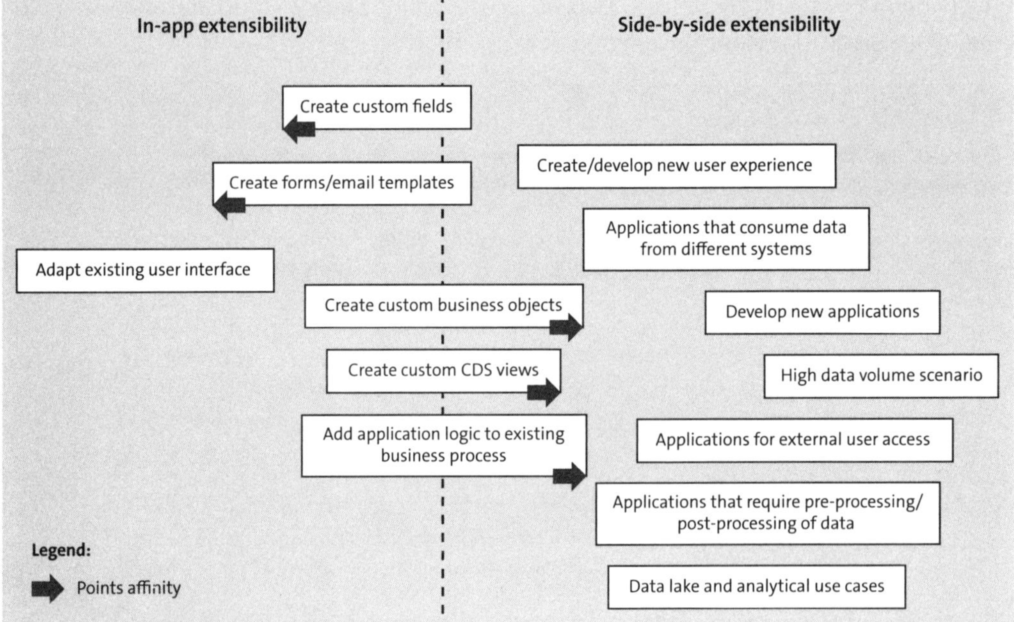

Figure 9.1 In-App Extensibility versus Side-by-Side Extensibility

9.2   SAP Cloud Platform's Extension Capabilities

SAP Cloud Platform is the go-to cloud-native extension framework for intelligent enterprises that allows you to discover and consume APIs and events from SAP's LOB solutions in order to develop decoupled extensions quickly in a flexible way. It also enables you to access SAP and third-party application APIs in a standardized way while developing your extensions.

The following are the high-level steps to connect SAP S/4HANA Cloud with SAP Cloud Platform to develop extensions:

1. SAP S/4HANA Cloud has multiple predefined communication scenarios for different use cases. You can use the Custom Communication Scenarios app in SAP S/4HANA Cloud to create communication scenarios if required.

2. Establish a connection between SAP S/4HANA Cloud with SAP Cloud Platform.

3. Make SAP S/4HANA Cloud APIs and events accessible within SAP Cloud Platform subaccounts.

Let's now look at these steps in detail and see how to set up and configure them:

1. Inside your SAP Cloud Platform Global Account, navigate to **System Landscape • Systems**.

2. Click on **Register System**, and in the pop-up screen, provide the **System Name** and **Type**. Click **Register**.

3. Copy the integration token generated for the SAP system; the status of the system registration is **Pending**.

4. In your SAP S/4HANA Cloud system, navigate to **Home • Communication Management • Display SAP Cloud Platform Extensions**.

5. Click **New** and provide the copied integration token under **Integration Token** and a **Description**. Click **Save**. The system status is set to **Enabled**, as shown in Figure 9.2.

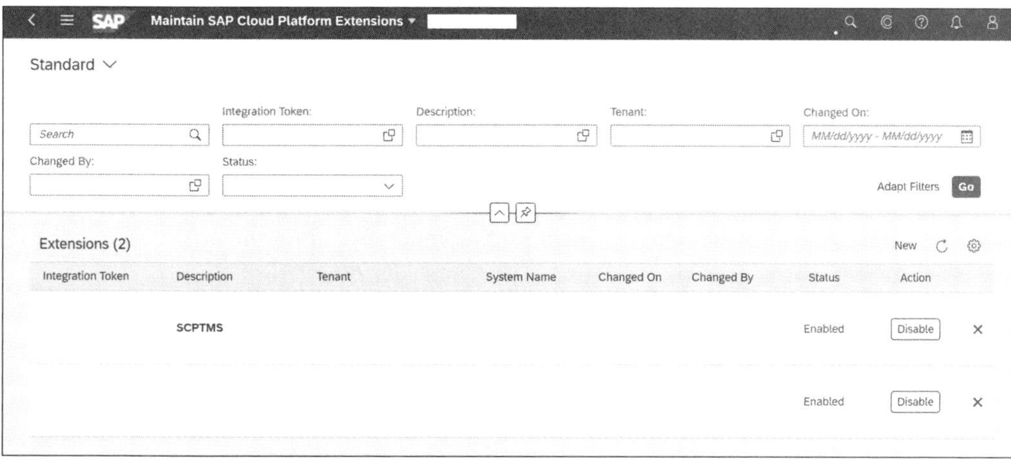

Figure 9.2 Maintain SAP Cloud Platform Extensions in SAP S/4HANA Cloud

6. Navigate back to the SAP Cloud Platform global account in **System Landscape • Systems**. The status of the registration changes to **Registered**, as shown in Figure 9.3, which means that the automated system process has been successfully completed.

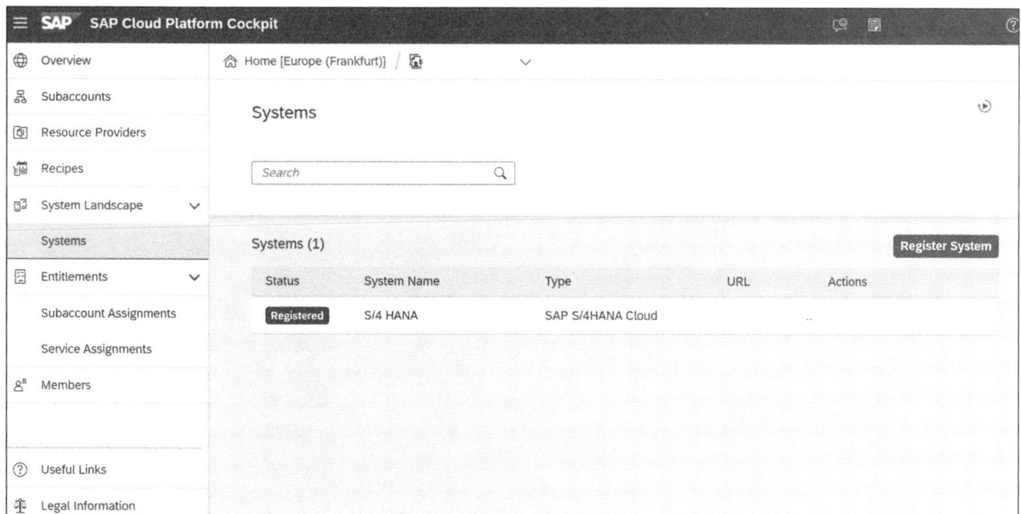

Figure 9.3 Register SAP System in SAP Cloud Platform

7. After registering the systems, you can create a formation to assign different SAP systems that you want to extend in the context of the same business case.

8. For this, navigate to **System Landscape • Formations** and choose **Create Formation**.

9. On the **Create Formation** screen, provide a unique **Name**, select a **Subaccount** (previously created), and select all systems to be assigned to this formation. Choose **Create**.

> **Note**
>
> At the time of writing this book, the formations capability is released only in SAP Cloud Platform Enterprise Agreement-based global accounts. You can edit formations later to change system assignment as well as edit the subaccount later. For this, navigate inside formation and choose **Assign Systems** or **Edit Subaccount**

10. Navigate to **Entitlements • Subaccount Assignments**. Enter the subaccount name to which you want to add this service plan and click **Go**.

11. Click **Configure Entitlements**, then choose **Add Service Plan**.

12. In the pop-up screen, select the **SAP S/4HANA Cloud Extensibility** service and choose the **System Name** previously created as shown in Figure 9.4. The following plans are available for the service:

    — **messaging**
      Provides access to events from SAP S/4HANA Cloud.

    — **api-access**
      Provides access to SAP S/4HANA Cloud APIs. Supports both predefined and custom communication scenarios for SAP S/4HANA Cloud.

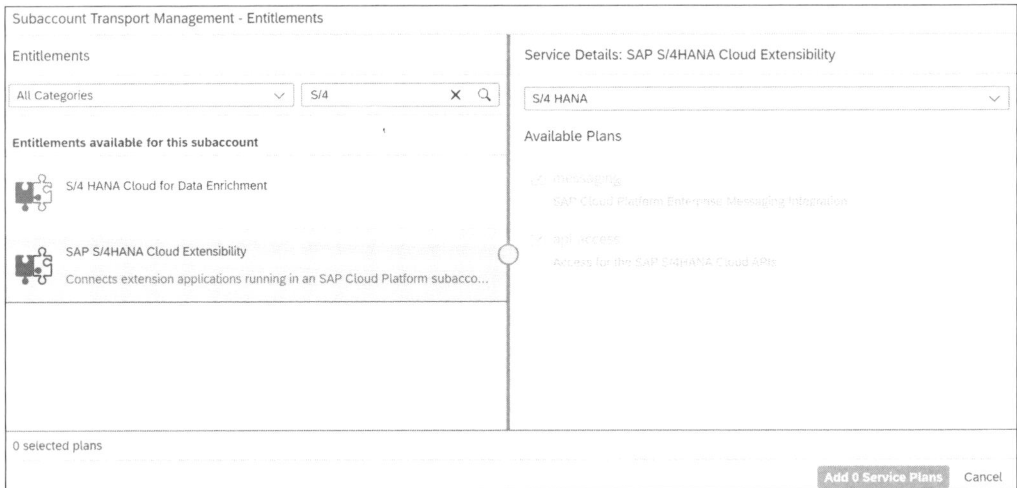

Figure 9.4 SAP S/4HANA Cloud Extensibility Service Plan

13. Select both plans and click **Add Service Plans**. Click **Save**.

> **Note**
>
> SAP S/4HANA Cloud API definitions are published to the SAP API Business Hub at *https://api.sap.com/package/SAPS4HANACloud?section=Artifacts*.
>
> SAP S/4HANA Cloud business events are available on the SAP API Business Hub at *https://api.sap.com/package/SAPS4HANACloudBusinessEvents?section=Artifacts*.

Exposed APIs and events can be consumed within SAP Cloud Platform to develop further business extension applications. SAP Cloud Platform provides the following runtime capabilities, which can consume APIs and events to develop extensions. Details of these runtimes are covered in depth in subsequent sections:

- SAP Cloud Platform Application Runtime
- SAP Cloud Platform Serverless Runtime
- SAP Cloud Platform, ABAP environment
- SAP Cloud Platform, Kyma runtime

SAP Cloud Platform's extension capabilities can be visualized as shown in Figure 9.5.

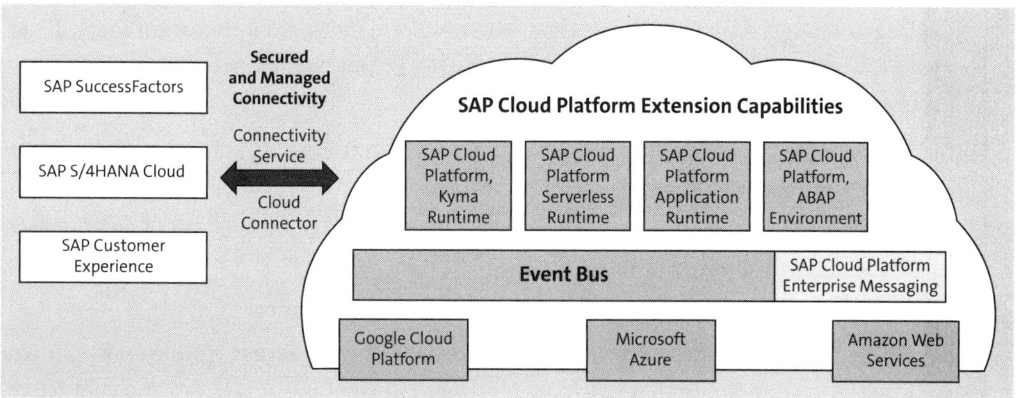

Figure 9.5 SAP Cloud Platform's Extension Capabilities

> **Note**
>
> At the time of writing this book, SAP Cloud Platform's extension capabilities are not yet supported for use with SAP S/4HANA on-premise, but the runtimes can be used to develop applications.
>
> Therefore, for SAP S/4HANA, you can still expose the services through the SAP Cloud Platform Connectivity's cloud connector, which can be consumed from applications developed in any of the runtimes listed earlier through SAP Cloud Platform Destination. These configurations were explained in Chapter 5, Section 5.2.

9.3   SAP Cloud Platform Runtimes

As mentioned in the previous section, in this section we'll discuss how to use the various runtimes in SAP Cloud Platform to develop applications and extensions:

- SAP Cloud Platform Application Runtime
- SAP Cloud Platform Serverless Runtime
- SAP Cloud Platform, ABAP environment
- SAP Cloud Platform, Kyma runtime

9.3.1   SAP Cloud Platform Application Runtime

You can develop microservice-based applications on SAP Cloud Platform using SAP Cloud Platform Application Runtime. This service allows you to bring your buildpacks into the Cloud Foundry environment and develop applications in different programming languages that your organization is comfortable using. In this section, we'll look at how you can develop an application using SAP Cloud Platform Application Runtime consuming SAP S/4HANA Cloud APIs set up using SAP Cloud Platform, as described in the previous section.

Before you begin, you must have assigned SAP Cloud Platform Application Runtime to your Cloud Foundry environment subaccount. Next, follow these steps to perform your setup:

1. Inside your Cloud Foundry subaccount, navigate to **Spaces** and choose the space where you want to develop your application.

2. Inside that space, navigate to **Services • Service Marketplace**. Choose the **SAP S/4HANA Cloud Extensibility** service tile.

3. Choose the tile and navigate to **Instances**. Click **New Instance**.

4. In the **Create Instance** pop-up screen, choose **Plan: api_access** from the dropdown and the **System Name** that you have registered in SAP Cloud Platform and click **Next**.

5. In next screen, **Specify Parameters**, you need to provide a JSON file for the communication scenario/API that you want to call from SAP S/4HANA Cloud. The JSON file to be provided has the parameters defined in Table 9.2.

Parameter	Required	Description
systemName	Yes	Name of the system registered in SAP Cloud Platform.
communication-Arrangement	Yes	Represents a communication arrangement in SAP S/4HANA Cloud.

Table 9.2 JSON Attributes for SAP S/4HANA Cloud Extensibility Service Instance

Parameter	Required	Description
communication-ArrangementName	Yes	Property within communicationArrangement. Name of the communication arrangement that will be created in SAP S/4HANA Cloud.
scenarioId	Yes	Property within communicationArrangement. ID of SAP S/4HANA Cloud communication scenario. Example: SAP_COM_<number>.
inboundAuthentication	Yes, if there is no outboundAuthentication defined	Property within communicationArrangement. Authentication type for SAP S/4HANA Cloud API access. Two types of authentication are supported: ■ Basic authentication with the value: BasicAuthentication ■ SAML bearer assertion authentication with the value: OAuth2SAMLBearerAssertion
outboundAuthentication	Yes, if no inboundAuthentication is defined	Property within communicationArrangement. Type of authentication used by SAP S/4HANA Cloud to call SAP Cloud Platform. The following methods are supported: ■ Basic authentication with the value: BasicAuthentication ■ OAuth 2.0 client credentials with the value: OAuth2ClientCredentials ■ No authentication
communicationSystem	No	Property within communicationArrangement. This is the communication system view of communication arrangement.

Table 9.2 JSON Attributes for SAP S/4HANA Cloud Extensibility Service Instance (Cont.)

Parameter	Required	Description
communication-SystemHostName	Yes	Property within communicationSystem. URL of remote system hosting the APIs that will be consumed in case the scenario contains outbound communication. This is the URL of the extension application running on SAP Cloud Platform.
port	No	Property within communicationSystem. Port for outbound calls to remote system. If not specified, default value is 443.
oAuthAuthEndPoint	No	Property within communicationSystem. OAuth authorization end point of the remote OAuth service for outbound communication.
oAuthTokenEndpoint	No	Property within communicationSystem. This is oAuth token endpoint used for outbound communication.
outboundCommunica-tionUser	No	Property within communicationSystem. Communication user used for outbound communication.
username	Yes	Property within outboundCommunicationUser. Username of the communication user.
Password	Yes	Property within outboundCommunicationUser. Password of the communication user.
outboundServices	No	Property within communicationArrangement. List of outbound service objects. You get this list from outbound services in SAP S/4HANA Cloud communication arrangement view.
outboundService	No	Property within outboundServices. Specific outbound service object.

Table 9.2 JSON Attributes for SAP S/4HANA Cloud Extensibility Service Instance (Cont.)

Parameter	Required	Description
name	Yes	Property within outboundService. Name of the outbound service. It must be the exact name displayed in SAP S/4HANA Cloud.
urlPath	No	Property within outboundService. This is a Path field in SAP S/4HANA Cloud.
isServiceActive	No	Property within outboundService This is the **Service Status** checkbox in SAP S/4HANA Cloud.
attributes	No	Property within outboundService. List of attribute objects, equivalent to the **Additional Properties** section of the outbound service.
attribute	No	Property within attributes. Specific attribute object.
name	Yes	Property within attribute. Name of additional attribute of outbound service. This is technical property name in the display communication scenarios in SAP S/4HANA Cloud.
value	Yes	Property within attribute. Value of additional attribute of outbound service.
attributes	No	Property within communicationArrangement. List of attribute objects; equivalent to the **Additional Properties** section of the communication arrangement in SAP S/4HANA Cloud.
attribute	No	Property within attributes. A specific attribute object that is an additional property in the communication arrangement.

Table 9.2 JSON Attributes for SAP S/4HANA Cloud Extensibility Service Instance (Cont.)

Parameter	Required	Description
name	Yes	Property within attribute. It's equivalent to the technical property name in the properties table in the Display Communication Scenarios app in SAP S/4HANA Cloud.
value	Yes	Property within attribute. Value of an additional property of the communication arrangement.

Table 9.2 JSON Attributes for SAP S/4HANA Cloud Extensibility Service Instance (Cont.)

Listing 9.1 provides a sample JSON file to configure API access for business partners from SAP S/4HANA Cloud. Please note that this example uses BasicAuthentication.

```
{
    "systemName": "<<System Name Provided in the System>>",
    "communicationArrangement": {
        "communicationArrangementName": "<<Communication Arrangement Name>>",
        "scenarioId": "SAP_COM_0008", //ScenarioId for Business Partner
        "inboundAuthentication": "BasicAuthentication",
        "outboundAuthentication": "BasicAuthentication",
        "outboundServices": [
            {
                "name": "Replicate Customers from S/4 System to Client",
                "isServiceActive": false
            },
            {
                "name": "Replicate Suppliers from S/4 System to Client",
                "isServiceActive": false
            },
            {
                "name": "Replicate Company Addresses from S/4 System to
Client",
                "isServiceActive": false
            },
            {
                "name": "Replicate Workplace Addresses from S/4 System to
Client",
                "isServiceActive": false
            },
            {
```

```
                        "name": "Replicate Personal Addresses from S/4 System to
    Client",
                        "isServiceActive": false
                },
                {
                        "name": "Business Partner - Replicate from SAP S/4HANA Cloud
    to Client",
                        "isServiceActive": false
                },
                {
                        "name": "Business Partner Relationship - Replicate from SAP
    S/4HANA Cloud to Client",
                        "isServiceActive": false
                },
                {
                        "name": "Business Partner - Send Confirmation from SAP S/4HANA
    Cloud to Client",
                        "isServiceActive": false
                },
                {
                        "name": "BP Relationship - Send Confirmation from SAP S/4HANA
    Cloud to Client",
                        "isServiceActive": false
                }
            ],
            "communicationSystem": {
                "communicationSystemHostname": "default.com",
                "outboundCommunicationUser": {
                    "username": "DefaultUser",
                    "password": "DefaultPassword"
                }
            }
        }
    }
}
```

Listing 9.1 SAP Business Partner Service Instance in SAP Cloud Platform

6. Click **Next** and go through the next steps and provide an **Instance Name**. Click **Finish**. A service instance is created.

7. Inside your subaccount, navigate to **Connectivity • Destinations** and you will find a destination service created with the same name as service instance name provided, as shown in Figure 9.6.

8. You can configure the authentication type with your SAP S/4HANA Cloud system further here. The following authentication options are supported with SAP S/4HANA Cloud:

- Basic authentication
- Principal propagation with OAuth 2.0 SAML bearer assertion

> **Note**
>
> The authentication type supported for each scenario is defined in the communication scenario.

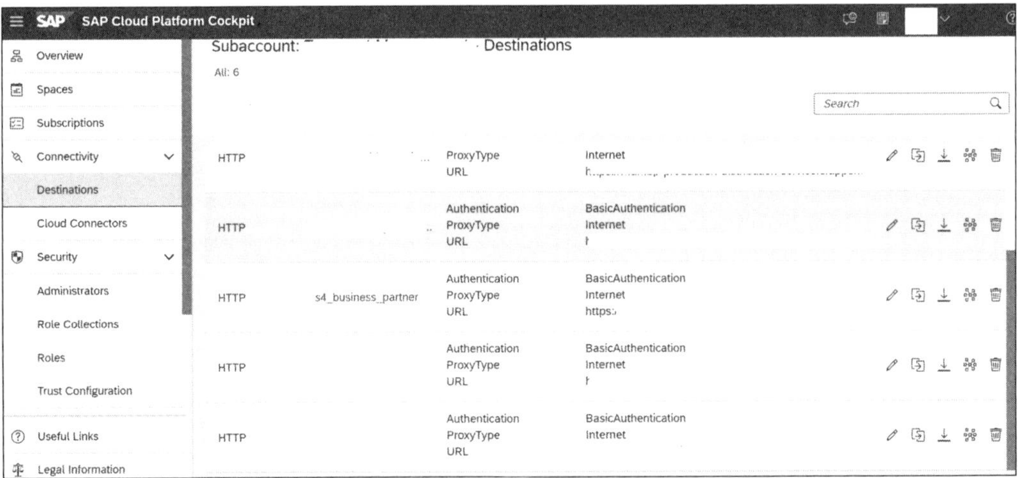

Figure 9.6 Destination Service in Subaccount with Service Instance Name

The next step is to create a Cloud Foundry-based application, which can be bound to the service instance created earlier and can use the destination service that has been automatically created. For this, use SAP Cloud Platform Business Application Studio and SAP Cloud Application Programming Model.

SAP Cloud Application Programming Model is an open, yet guided, framework of languages, libraries, and tools for building enterprise-grade services and applications. It guides developers through proven best practices and a great wealth of out-of-the-box solutions for recurring tasks and support for SAP Fiori and SAP HANA, thus helping organizations develop at an accelerated pace while still managing enterprise readiness.

> **Tip**
>
> You can also combine SAP Cloud Application Programming Model-based services with alternative UI or database technologies because SAP Cloud Application Programming Model's core components are explicitly designed for you to integrate with alternatives.

Primary building blocks of SAP Cloud Application Programming Model applications are as follows:

- **Core data services (CDS)**
  CDS serves as a modeling language—for example, for capturing domain models and service definitions. It's a family of domain-specific languages and corresponding notations.

- **Service SDKs**
  The SAP Cloud Application Programming Model has libraries for Java and Node.js to provide and consume services through synchronous and asynchronous APIs. Service SDKs help to provide out-of-the-box integration to lower platform-level services such as authentication and credential flows in an abstracted manner, which helps avoid any hard wirings.

- **Development tools**
  Although using SAP Cloud Application Programming Model doesn't require any specific editor, dedicated support for SAP Cloud Application Programming Model is provided through SAP Cloud Platform Business Application Studio, as well as through plug-ins for Visual Studio Code, Eclipse, and SAP Web IDE (full stack).

To begin the creation of your application, follow these steps:

1. Navigate inside your Cloud Foundry environment subaccount to **Subscriptions** and select **SAP Business Application Studio**.

2. Click the **Subscribe** button (if the service is not already subscribed).

3. Click **Manage Roles** and assign role templates from SAP Cloud Platform Business Application Studio to your role collections. Table 9.3 lists the role templates with SAP Cloud Platform Business Application Studio.

Role Template	Description
`Business_Application_Studio_Administrator`	Provides users access to manage administrative rights and manage user data
`Business_Application_Studio_Developer`	Provides users access to load and develop applications

Table 9.3 SAP Cloud Platform Business Application Studio Role Templates

4. If role collections do not exist, you can create one by navigating to **Security • Role Collections** inside your Cloud Foundry environment subaccount.

5. Assign the role collection to users. Navigate to **Security • Trust Configurations**, choose your identity provider, and enter a user ID to assign role collection. For SAP's default identity provider, enter your email address as the user ID.

6. Click **Go to Application** to launch SAP Cloud Platform Business Application Studio.

7. Click **Create Dev Space**.

8. Enter a **Dev Space name** and select the **SAP Cloud Business Application** radio button. Select other development tools you need in the space as shown in Figure 9.7.

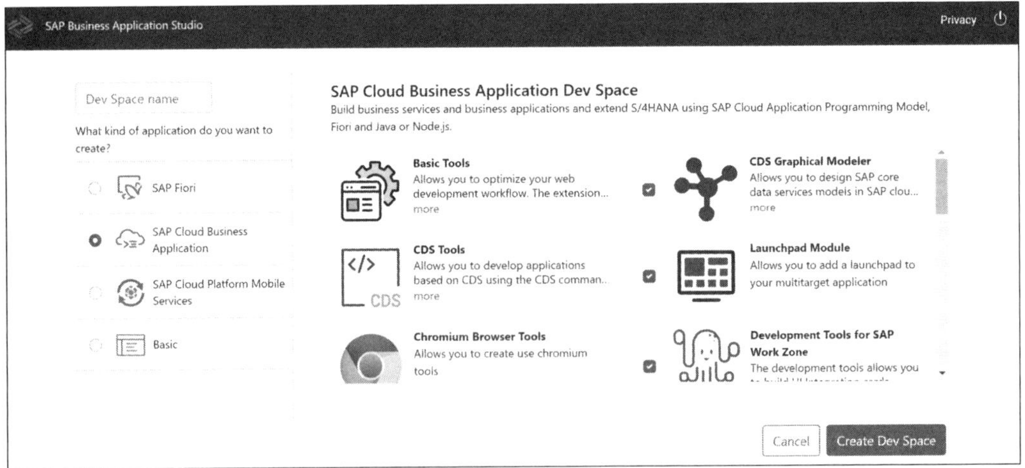

Figure 9.7 SAP Cloud Platform Business Application Studio: SAP Cloud Business Application

9. Click **Create Dev Space** and a new dev space is created in SAP Cloud Platform Business Application Studio.

Tip

You can set the organization and space for SAP Cloud Platform Business Application Studio. To do so, follow these steps:

1. As shown in Figure 9.8, click the **Cloud Foundry** icon ❶, followed by the **Login** icon ❷.

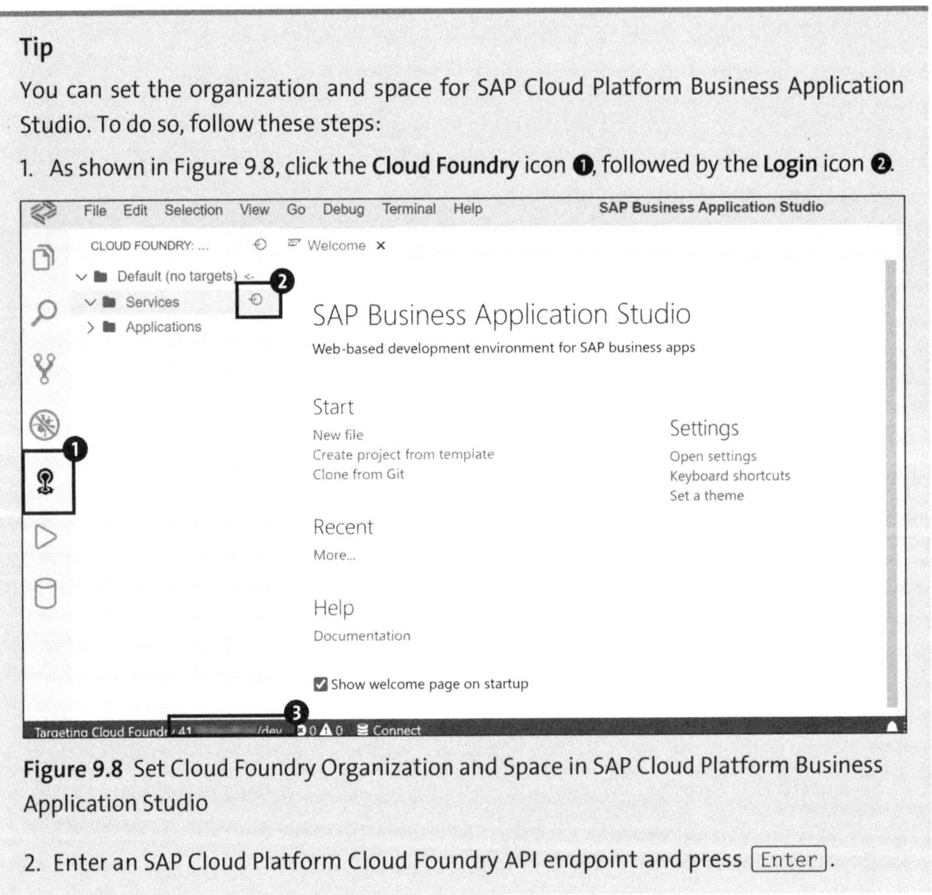

Figure 9.8 Set Cloud Foundry Organization and Space in SAP Cloud Platform Business Application Studio

2. Enter an SAP Cloud Platform Cloud Foundry API endpoint and press Enter.

3. Enter your user name and password, select your organization, and click **Enter**.

4. Then select the space information and click **Enter**.

This will set the organization and space in your SAP Cloud Platform Business Application Studio, and they will be shown as in Figure 9.8 ❸.

10. Select the space created, and on the **Welcome** screen, select **New Project from Template**. Then select **@sap/cap Project** as shown in Figure 9.9. Click **Next**.

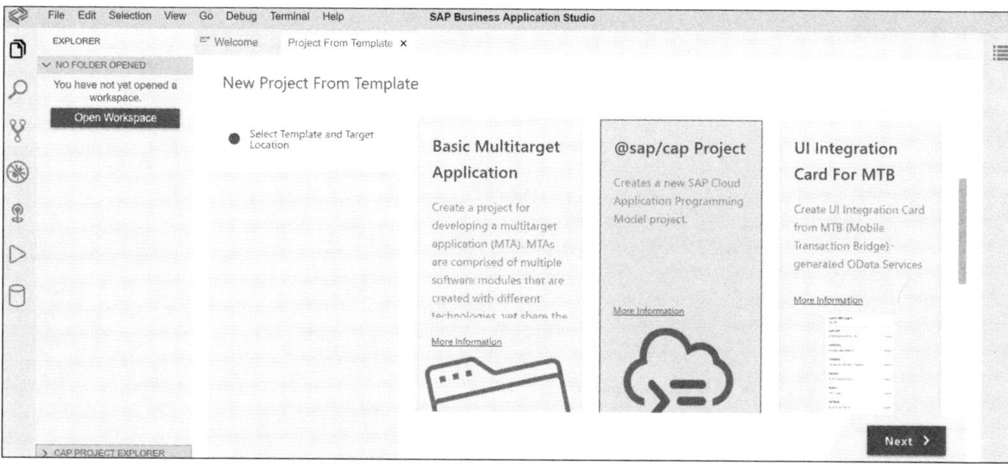

Figure 9.9 SAP Cloud Application Programming Model Project Template

11. Enter a project name and select the required features as shown in Figure 9.10 for the project, then click **Next**.

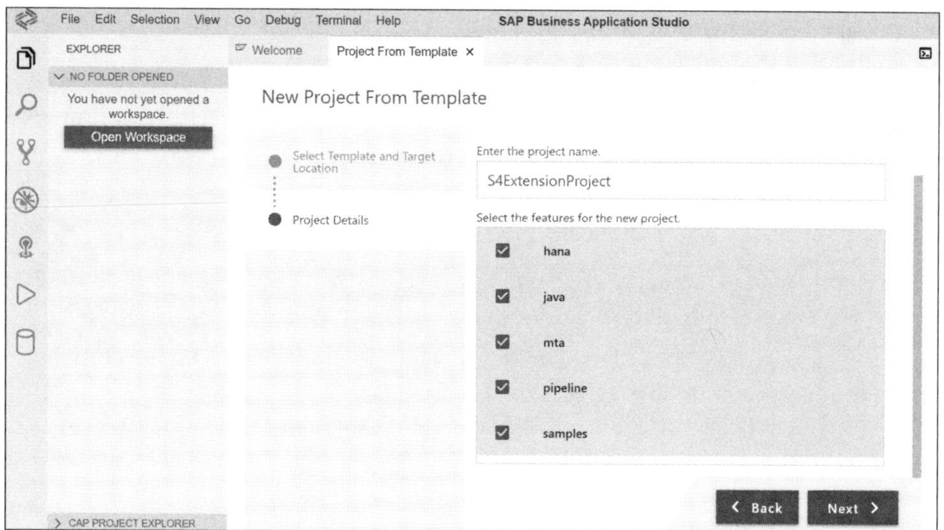

Figure 9.10 Select Features for Project

12. This will create a project template with folders based on features you have selected, as shown in Figure 9.11. From here, you can further develop/enhance your application.

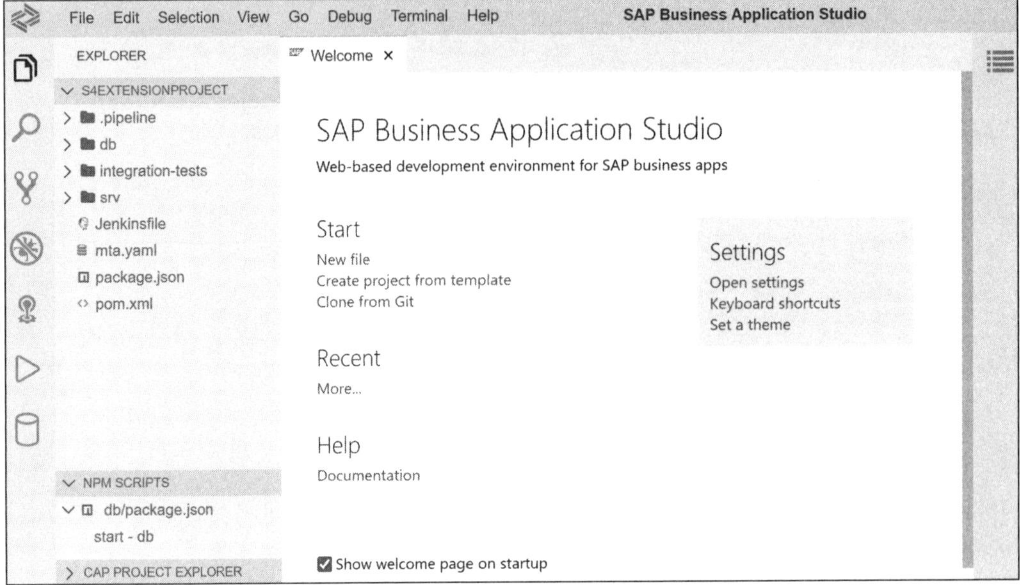

Figure 9.11 Project Explorer

13. To access the SAP S/4HANA Cloud APIs, you need to bind your application to a service instance of the **SAP S/4HANA Cloud Extensibility** service that was created as described previously.

14. Thus, **SAP S/4HANA Cloud Extensibility** service API details are available through the `VCAP_SERVICES` environment variable of the application. This can be used from your application through cloud SDKs.

Figure 9.12 summarizes the high-level overview of the architecture setup and flow performed in this section using SAP Cloud Platform.

> **Note**
>
> You can also create this same extension by activating the messaging plan of the **SAP S/4HANA Cloud Extensibility** service.
>
> We'll cover how this plan can be set up and how to develop extensions using SAP Cloud Platform Serverless Runtime ahead.

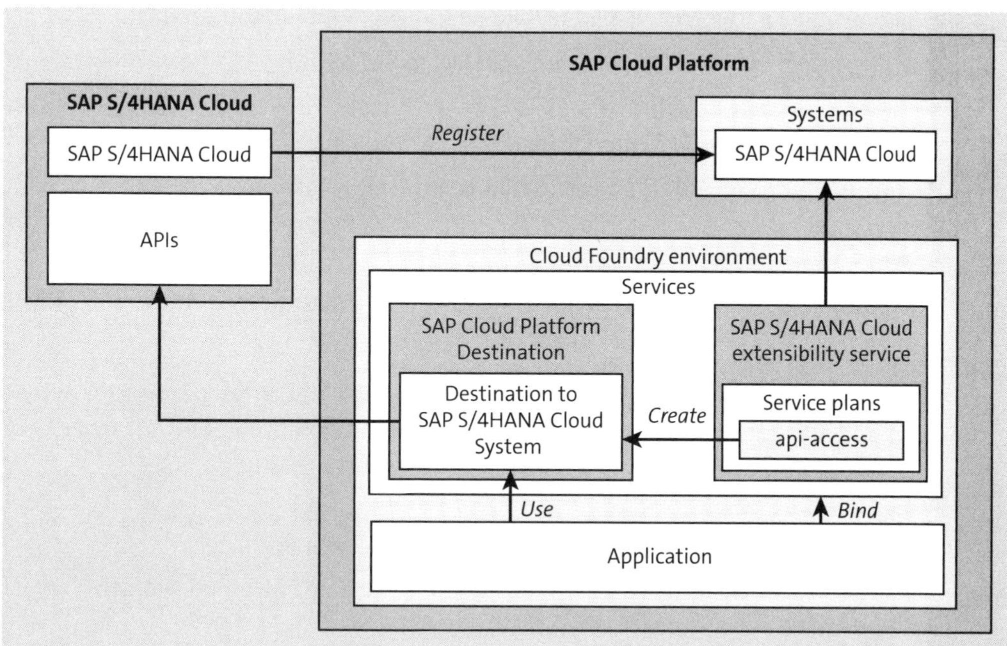

Figure 9.12 High-Level Overview of System Setup Using SAP Cloud Platform

9.3.2   SAP Cloud Platform Serverless Runtime

SAP Cloud Platform Serverless Runtime helps you build, run, and manage serverless applications that extend your SAP S/4HANA system and react to business events. It also provides the capability to provision OData services from your SAP systems.

In a SAP Cloud Platform Serverless Runtime-based application, you get the following advantages:

- You can focus on application logic and not worry about scaling as this is the responsibility of the cloud provider.

- Cost is based on the frequency of execution of business logic, so the lower the execution, the lower the cost of running applications.

Thus, this service enables you to run your cloud-native extensions in a fully managed serverless environment that's decoupled from your digital core and can be triggered on demand. You also have the capability to develop a full-fledged user interface to build and manage your extensions and to use the OData provisioning feature within this service to connect your SAP systems in order to access highly available, scalable, and resilient OData APIs.

Functions can be triggered using the following trigger types in this service:

- HTTP, which executes functions on an HTTP request.

- Timer, which executes functions based on a schedule. The schedule can be based on time zones and can be written using CRON expressions.

- AMQP, which executes functions based on incoming links, outgoing links, and connection rules for SAP Cloud Platform Enterprise Messaging using the AMQP 1.0 over WebSocket messaging protocol.

- CloudEvents, which executes functions based on rules defined in the subscribed CloudEvents.

Let's now look at how to quickly set up this runtime and how to build serverless extensions and applications. We'll also discuss how to use events from SAP S/4HANA Cloud to trigger the functions.

Creating an SAP S/4HANA Extensibility Messaging Service Instance

Before you begin, there are two prerequisites for setting up SAP Cloud Platform Serverless Runtime:

1. You must have SAP Cloud Platform Enterprise Messaging assigned and configured for your Cloud Foundry environment subaccount, as explained in Chapter 5, Section 5.2. This is required to receive events from SAP S/4HANA Cloud.

2. The messaging service plan of the **SAP S/4HANA Cloud Extensibility** service must be assigned to your Cloud Foundry environment subaccount from the **Entitlements** area. This service plan connects the SAP S/4HANA Cloud tenant to the SAP Cloud Platform Enterprise Messaging service in the same subaccount.

Now you're ready to start your setup by creating a new instance as follows:

1. Inside your space, navigate to **Services** • **Service Marketplace** and choose the **SAP S/4HANA Cloud Extensibility** tile.

2. Navigate to **Instances** and click **New Instance**.

3. In the **Create Instance** pop-up, select **Plan: messaging** from the dropdown and set the **System Name** as the system name provided in the configuration in SAP Cloud Platform. Click **Next**.

4. On the next screen, provide a JSON file to configure the service. Table 9.4 provides the mandatory JSON parameters required for configuration of the service. Click **Next**.

Parameter	Description/Value
`systemName`	Name of system provided during registration in SAP Cloud Platform.

Table 9.4 SAP S/4HANA Cloud Extensibility Messaging Service Plan JSON Parameters

Parameter	Description/Value
emClientId	ID for enterprise messaging client. Has maximum length of 4. This generates the communication arrangement name, channel name, emname (enterprise messaging name), and namespace when these parameters are not explicitly provided as follows: ■ *Communication arrangement* is the name of the communication arrangement for the SAP S/4HANA Cloud tenant and has the following default value: SAP_CLOUD_PLATFORM_XF_<emClientId>. ■ *Channel name* is the name of the communication channel and has the following default value: SAP_CP_XF_<emClientId>. Channel name must be unique in the SAP S/4HANA Cloud tenant. *Topic space* is the topic that events should use and is the identifier for the events that originate from same source. It contains exactly three segments, like a/b/c Default value: sap/S4HANAOD/<emClientId>. This value is used in the next step while configuring the SAP Cloud Platform Enterprise Messaging namespace.

Table 9.4 SAP S/4HANA Cloud Extensibility Messaging Service Plan JSON Parameters (Cont.)

Sample JSON for the configuration is provided in Listing 9.2.

```
{
    "emClientId": "<enterprise_messaging_client_id>",
    "systemName": "<System_Name_in_Extension_Factory>"
}
```

Listing 9.2 JSON Sample for SAP S/4HANA Cloud Messaging Service Plan Configuration

5. On the next screen, provide an **Instance Name** and click **Finish**. A new service instance is created.

6. You'll now see a message client created in SAP Cloud Platform Enterprise Messaging with the name emClientId and namespace sap/S4HANAOD/<emClientId>. To confirm this, navigate to **Subscriptions** inside your Cloud Foundry subaccount, choose the **Enterprise Messaging** tile, and click **Go to Application**.

7. In SAP S/4HANA Cloud, under **Communication Management • Communication Arrangements**, you'll see the communication arrangement created with the name SAP_CLOUD_PLATFORM_XF_<emClientId> and all the necessary settings set, including **Channel Name** set to SAP_CP_XF_<emClientId> and **Topic Space** set to sap/S4HANAOD/<emClientId>.

8. Now enable events that should be sent to SAP Cloud Platform. Navigate inside SAP S/4HANA Cloud to **Implementation Cockpit • Manage Your Solution**, and perform these steps:

   – Open the service and navigate to **Configure Your Solution**.

   – In the filter, enter "Messaging". From the results, select **Event Handling**.

- Go to **Maintain Event Topics** and click **Configure**.
- Select the **Channel** created earlier and confirm the selection.
- On the next screen, select different topics which should publish the message to SAP Cloud Platform. For this, choose **New Entries** and select the topics that you want to add to the channel. Choose **Save** and then choose **Exit**.

9. Navigate to **Subscriptions**, choose the **Enterprise Messaging** tile, and click on **Go to Application**. Navigate inside the messaging client created earlier and click the **Events** tab. The events selected in the previous step will be displayed.

10. Now set up an SAP Cloud Platform Enterprise Messaging instance for the application, which can connect to the SAP Cloud Platform Enterprise Messaging client created by SAP Cloud Platform. This helps enable a single event bus for applications that can connect to different backend systems.

11. Navigate inside your space to **Services • Service Marketplace** and select the **Enterprise Messaging** tile.

12. Select **Instances** and click **New Instance**.

13. In the **Create New Instance** pop-up, select **Plan** as the default and click **Next**.

14. For **Specify Parameters**, you need to provide JSON configuration parameters as shown in Listing 9.3. Click **Next**.

```
{
    "emname": "<message_client_name>",
    "namespace": "sap/<message_client_name>/<uniqueID>",
    "options": {
        "management": true,
        "messagingrest": true,
        "messaging": true
    },
    "rules": {
        "queueRules": {
            "outboundFilter": [
                "${namespace}/#",
                "sap/S4HANAOD/<emClientId>/#"
            ]
        },
        "topicRules": {
            "outboundFilter": [
                "${namespace}/#",
                "sap/S4HANAOD/<emClientId>/#"
            ]
        }
    }
}
```

Listing 9.3 JSON Parameters for SAP Cloud Platform Enterprise Messaging Instance

15. On the next screen, you can optionally bind an application from the dropdown if there's an application available. Click **Next**.

16. On the next screen, provide emname as the **Instance Name**, as in the JSON in Listing 9.3.

17. A message client is created in SAP Cloud Platform Enterprise Messaging where you'll create the queue and queue subscriptions, as explained in Chapter 5, Section 5.2.

Set Up SAP Cloud Platform Serverless Runtime

Let's now discuss how to set up and configure SAP Cloud Platform Serverless Runtime, which can be used to create functions and be triggered on events configured earlier:

1. Navigate to **Services • Service Marketplace** inside your space within the subaccount and choose the **Serverless Runtime** tile.

2. Navigate to **Instances** and click **New Instance**.

3. In the **Create Instance** pop-up, select the **Plan** and click **Next**.

4. Specify the JSON parameters with a JSON file, as provided in Listing 9.4. To create extensions and to provision OData services, set these parameters to true. Click **Next**.

```
{
    "extensions": true,
    "odp": true
}
```

Listing 9.4 JSON Parameters for SAP Cloud Platform Serverless Runtime

5. On the next screen, optionally choose an application to bind this service from the dropdown. Click **Next**.

6. Provide an **Instance Name** and click **Finish**. A service instance for the service is created.

7. In your subaccount, navigate to **Subscriptions** and choose the **Extension Center** tile

8. Next, choose **Subscribe**.

9. Now you need to assign roles to users. Click **Manage Roles** and you'll see the role templates in the Extension Center, as shown in Table 9.5.

Role Template	Description
FunctionsManage	Provides authorization to users to create, edit, and delete extensions, functions, triggers, configmaps, and secrets.
FunctionsRead	Provides authorization to users to view extensions, functions, triggers, configmaps, and secrets.

Table 9.5 Role Templates in Extension Center Service

Role Template	Description
ODPAPIAccess	Provides authorization to users to make API calls to the SAP backend, proxied by OData provisioning.
ODPManage	Provides authorization to users to view and register services with OData provisioning.

Table 9.5 Role Templates in Extension Center Service (Cont.)

10. If a role collection is already created, add the role template to it. Otherwise, navigate to **Security • Role Collections** and create a new role collection. Click the role collection and add the role templates from the service.

11. Now provide users with access to the role collection. For this, navigate to **Security • Trust Configuration**. Choose the identity provider and enter the user ID. Click **Show Assignments**, then click **Assign Role Collection** to assign the role collection to the user.

12. Click **Go to Application** inside the Extension Center. This will launch the Extension Center home page, as shown in Figure 9.13.

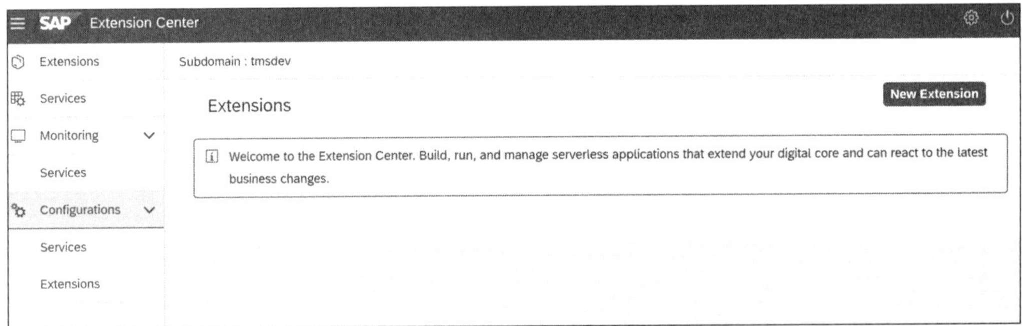

Figure 9.13 SAP Cloud Platform Extension Center Home Screen

13. To create new extensions, navigate to **Extensions** and click **New Extension**.

14. In the **Create Extension** pop-up, choose **Template with a Function and a Trigger** and then choose **Step 2**.

15. On the next screen, provide an **Extension Name** and **Runtime**. The runtimes supported are nodejs10 and nodejs8. Click **Step 3**.

16. Now provide a **Function Name** and click **Step 4**.

17. Enter a **Trigger Name** and **Trigger Type** and click **Step 5**.

18. Review the details and click **Create**. The function created will open as shown in Figure 9.14.

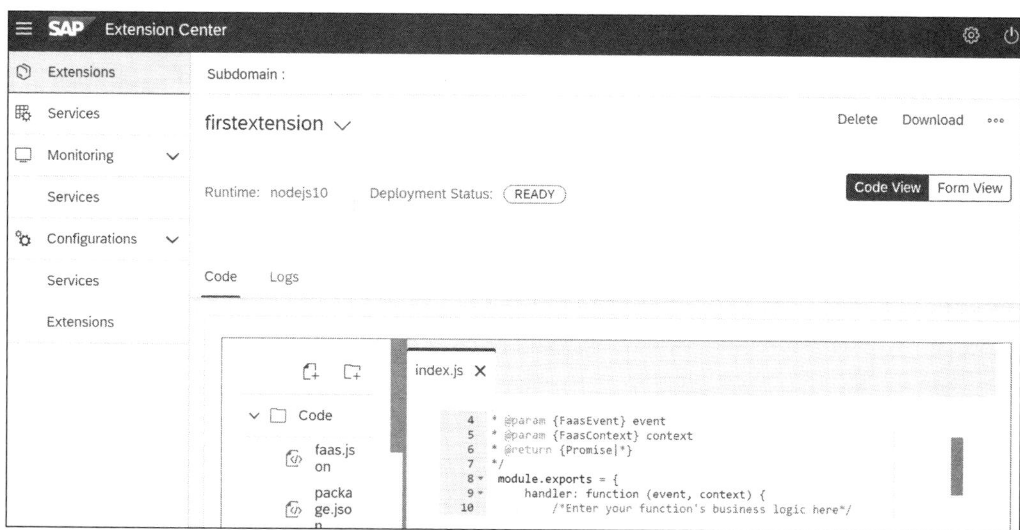

Figure 9.14 Extension Center Code Editor

A SAP Cloud Platform Serverless Runtime extension project has the structure defined in Table 9.6.

File or Folder Name	Description
faas.json	This is the configuration file for Extension Center extensions.
	This is under the root folder of the extension and contains all functions, triggers, secrets, and configurations used in the extension.
	This file contains following information in JSON format:
	■ `project`: name of the project.
	■ `version`: current version of the project.
	■ `runtime`: runtime used in the function.
	■ `library`: relative path to folder within project which contain function's source code.
	■ `secrets`: defines a list of secrets in the project. Secrets are used to input configuration data for functions. A collection of secrets is used to store sensitive information such as credentials. Each secret has JSON format and has a source-relative path to a folder where secrets are stored.
	■ `configs`: defines `ConfigMaps` in the project. Add `ConfigMaps` to provide configurations like settings for storing unencrypted configuration information that should not be hard-coded. Has the same structure as secrets.
	■ `functions`: defines all functions in the project. Each function is defined as a JSON object with reference to a module, handler, timeout, secrets, and configs.
	■ `triggers`: defines all triggers in the project.

Table 9.6 Project Structure for Extension Center Extension

File or Folder Name	Description
lib	This folder contains the source code of all functions in the extension.
data	This folder defines all the secret configurations in the extension.
deploy	This folder contains the value for the properties of the secret to be used during deployment.

Table 9.6 Project Structure for Extension Center Extension (Cont.)

19. Navigate to **Configurations • Extensions** and choose **New Credentials**.

20. In the **Create Extension Credentials** pop-up, provide the **Service Instance**, **Service Key**, and **Credentials** for the SAP Cloud Platform Enterprise Messaging service type created previously. Choose **Create**.

21. Navigate to **Extensions** and choose the extension already created, then switch to **Form View**. The following tabs are visible in the UI:

    - **Functions**: Here you can view a list of functions present in the extension, including name, module, handler, and timeout details of functions within this extension. You can click **Add Function** to add a new function to this extension.

    - **Triggers**: Here you can provide details of triggers in the extension. Choose **Add Trigger** to add new triggers. From the **Add Trigger** dropdown, you can create following trigger types: **HTTP**, **Time**, **AMQP**, or **Cloud Events**. Then follow these steps:

        - To add SAP Cloud Platform Enterprise Messaging as a trigger type to listen to the queues, make AMQP the trigger type.

        - In the **Add AMQP Trigger to the Extension** pop-up, provide a **Trigger Name** and choose **Step 2**.

        - On the next screen, add a **Link Name** and **Source Address** for the incoming queue name of the SAP Cloud Platform Enterprise Messaging service to which events are published. Select **+** and choose **Step 3**.

        - Optionally, provide a queue or topic name to which you want to send messages and select **+**. Choose **Step 4**.

        - Provide rules for invoking functions and optionally define filters on events. Select **+** and choose **Step 5**.

        - On the **Credentials** screen, you can either **Provide New Credentials** and provide the **AMQP URI, Token Endpoint, Client ID**, and **Client Secret** from the services instance for SAP Cloud Platform Enterprise Messaging or choose **Use Existing Credential Store** and add a credential store created earlier.

> **Note**
>
> You can also configure the trigger from the SAP Cloud Platform Enterprise Messaging queue using other triggers supported by the Extension Center.

- **Secrets**: Here you can view lists of secrets present in extensions.
- **Configs**: Here you can view lists of `ConfigMaps` present in extensions.
- **Logs**: Here you can view execution logs of functions in extensions.

Choose **Save and deploy** to deploy the extension.

> **Note**
>
> You can create function extensions using SAP Cloud Platform Business Application Studio as well. You also can trigger functions based on APIs, with an HTTP trigger type.
>
> To use the OData provisioning feature of Extension Center, follow these steps:
>
> 1. Navigate to **Services** and choose **Register**.
> 2. Select **Destination** and look for the **Service Name**. Choose **Register**.

9.3.3   SAP Cloud Platform, ABAP Environment

SAP Cloud Platform, ABAP environment contains standard technology components familiar from the standalone ABAP application server. Thus, ABAP developers can make use of their existing ABAP know-how to develop and run applications in SAP Cloud Platform. However, note that ABAP in SAP Cloud Platform introduces a subset of the ABAP language optimized for the cloud and hence excludes statements that aren't compliant with secure cloud operations, such as direct access to the file system.

This stack supports a new RESTful programming model, including SAP Fiori and CDS, and services and APIs are offered according to a whitelisting approach. The usage of whitelisted APIs ensures secure and stable upgrades of the underlying platform, and usage of SAP objects that aren't whitelisted will lead to syntax errors.

In addition, this environment gives you the features and flexibility you need for your future-ready journey by letting you move your existing on-premise ABAP code to the cloud in a decoupled way. This can be done with the Custom Code Migration app in the ABAP environment, which helps analyze your custom code that needs to be migrated from an SAP Business Suite system to SAP S/4HANA (on-premise) and migrate it to the ABAP environment. This app also supports identifying unused custom objects based on collected usage data, thus helping to simplify your move to SAP S/4HANA and make your landscape future ready. However, existing code cannot be *lifted and shifted* to the cloud as-is, as your existing custom code may have dependencies that would cause it to break.

> **Note**
>
> Another way to help you analyze code is the new cloud readiness checks provided with the Code Inspector in ABAP Test Cockpit. Check variant `SAP_CP_READINESS_REMOTE` is available with `SAP_BASIS` 7.52 and helps migrate your on-premise ABAP code to SAP

Cloud Platform, ABAP environment. The checks detect incompatibilities in the use of unsupported objects, unreleased objects, and language elements not supported in SAP Cloud Platform, ABAP environment.

The following SAP Notes are published to help you analyze your current on-premise system's readiness to be moved to the SAP Cloud Platform, ABAP environment:

- SAP Note 2682626: Code Inspector Check for Restricted Language Scope Version
- SAP Note 2684665: Custom Code Checks for SAP Cloud Platform ABAP Environment
- SAP Note 2830799: Custom Code Checks for SAP Cloud Platform ABAP Environment (2)

A third way to analyze your custom code and keep your core clean is to use the Custom Code Migration app in the SAP S/4HANA system itself.

Let's now see how to set up SAP Cloud Platform, ABAP environment. Before you begin, you must have met the following prerequisites:

- The ABAP environment is assigned to your Cloud Foundry environment subaccount.
- The SAP Cloud Platform Application Runtime and SAP Cloud Platform Destination quotas are assigned to your Cloud Foundry environment subaccount.

For the setup itself, follow these steps:

1. Navigate inside your Cloud Foundry environment subaccount to **Subscriptions**.
2. Choose the **Web Access for ABAP** tile and click **Subscribe**.
3. Navigate to your space and then select **Services • Service Marketplace** and select **ABAP System**.
4. Navigate to **Instances** and select **New Instance**.
5. In the Create Instance wizard, select the plan and click **Next**.
6. On the next screen, in **Specify Parameters (Optional)**, you need to provide parameters in JSON format as explained in Table 9.7.

Parameter	Description
admin_email	Initial user of the system with administrator rights.
description	A meaningful description of the system. This parameter is optional.

Table 9.7 JSON Parameter Descriptions for SAP Cloud Platform, ABAP Environment

Parameter	Description
is_development_allowed	Values: true or false. Specifies whether development objects in the system can be changed. You specify the value as true for the development system; for the production instance, the value is false because changes only are imported.
size_of_runtime	Specifies the abap/abap_compute_unit quota plan. ABAP compute units are represented in numbers of 16 GB blocks. So size_of_runtime = 1 would mean 16 GB of RAM for compute units. Available configurations are 1, 2, 4, 6, or 8.
size_of_persistency	Specifies the abap/hana_compute_unit quota plan, with the SAP HANA compute unit represented in a number of 16 GB blocks. Minimum value is 4 (64 GB). The following configurations are available: 4, 8, or 16.
sapsystemname	This is an optional element. This specifies a three-character name for the system.

Table 9.7 JSON Parameter Descriptions for SAP Cloud Platform, ABAP Environment (Cont.)

A JSON example is shown in Listing 9.5.

```
{
    "admin_email": "<email address of admin>",
    "description": "<optional - description for system> ",
    "is_development_allowed": <true/false>,
    "size_of_runtime": "1",
    "size_of_persistency": "4",
    "sapsystemname": "<optional, three-character name system_name>"
}
```

Listing 9.5 JSON Example for SAP Cloud Platform, ABAP Environment

7. Click **Next**, navigate to a screen to provide an **Instance Name**, and select **Finish**.

8. The system is created (which may take some time) and an email is triggered to the administrator specified in admin_email.

9. Click the created ABAP system instance, navigate to **Service Keys**, and choose **Create Service Key**. In the **Create Service Key** pop-up, provide a **Name** and choose **Save**. A service key is created for the ABAP service. Copy this service key or download it; you'll use it later to connect to this service from Eclipse.

10. Navigate to **Referencing Apps** and choose **Open Dashboard**. The SAP Fiori launchpad for SAP Cloud Platform, ABAP environment is launched as shown in Figure 9.15.

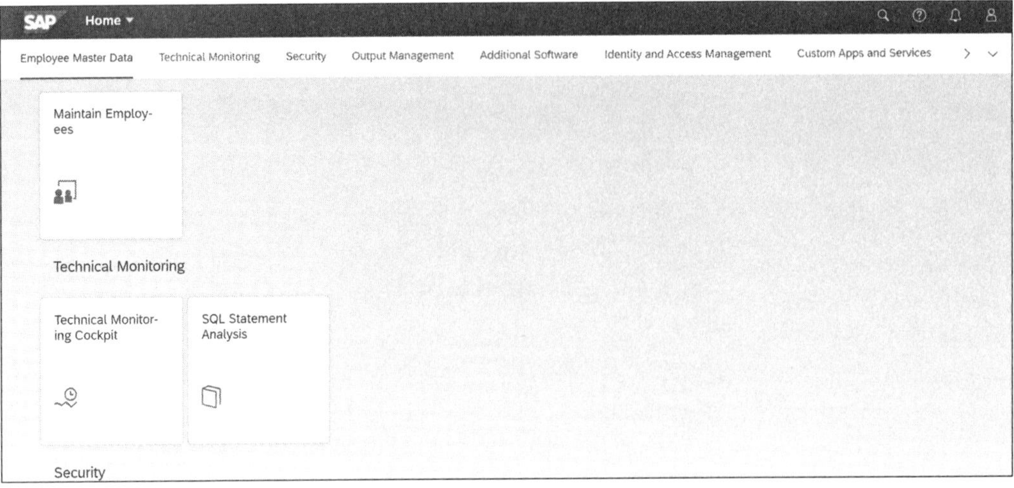

Figure 9.15 SAP Cloud Platform, ABAP Environment Launchpad

Next you'll add other users to the environment in different roles. To do this, follow these steps:

1. Create the required roles. For this, navigate to the Maintain Business Roles app from the SAP Cloud Platform, ABAP environment home page.

2. Choose **Create from Template** and select a template already provided by the ABAP service. Enter a **New Business Role ID** and **Description** (optional). Choose **Ok**. Table 9.8 lists the default templates provided by SAP Cloud Platform, ABAP environment.

ID	Description
SAP_BR_ADMINISTRATOR	Administrator: allows access rights for user management, communication management, security, employee master data management, lifecycle management of software components, installing additional software, technical monitoring, and output management functionalities
SAP_BR_BPC_EXPERT	Configuration expert—business process configuration: allows authorization for business configuration customization functionality
SAP_BR_DEVELOPER	Developer: provides authorization for development capabilities
SAP_BR_IT_PROJECT_MANAGER	Project manager—IT: provides authorization for configuring custom code migration

Table 9.8 SAP Cloud Platform, ABAP Environment Business Role Templates

3. Navigate to the **Assigned Business Catalogs** tab inside the new role created to add or remove business catalogs required for this role.

4. Click **Maintain Restrictions** to set **Unrestricted** or **No Access** for write, read, and value help functions as required and navigate back.

5. Choose **Save** to save the role.

6. Next, navigate to the Maintain Employees app, choose **New**, and provide employee data information. Choose **Save**.

7. Navigate to the Maintain Business Users app and choose **New**. Select the employee maintained and select **OK** on the confirmation screen.

8. On the new screen, navigate to the **Assigned Business Role** tab, click **Add**, choose **Business Roles**, and click **Ok**. Choose **Save**.

The process explained here walks you through the steps of creating users and assigning necessary roles to them. The next step is to set up development environment.

> **Note**
>
> Before proceeding with the following steps for setting up development rights and access, make sure that you have business user created and assigned to the developer role as explained earlier.

Follow these steps to set up SAP Cloud Platform, ABAP environment based on Eclipse:

1. Inside the SAP Cloud Platform cockpit, navigate to **Useful Links • Tools**.

2. In the new web page that opens, select the **ABAP** tab and install the software components as explained. After setup, you'll see the required ABAP tools in Eclipse as shown in Figure 9.16.

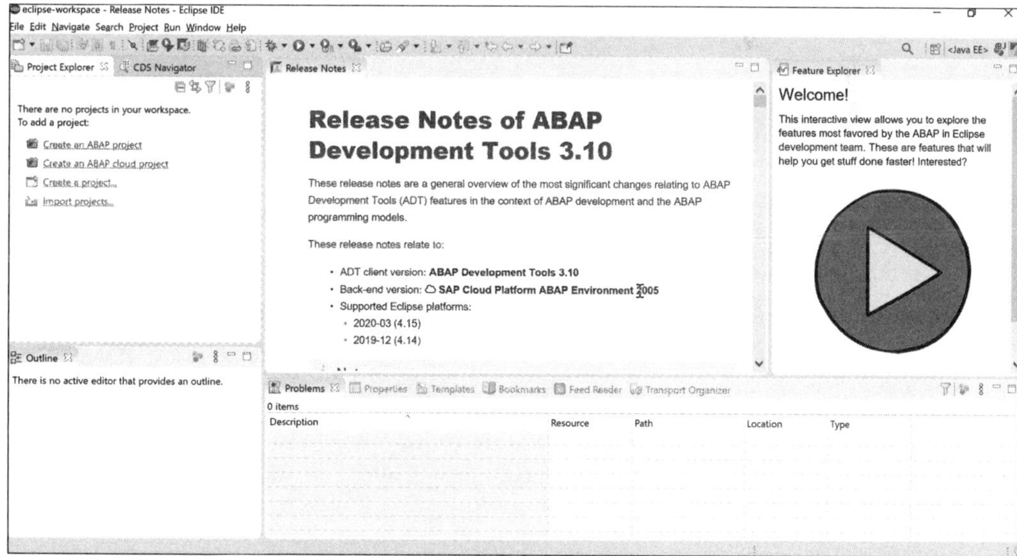

Figure 9.16 SAP Cloud Platform ABAP Development Environment

3. To connect Eclipse to SAP Cloud Platform, ABAP environment, click **File • New • Project**. Select **ABAP • ABAP Cloud Project** and click **Next**.

4. Select the **Service Key** radio button and click **Next**.

5. Import the service key downloaded previously from the SAP Cloud Platform cockpit and click **Next**.

6. Log on in the next screen using your email address and password for your Cloud Foundry account. Note that single sign-on is not permitted.

7. On the **Service Instance Connection** page, you'll see **Service Instance URL, Email, User ID, SAP System ID, Client**, and **Language** details. Select **Next**. Provide a **Project Name**, select **Favorite Package**, and click **Finish**.

8. You are now ready to create your ABAP project inside the package and can call the SAP S/4HANA services through SAP Cloud Platform Destination. The services created using the ABAP development environment are exposed by bundling them into the communication scenario. The communication user, communication system, and communication arrangement are created from the respective apps under the communication management catalog in the ABAP service home page.

> **Note**
>
> SAP Cloud Platform also provides an easy and automated way to create and set up the ABAP environment through recipes. For this, inside your global account, navigate to **Recipes** and click **Start Recipe** for the **Prepare an Account for ABAP Development** recipe.
>
> Access to the Custom Code Migration app in SAP Cloud Platform, ABAP environment requires your user to be assigned to the SAP_BR_IT_PROJECT_MANAGER role. You can find more details about setting this up at *http://s-prs.co/v515732*.

9.3.4   SAP Cloud Platform, Kyma Runtime

SAP Cloud Platform, Kyma runtime is a fully managed Kubernetes runtime-based service in SAP Cloud Platform for developing applications and extensions. This runtime helps customers build applications and extensions by using serverless functions and combining them with containerized microservices. SAP Cloud Platform, Kyma runtime is available across all cloud providers, thus helping to avoid hyperscaler lock-in while developing Kubernetes-based applications with a highly scalable infrastructure.

SAP Cloud Platform, Kyma runtime has the following in-built capabilities:

- Service mesh makes it easy for cloud-native developers to run their workloads by providing a service mesh based on Istio, which allows service-to-service communication and proxying.

- The API gateway allows you to consume extensions and applications from the outside world.

- SAP Cloud Platform, Kyma runtime provides an event bus so that you can build event-based and API-based extensions.

- The serverless engine allows you to build serverless functions to combine with microservices.

- The service catalog allows for easy instantiation and consumption of services made available to the runtime.

SAP Cloud Platform, Kyma runtime is based on open standards and allows you to add custom Docker images to applications. Thus, it becomes easy to upskill your developers quickly, reducing time to market for a cloud-native solution. It also contains SAP solution-specific connectivity, which makes it an ideal runtime to create extensions and applications that integrate with SAP and non-SAP solutions. In the next section, we'll look at how to set up and configure SAP Cloud Platform, Kyma runtime.

You should be aware that SAP Cloud Platform, Kyma runtime is available only in a CPEA-based licensing agreement. Before you begin, check for availability of the service in your region/data center. To start your setup, follow these steps:

1. Inside your SAP Cloud Platform global account, create a subaccount in a hyperscaler data center where SAP Cloud Platform, Kyma runtime is available.

2. Inside the subaccount, on the **Overview** screen, click **Enable Kyma**. In the pop-up screen, provide a **Cluster Name**, **Description** (optional), **Provider**, and **Region** and choose **Create**.

> **Note**
>
> You can enable Kyma inside the subaccount where you have already enabled Cloud Foundry.

3. This starts creation of the SAP Cloud Platform, Kyma runtime environment. Once it's provisioned, you will get a link to the dashboard for the Kyma environment.

4. Before accessing the Kyma dashboard, you need to provide required roles to the user. For this, navigate to **Security • Role Collections** and click **New Role Collection** to create a new role collection (choose an existing role collection if one already exists).

5. Click the role collection and click **Add Role** to add Kyma-specific roles. Kyma provides the roles in Table 9.9.

Role Name	Description
KymaRuntimeNamespaceAdmin	Kyma administrator role
KymaRuntimeNamespaceDeveloper	Kyma developer role

Table 9.9 SAP Cloud Platform, Kyma Runtime Roles

6. Navigate to **Overview** in your subaccount and click **Link to Dashboard** to launch the Kyma home screen as shown in Figure 9.17.

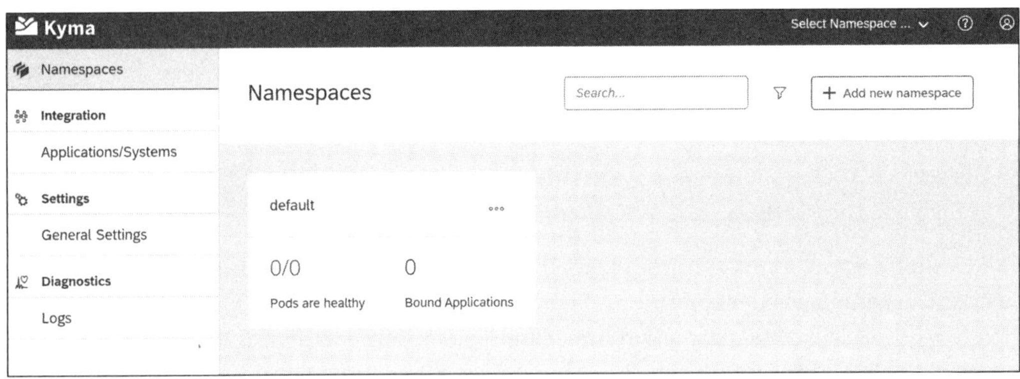

Figure 9.17 SAP Cloud Platform, Kyma Runtime Home Screen

7. Start your extension or application development by navigating inside a namespace.

9.3.5   Decision Matrix

In the previous sections, we discussed the various runtimes that are available with SAP Cloud Platform and how you can use them to create new extensions and applications in a quick and easy way, thus helping to reduce your time to market and increase your return on investment.

In this section, we offer a simple decision matrix you can use as guidance to select the runtime you need to use for an extension or application's development. Table 9.10 provides guidelines on decision criteria for these runtimes that you can use while developing your applications and extensions.

Decision Criteria	SAP Cloud Platform Application Runtime	SAP Cloud Platform Serverless Runtime	SAP Cloud Platform, ABAP Environment	SAP Cloud Platform, Kyma Runtime
Application type	Microservices-based applications and extensions	Serverless applications and extensions	ABAP RESTful applications and extensions	Microservices-based, serverless, and docker-based applications and extensions
Developer skillset	Cloud-native developers in Java/Node.js or any major programming language	Serverless function developers in Node.js	ABAP experience along with CDS knowledge	Cloud-native and Kubernetes- and/or Docker-based application development knowledge; Node.js knowledge for serverless functions

Table 9.10 Decision Matrix with Development Runtimes

Decision Criteria	SAP Cloud Platform Application Runtime	SAP Cloud Platform Serverless Runtime	SAP Cloud Platform, ABAP Environment	SAP Cloud Platform, Kyma Runtime
Development framework	SAP Cloud Application Programming Model framework available to help accelerate the extension development, though not mandatory	Fully managed serverless development capability with no specific configuration requirement	Capabilities to analyze your existing custom ABAP code and migrate it to the cloud, thus accelerating your move to SAP S/4HANA	Capability to deploy docker-based applications and out-of-the-box consumption of SAP and hyperscaler services
Price metered	Metered by application runtime used for application development	Metered on execution of functions in a pay-per-use model	Metered for ABAP server deployment	Metered on number of nodes and storage deployed

Table 9.10 Decision Matrix with Development Runtimes (Cont.)

9.4   Integrating Applications with Hyperscaler Services

Organizational landscapes are often complex and have systems and services from hyperscalers in their infrastructure. Such a blended organizational landscape is complex to understand and leads to overhead, which reduces speed, agility, and flexibility in your transformation journey. These various SAP and non-SAP landscapes and the data flowing through these systems need to be brought together to bring maximum value to your business while building extensions and applications. At the same time, they should provide a way to reduce overhead and offer speed and agility to transform based on business needs. SAP Cloud Platform offers capabilities to configure, integrate, and provision with hyperscaler services in a Cloud Foundry or Kyma environment.

> **Service Manager**
>
> In addition to the capabilities discussed in the following sections, you can also integrate SAP Cloud Platform with services and applications hosted in Kubernetes clusters through service manager functionality.
>
> The setup and configuration for this process is explained in Chapter 5, Section 5.5 and is worth reviewing in this context.

Note

You need to own the relevant hyperscaler resources, and this functionality is subject to the availability of the supported non-SAP vendors in your country or region.

In the following sections, we'll discuss different capabilities available in SAP Cloud Platform to integrate with services from hyperscalers. You will select the option based on your scenario and what each option offers.

9.4.1   Resource Provider

SAP Cloud Platform provides integration with hyperscaler services through configuring a new resource provider. The setup and configuration for this is as follows:

1. In your global account, navigate to **Resource Provider** and select **New Provider**.

2. In the **New Resource Provider** pop-up, select a provider from list of supported non-SAP vendors.

3. Enter a **Display Name** for the provider.

4. Enter a **Technical Name** for the provider. This name is used by developers as a parameter when creating a service instance from this provider. You can create more than one instance of a given resource provider, each with its unique configuration parameters, so the technical name should be unique.

5. Choose either **Manually Enter Provider Configuration Parameters** or **Add Provider Configuration Properties by JSON File**.

Note

Depending on the supported vendor, the configuration parameters may be different. You get these configuration parameters from the vendor system with the required user rights.

6. Choose **Create**.

7. After creating a new resource provider, the supported services are added as entitlements in your global account. This can be assigned to your subaccounts from the **Entitlements** section inside your global account.

Note

You can also integrate PostgreSQL on SAP Cloud Platform with your application by creating a user-provided service instance, as discussed in Chapter 5, Section 5.8.1.

9.4.2   Service Broker

SAP Cloud Platform also offers this integration through a service broker based on the Open Service Broker API specification. The broker provides easy integration of multiple hyperscaler services directly into the application platform. These brokers are open source and provided by other vendors. A service broker from the hyperscaler is deployed to your space and is called natively when a developer wants to create a service instance of a hyperscaler service exposed through the service broker. A high-level overview of the steps required for this configuration is shown in Figure 9.18.

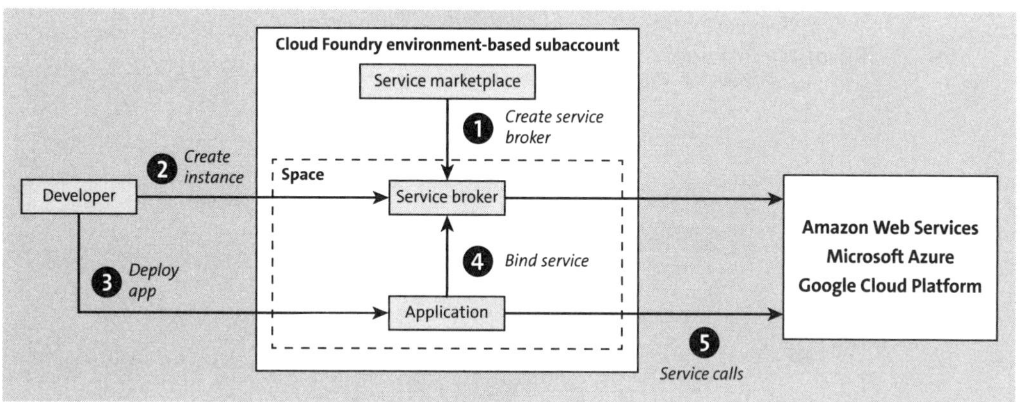

Figure 9.18 Service Broker Integration with Hyperscaler Service

Once the setup with the service broker is instantiated, the services from hyperscalers are visible under **Services • Service Marketplace** in your space, as shown in Figure 9.19 for Microsoft Azure services integration.

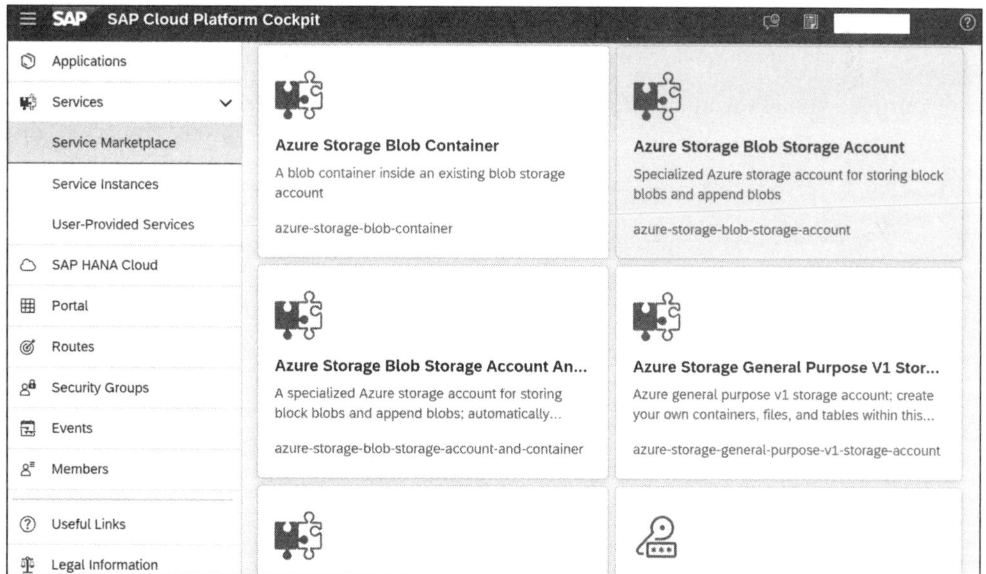

Figure 9.19 Hyperscaler Services in SAP Cloud Platform Service Marketplace

Figure 9.20 shows the services from Azure that can be integrated with SAP Cloud Platform, which are exposed through the Microsoft Azure service broker.

> **Note**
>
> You can find details about configuring this integration with Azure and setting up Azure through a service broker at *http://s-prs.co/v515733*.

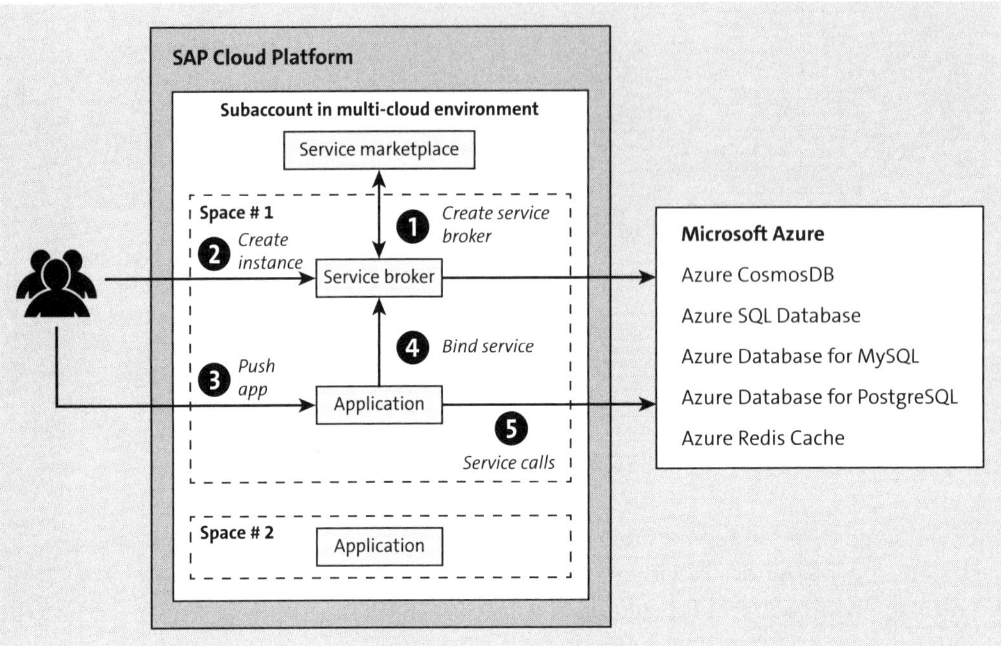

Figure 9.20 Azure Services Integration through Service Broker

Figure 9.21 shows the integration with AWS service broker and shows the services which are exposed through this service broker.

> **Note**
>
> More information about the deployment of a service broker for AWS and configuration of AWS can be found at *http://s-prs.co/v515734*.

Figure 9.22 shows services from the Google Cloud Platform. These services are exposed through the Google Cloud Platform's service broker.

> **Note**
>
> More information on the service broker configuration for the Google Cloud Platform and settings in the platform to enable this scenario can be found at *http://s-prs.co/v515735*.

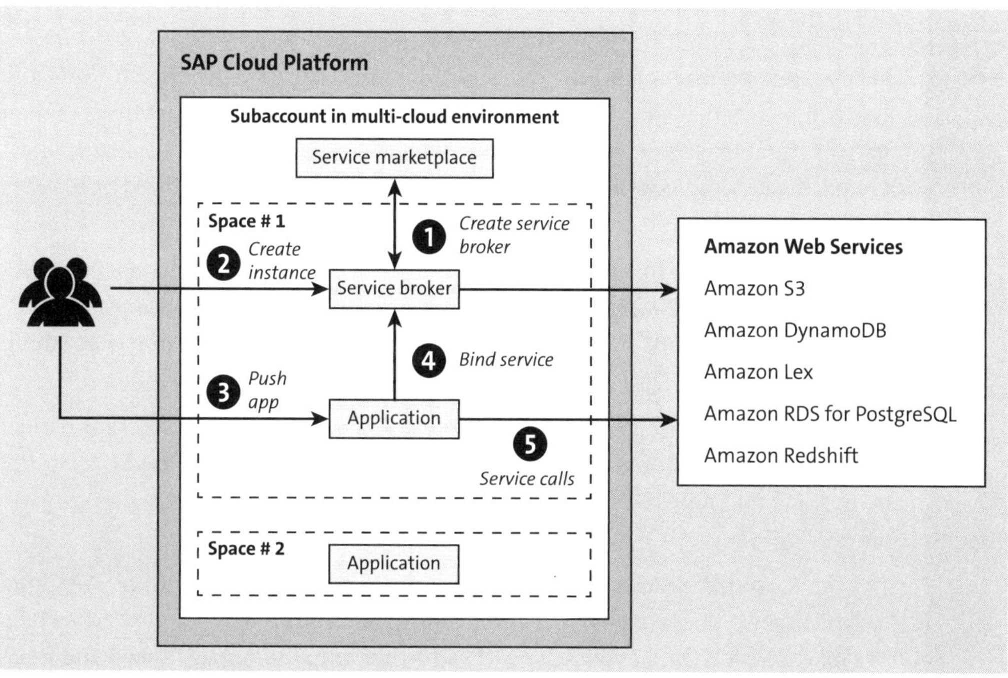

Figure 9.21 AWS Integration through Service Broker

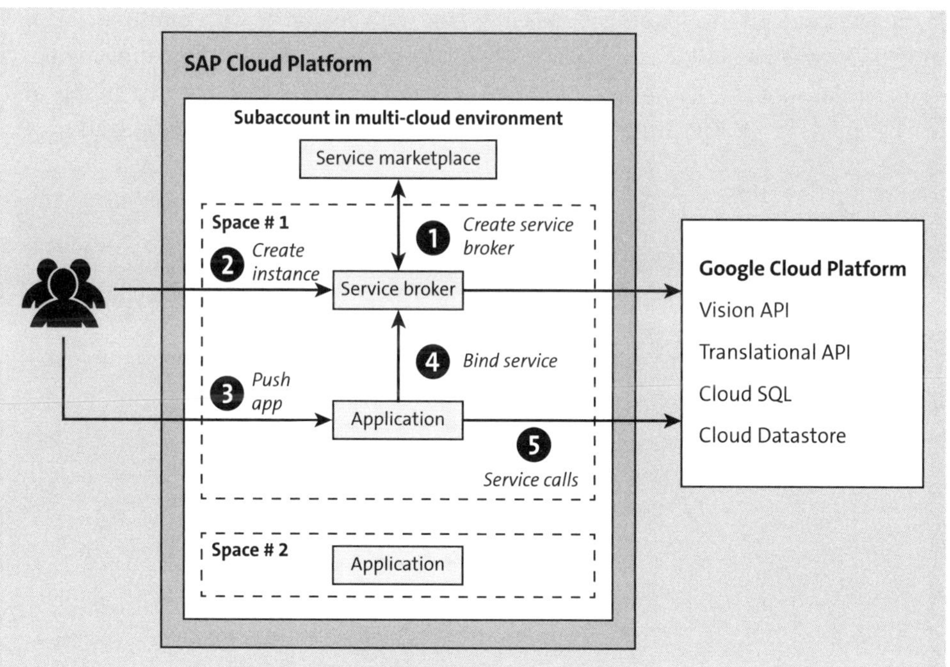

Figure 9.22 Google Cloud Platform Services Integration through Service Broker

9.5   Summary

In this chapter, we discussed how organizations can customize their solutions and develop completely new solutions to differentiate themselves from competitors in their industry and market segment. We discussed why it's important for a business to meet such requirements and how IT and business can collaborate to realize these needs with great speed and agility.

We then further dove into in-app extensions and side-by-side extensions using SAP S/4HANA. We looked at various scenarios and discussed when and where it's advisable to create these extensions from an architectural standards, best practices standpoint.

Because creating extensions is quite important to future proof your business and meet changing market conditions, we saw how SAP Cloud Platform can be used to simplify the extension build process, which helps improve speed and flexibility in your organization.

Further, we looked at different runtimes and capabilities available within SAP Cloud Platform. These can be used based on your business requirements and skills to help you create extensions and applications, thus further amplifying your speed and flexibility.

Finally, we discussed how SAP Cloud Platform can help simplify application and extension development in a heterogenous landscape for a customer with multiple SAP and non-SAP systems. SAP Cloud Platform helps to greatly simplify the connectivity to hyperscaler services. This helps bring SAP and non-SAP data together in a simple and easily consumable way, bringing tremendous business value to your organization.

Chapter 10
Continuous Integration and Delivery

In previous chapters, we noted that business requirements change quite often and that your enterprise system needs to be agile and flexible to deliver continuous innovations. In this chapter, we'll look at various services on SAP Cloud Platform that can aid your organization, helping to deliver continuous innovations to business in a quick, agile, and flexible way.

In previous chapters, we discussed building your applications, extensions, and integrations in your landscape to become a future-ready intelligent enterprise. In this chapter, we'll investigate how these innovations can be brought to customers on a continuous basis with an agile and flexible approach. Your organization needs a well-defined continuous integration and development plan, which can be adapted to your business innovation needs to keep the innovation pace.

Thus, a comprehensive continuous integration and development plan is essential in the in process to build applications in the cloud and bring continuous innovation to end users.

The purpose of this chapter is to introduce you to the challenges of building applications in the cloud, plus strategy and guidance to build a continuous integration and delivery (CI/CD) strategy while building applications in SAP Cloud Platform. We will also introduce various CI/CD toolsets available in SAP Cloud Platform to build applications and extensions, which can help you bring innovations to your end users with speed and agility.

10.1 Continuous Integration and Delivery Overview

Business requirements and customer expectations change frequently, and application development in the cloud needs to match these fast innovation cycles. Many cloud applications have a nearly continuous delivery mode to meet this challenge. Software is developed in shorter cycles, thus enabling fast innovation and quick feedback from end users, all the while making sure that end users are not impacted or that there is no downtime for the software.

Let's look at advantages for companies to deploy changes to production multiple times per week/day, as follows:

- Smaller changes means less risk.
- More frequent change means a closer feedback loop, helping pinpoint where and when a change has introduced a bug.
- You gain more immediate feedback from customers.
- You can maximize your cloud capabilities and offerings.

Hence, the application development moves from a *never change a running system* paradigm to a *change as often as necessary* paradigm, which has to be enabled and complemented by a mature continuous integration and delivery foundation, including technologies and processes to get changes of all types—new features, configuration changes, bug fixes, beta solutions—into production or in the hands of end users safely and quickly in a sustainable way.

Continuous integration and delivery is not just a technical change; it involves cultural transformation. A mature CI/CD cycle requires processes and tools to iterate the code changes in small steps and transfer them through a pipeline of various testing phases to production. This is commonly referred to as the *deployment pipeline*, and it leads to delivering the system without disruption to end users. Now let's look at what it takes to build a mature continuous integration and delivery cycle. Ahead, we'll provide an initial look at the products and tools SAP provides in this space.

10.1.1   Continuous Integration

Continuous integration is the practice of building software continuously by integrating the code changes frequently into a common source code repository and then validating and testing the changes automatically. This enables you to detect errors as quickly and as early as possible in the development lifecycle.

A basic flow for continuous integration steps can be described as follows (as shown in Figure 10.1):

❶ Developer writes the code and tests it

❷ Developer pushes the code changes into a central source code management system

❸ The central source code management system triggers the continuous integration server

❹ The continuous integration server runs the automated build and tests

❺ The CI server sends feedback to the developer

Over time, all these processes happen in shorter cycles.

Implementing continuous integration in an organization requires process changes tied to how you develop software, in addition to the technical aspects described previously. There are multiple well-established agile software engineering methodologies, like scrums and Kanban, which organizations are adopting to achieve this.

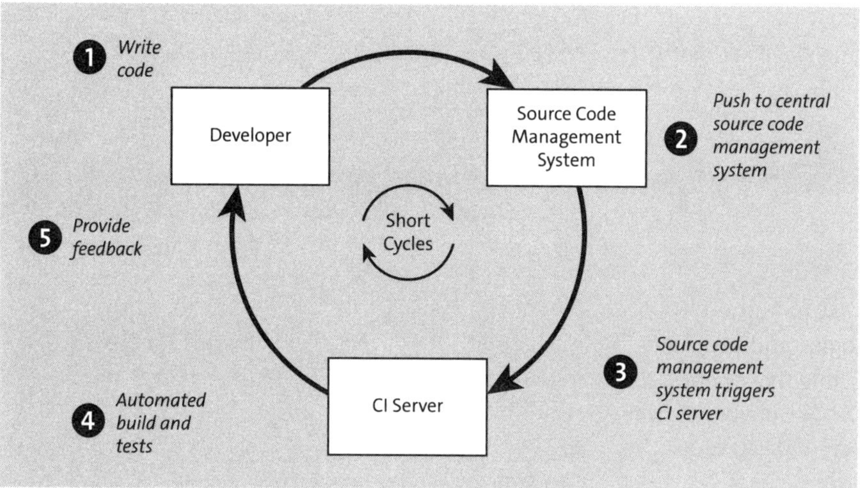

Figure 10.1 Continuous Integration Basic Flow

Continuous integration is also tied to a well-defined change management process: documenting bug fixes/new feature requests → impact analysis → approve/deny cycle → implementing the change → feedback received from continuous integration process. Thus, a process is established to track, manage, and detect these changes early. These process elaborations are outside the scope of this book.

10.1.2   Continuous Delivery

Continuous delivery is the process of taking the implemented changes as part of a continuous integration cycle to end customers as quickly as possible without any possible disruption. The goal of the continuous delivery cycle is to bring value to end users/customers as part of the delivery cycle, thus enabling a faster feedback cycle from the actual users of the software.

Figure 10.2 shows an example of a continuous delivery cycle in which the committed changes from a developer pass through the deployment pipeline of integration, performance, and acceptance cycles. The changes are tested rigorously for defined software qualities before getting delivered to an actual production system. For achieving a faster deployment lifecycle, it's imperative that these tests are automated as much as possible through this lifecycle, which creates a repeatable and reliable process for releasing software.

Continuous delivery emphasizes that every change through the pipeline meets the qualities to be deployed to the production environment at any time. Organizations want to implement some checks and balances before deployment to production, so based on this process, certain organizations decide to make deployment to production a manual activity instead of an automated process. Thus, from a process perspective, continuous delivery need not always ensure a continuous deployment, and faster deployment times require a high degree of automation and collaboration across the team, which leads to maturity of the DevOps lifecycle, as outlined in Figure 10.3.

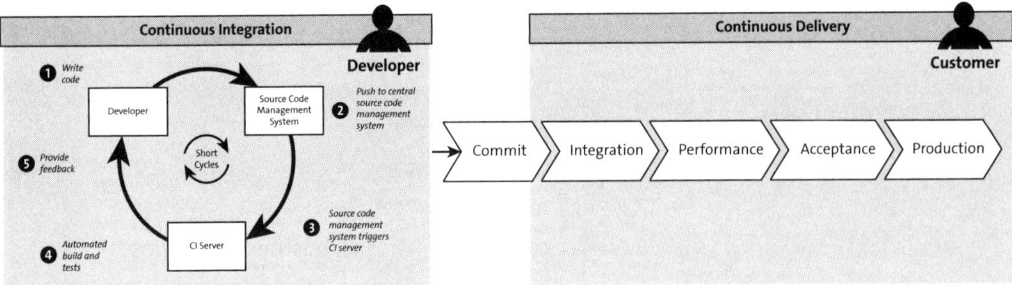

Figure 10.2 Continuous Delivery Cycle

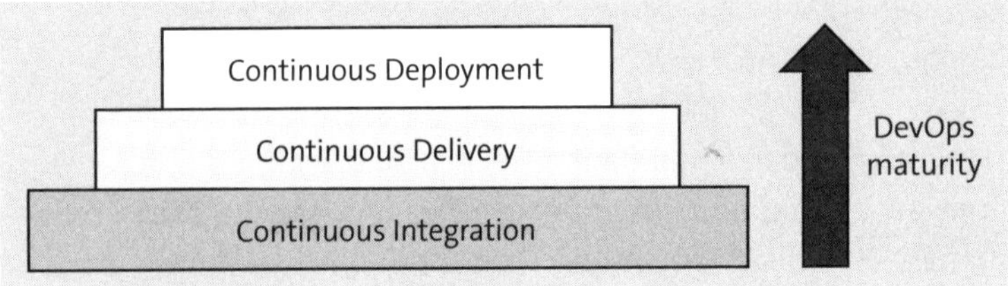

Figure 10.3 DevOps Maturity

> **Note**
>
> An important aspect of DevOps is operations through continuous monitoring, which we will talk about in Chapter 11.

Studies suggests that companies with mature DevOps cultures receive the following benefits:

- Deploy faster
- Have a closer feedback loop with customers
- Recover faster from failures
- Have high employee engagement and loyalty
- Have an overall lower TCO

> **Note**
>
> You can read more about SAP's best practices and principles for continuous integration and delivery at *http://s-prs.co/v515724*.

10.1.3   SAP Cloud Platform Services

SAP Cloud Platform has multiple services to help you with development and establishing a continuous integration and delivery process. In this section, we'll cover these tools at a high level.

In subsequent sections, we'll go into the details of some of the important services.

- **SAP Cloud Platform Business Application Studio**
  SAP Cloud Platform Business Application Studio is a development IDE based in a multicloud environment. It comes with a predefined set of development environments, called *dev spaces*, tailored for developing specific SAP business scenarios.

- **SAP Web IDE (full-stack)**
  SAP Web IDE full-stack is the development IDE service available on SAP Cloud Platform based in the Neo environment. It enables and helps developers to develop applications and extensions on SAP Cloud Platform.

> **Tip**
>
> At the time of writing this book, the SAP Cloud Platform Business Application Studio IDE is still evolving. We cover some of the scenarios in this chapter using SAP Web IDE full-stack.
>
> However, we recommend that you start your development in SAP Cloud Platform Business Application Studio. Then, if your particular development scenario isn't yet supported, consider using SAP Web IDE for development.

- **SAP Cloud Platform Git service**
  SAP Cloud Platform Git service is the source code management service.

- **Corporate Git service**
  Come companies have a Git repository in their on-premise landscape. SAP Web IDE allows you to connect and utilize your existing Git repository for source code management.

- **SAP Cloud Platform Feature Flags**
  SAP Cloud Platform Feature Flags is a service available in the Cloud Foundry environment in SAP Cloud Platform. The service helps enable or disable new features at application runtime, based on predefined rules or a release plan schedule.

- **SAP Cloud Platform Transport Management**
  SAP Cloud Platform Transport Management lets you manage transports between SAP Cloud Platform accounts in Neo and Cloud Foundry, such as DEV to TEST to PROD accounts. SAP Cloud Platform helps to transport the contents from development to a production environment in an automated way.

10.2   Building a DevOps Strategy and Plan

In the previous section, we talked about continuous integration and delivery and various SAP Cloud Platform services available to develop and manage changes. In this section, we'll guide you through how to bring these concepts into action in the SAP Cloud Platform environment, build your development and production strategy and environment, and outline the next steps.

10.2.1   Building Your Strategy and Landscape in SAP Cloud Platform

In this section, we'll talk about various considerations while developing and planning the right subaccount and spaces structure for application development. A few considerations are as follows:

- **Staged development**
  You'll want to create and separate between development, test, and production environments. You need to manage different environments separately due to security needs, member isolation, data separation, and so on.

- **End user needs/data regulation needs**
  Certain scenarios need data to be stored in a specific region because of legal reasons or to be closer to end users of a specific region.

- **Cost separation**
  Some organizations need to separate and monitor costs based on individual lines of business or projects. You have better flexibility to monitor the usage costs at the subaccount level.

> **Tip**
>
> In a consumption-based global account, you can export a usage report in which the data can be segregated at the individual subaccount level.
>
> With a directory, this data can be extracted/consolidated at the directory level.

You also want to consider additional criteria, such as your current landscape setup in your organization and your SAP S/4HANA environment setup. This has a certain advantage of being able to leverage existing CI/CD process in your organization, but we advise you to consider that typically on-premise and cloud systems follow different application development lifecycles.

Based on the considerations outlined thus far, Figure 10.4, Figure 10.5, and Figure 10.6 offer example scenarios for how to set up and design your landscape for SAP Cloud Platform application development.

In Figure 10.4, your landscape is designed to be staged according to subaccounts. Here you would have all your applications segregated by a subaccount tiering strategy. This is relevant for both Cloud Foundry and Neo runtime environments.

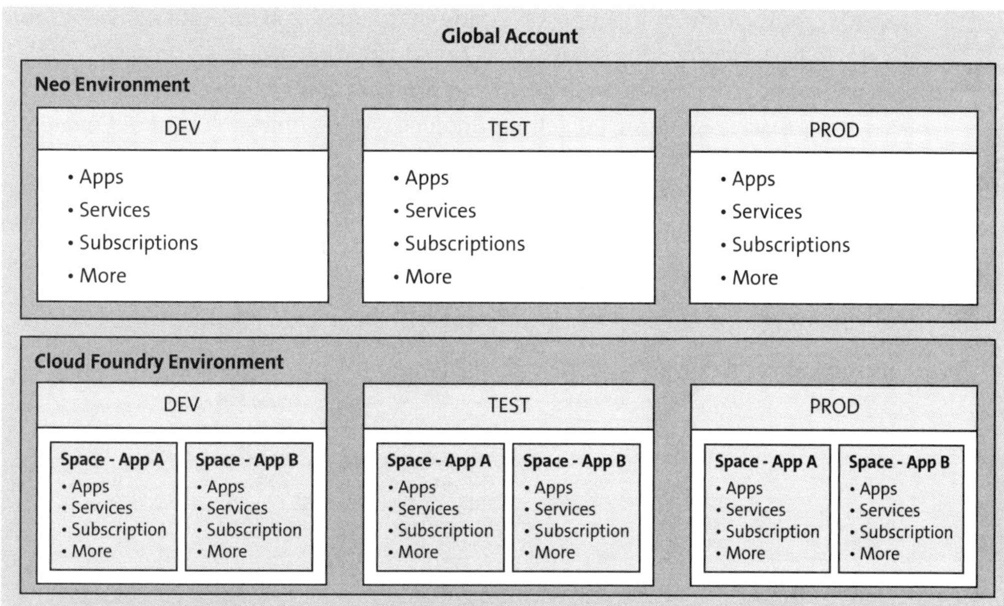

Figure 10.4 Scenario 1: Subaccount Setup Based on Staged Development

Figure 10.5 is another example of tiering your landscape. This case shows tiering your landscape based on individual lines of business in your organization. Such models allow you to segregate the cost of running/operating between different lines of business in an organization while taking care of any specific data-privacy needs for applications with sensitive information. In a Cloud Foundry environment, this model of segregation allows you to further use spaces to segregate between different tiers of DEV, TEST, and PROD, although this is not mandatory and can be done at the subaccount level too, like for Neo.

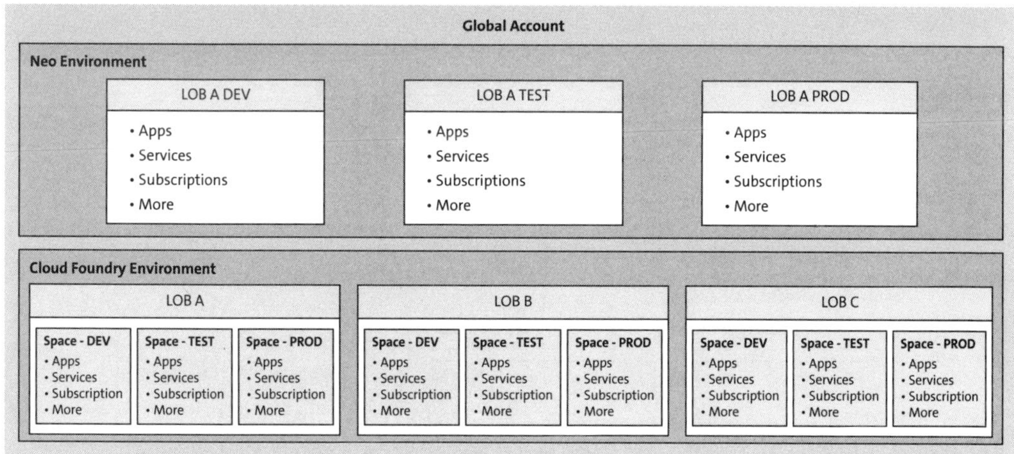

Figure 10.5 Scenario 2: Subaccount Setup for Lines of Business (Triggered by Cost Segregation or Data Privacy/Isolation per LOB)

Figure 10.6 shows another example of tiering your landscape for usage of SAP Cloud Platform. In this example, the landscape is tiered based on projects. Typical examples for such scenarios for certain projects in your organization are region-specific or have special data privacy regulation guidelines that need to be applied. This would mandate you to use a data center that is logically close to your end users and/or data centers that meet your data privacy regulations. Thus, you can segregate your landscape to meet the region-specific requirements based on projects.

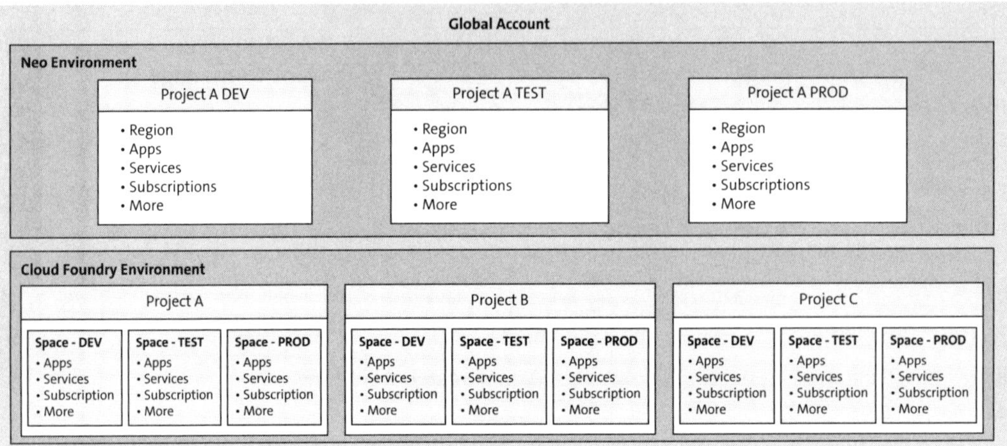

Figure 10.6 Scenario 3: Subaccount Setup Divided by Projects (Triggered by Region or Data Isolation Requirements)

Note

For all these scenarios, the subaccount can be further grouped together based on a directory concept, as shown in Figure 10.7.

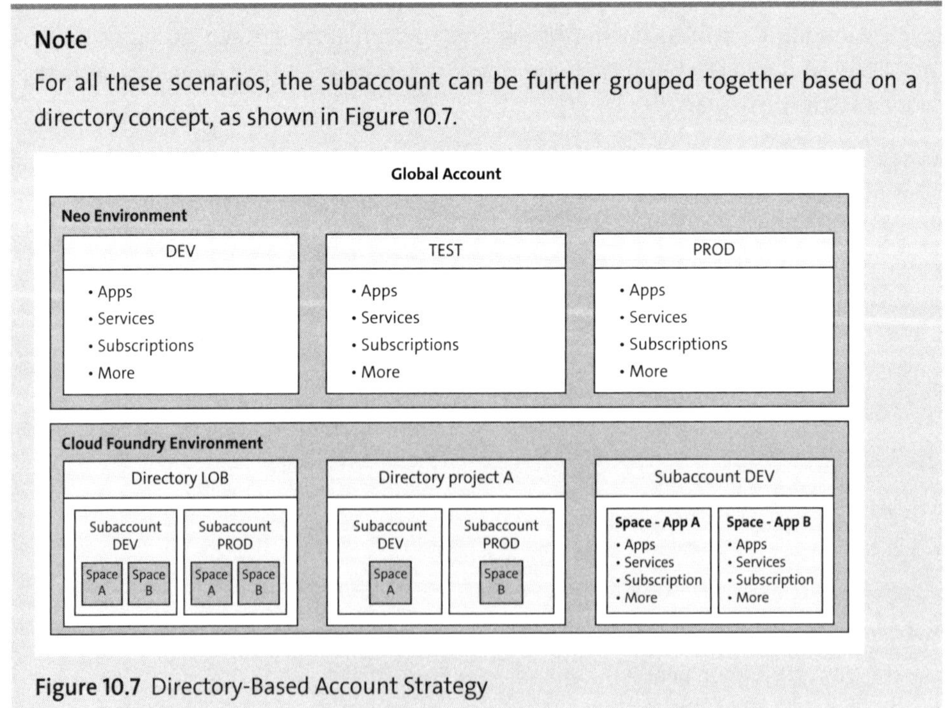

Figure 10.7 Directory-Based Account Strategy

10.2.2   Managing Roles and Responsibilities

In Chapter 8, Section 8.2, we covered various predefined roles available at the global account, subaccount, and space levels for SAP Cloud Platform. We covered custom platform role capabilities as well in that section.

In this section, we want to remind you of those roles and functions as you proceed to set up your landscape structure involving various subaccounts and spaces. Note that the quota plan is only relevant for Cloud Foundry, not for Neo.

In addition to assigning the right people in an organization the right roles, you can further define a quota plan for individual spaces within a subaccount to manage space level consumption.

Figure 10.8 shows how you can create quota plans for individual spaces. To create a quota plan, follow these steps:

1. Navigate to your Cloud Foundry subaccount.
2. Select **Cloud Foundry • Quota Plans**.
3. Click **New Plan**. This will open the pop-up shown in Figure 10.8.

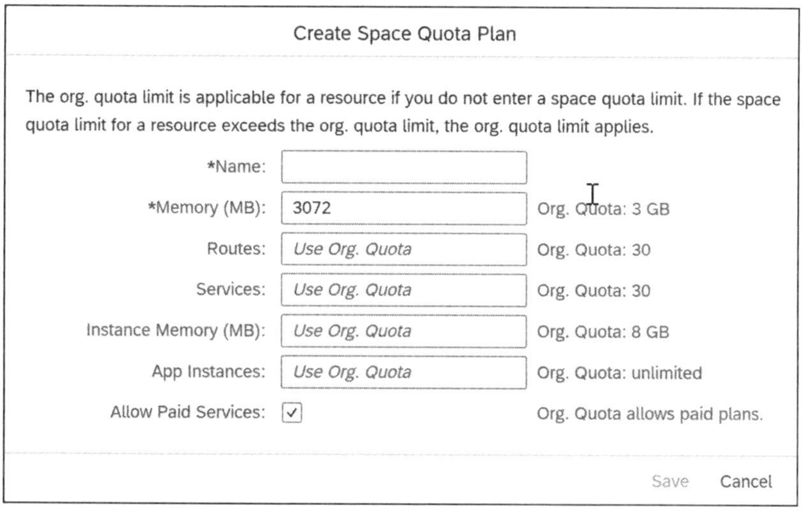

Create Space Quota Plan

The org. quota limit is applicable for a resource if you do not enter a space quota limit. If the space quota limit for a resource exceeds the org. quota limit, the org. quota limit applies.

*Name:		
*Memory (MB):	3072	Org. Quota: 3 GB
Routes:	Use Org. Quota	Org. Quota: 30
Services:	Use Org. Quota	Org. Quota: 30
Instance Memory (MB):	Use Org. Quota	Org. Quota: 8 GB
App Instances:	Use Org. Quota	Org. Quota: unlimited
Allow Paid Services:	✓	Org. Quota allows paid plans.

Save   Cancel

Figure 10.8 Create Space Quota Plan

10.3   SAP Cloud Platform Transport Management

SAP Cloud Platform Transport Management manages application transport and content transport within a landscape in SAP Cloud Platform. SAP Cloud Platform Transport Management is available in the SAP Cloud Platform Cloud Foundry environment and can be used to transport the following entities:

- Multitarget application archives (MTAs; *.mtar* files) between different cloud subaccounts

- Delivery units in SAP HANA extended application services, classic model (SAP HANA XS) between different SAP HANA instances that are assigned to cloud subaccounts

- Application content transported in an application-specific format between different cloud subaccounts (ZIP files)

In the following sections, we'll discuss how you can create multitarget applications using the SAP Cloud Platform development environment and package them for transport using SAP Cloud Platform Transport Management.

10.3.1 Multitarget Applications and Multitarget Application Archive

An MTA is logically a single application, consisting of multiple related and interdependent parts that are developed using different technologies or programming paradigms and designed to run on different target runtime environments, with a single consistent lifecycle. This approach of developing an application simplifies the application transport: applications have multiple interconnected parts as one single unit and you make them work together. SAP Cloud Platform wisely deploys the pieces simultaneously, depending on the technology they are based on. Thus, MTA is an SAP construct that allows you to combine modules into a single deployable unit for lifecycle maintenance.

In the following sections, we'll discuss creating an MTA and then building an MTA application.

Creating an MTA

SAP Web IDE provides templates as an easy way to create MTA applications. Before you begin, you must meet the following prerequisites:

- You have SAP Web IDE enabled and appropriate developer/administrator access rights given to users.

- You have activated the SAP Cloud Platform Business Application Development Tool add-on in SAP Web IDE.

- You have created an SAP Cloud Platform Business Application Studio project and selected the option to **Include Support for Continuous Delivery Pipeline of SAP Cloud SDK** as shown in Figure 10.9.

The project thus created is displayed in the workspace of the SAP Web IDE with a *mta.yaml* file created containing default modules supported by SAP Cloud Platform, as shown in Figure 10.10. You can then start building your application and make changes to different modules.

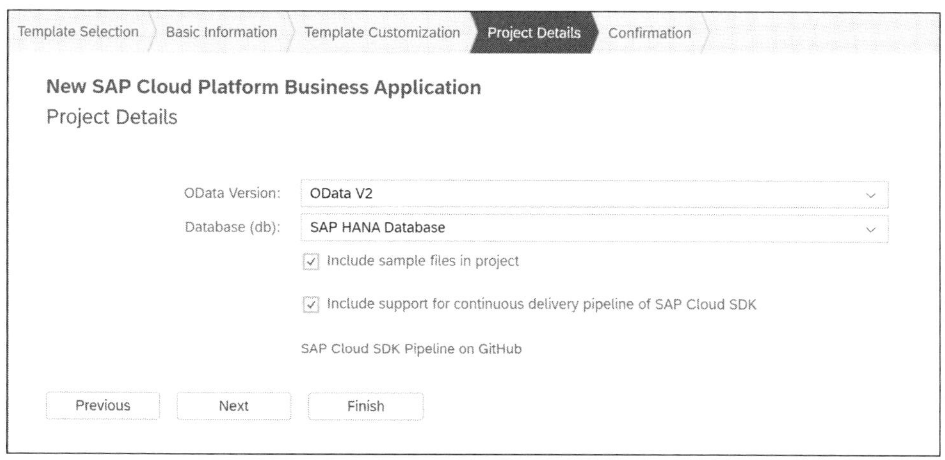

Figure 10.9 Business Application Project Details

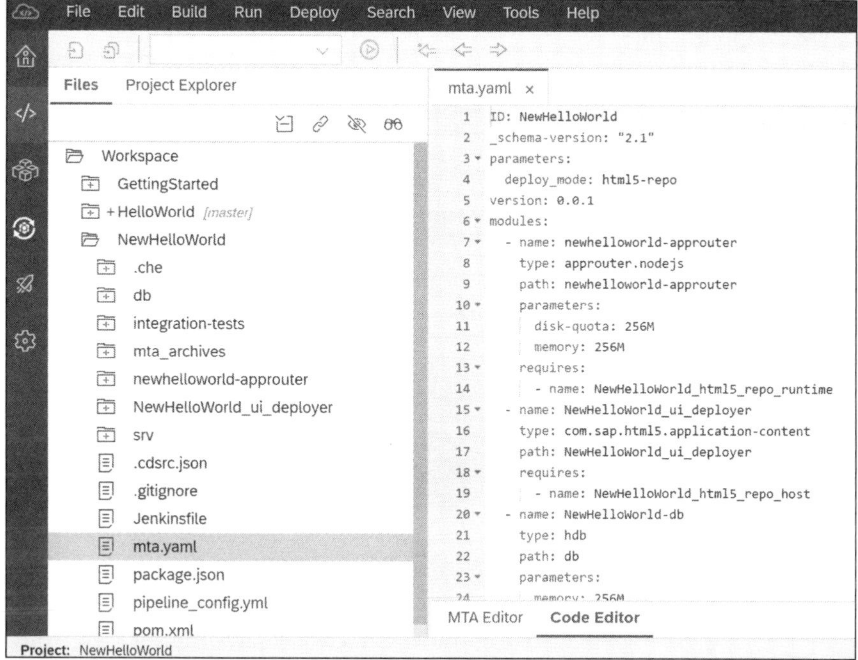

Figure 10.10 MTA Descriptor for Business Application Template

> **Note**
>
> The created *mta.yaml* project is only compliant with the Cloud Foundry runtime of SAP Cloud Platform because the selected **SAP Business Application Project** template is for Cloud Foundry.
>
> If you're creating a project for the Neo runtime, you need to select a template meant for the Neo runtime in SAP Web IDE.

> **Note**
>
> Building applications is explained in detail in other books on application development using SAP Cloud Platform—for example:
>
> - SAP Cloud Platform: Cloud-Native Development (2019, SAP PRESS): *https://www.sap-press.com/4713/*
> - Developing Application with the SAP Cloud Application Programming Model (2020, SAP PRESS): *https://www.sap-press.com/5152/*
> - Extending SAP S/4HANA: Side-by-Side Extensions with the SAP S/4HANA SDK (2020, SAP PRESS): *https://www.sap-press.com/4655/*

Once the application development is complete and tested, the next step is to build the project to generate an MTA archive (*.mtar* file), which is the deployable file for SAP Cloud Platform. Let's discuss how to do that now.

Building an MTA Application

You can build an MTA application as follows using SAP Web IDE:

1. For projects in SAP Web IDE and supported by SAP Web IDE, right-click **Projects • Build • Build with Cloud MTA Build Tool (recommended)** as shown in Figure 10.11.

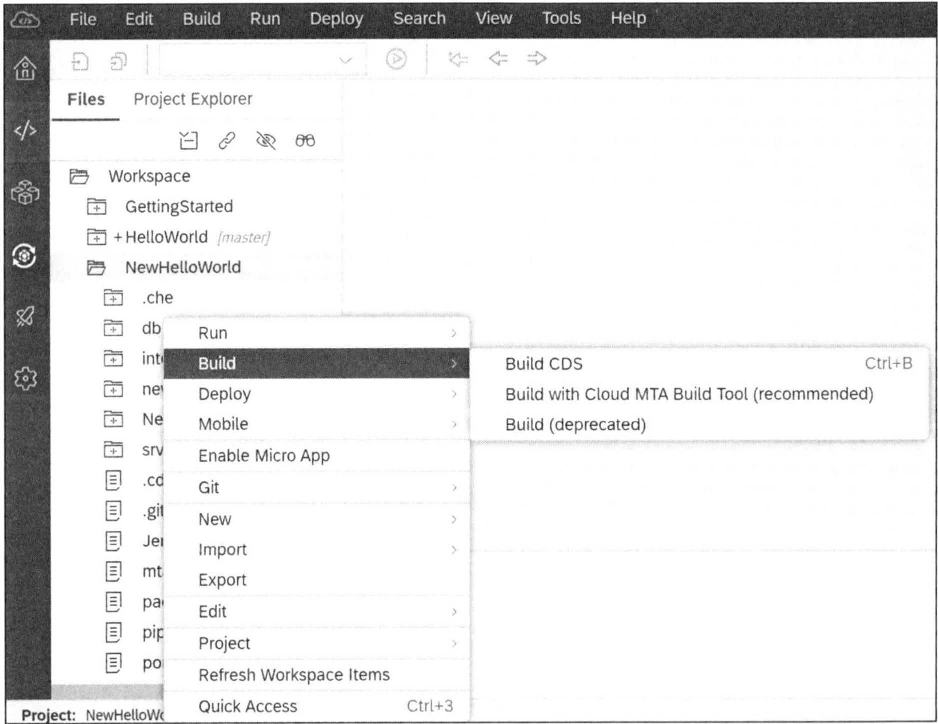

Figure 10.11 Build with Cloud MTA Build Tool

2. A successful build will create a file with the *.mtar* extension in the **mta_archives** folder under the project name, as shown in Figure 10.12.

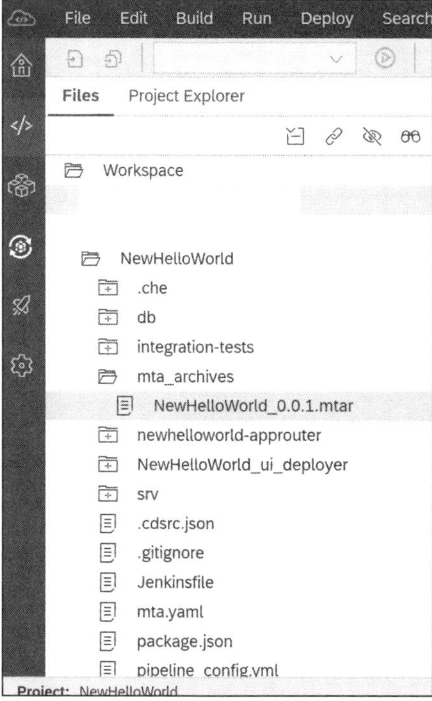

Figure 10.12 MTA Archive

You can deploy the generated *.mtar* file directly from SAP Web IDE to your Cloud Foundry development environment by selecting right-clicking the *.mtar* file and selecting **Deploy • Deploy to SAP Cloud Platform**, as shown in Figure 10.13.

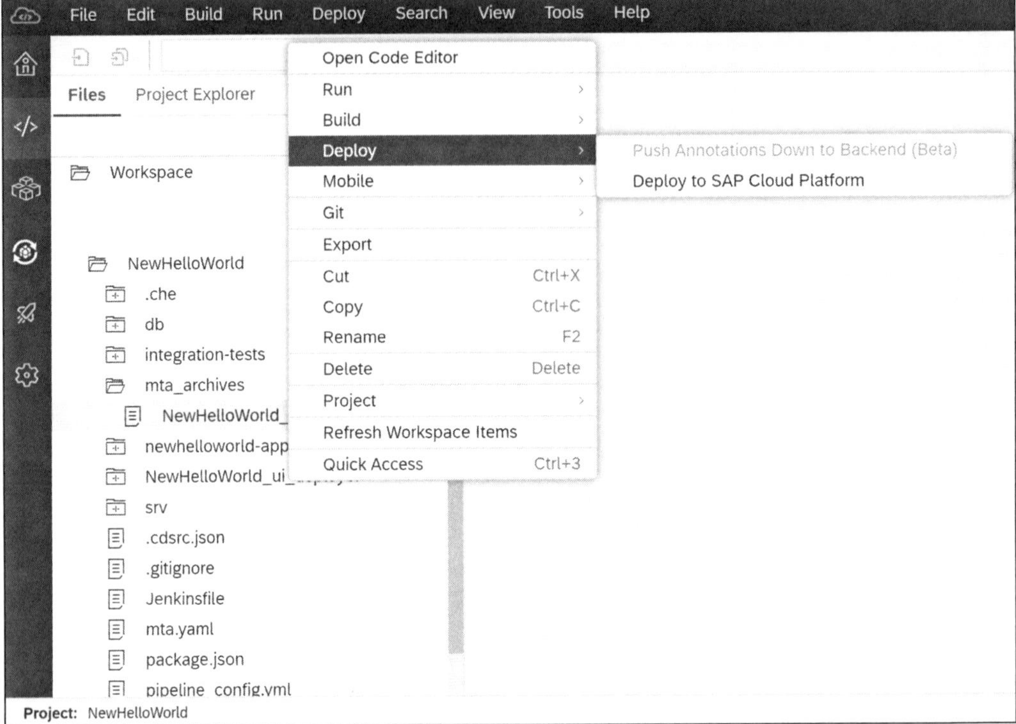

Figure 10.13 Deploy to SAP Cloud Platform

If you have used the MTA Archive tool to build your project, you can deploy the generated build file using console tools directly into the target environment.

10.3.2   Set Up SAP Cloud Platform Transport Management

In this section, we'll describe how the MTA archive build from the previous steps can be transported using SAP Cloud Platform Transport Management.

> **Note**
>
> SAP Cloud Platform Transport Management is available in the Cloud Foundry runtime. To use the service, you therefore need a subaccount created in the Cloud Foundry runtime.

In the following sections, we'll discuss how to activate and then configure SAP Cloud Platform Transport Management to transfer the contents.

Activate the Service

Before you can activate the service, you have must a subaccount with a space created in the Cloud Foundry runtime in SAP Cloud Platform. To begin the activation, follow these steps:

1. From your SAP Cloud Platform global account, choose **Entitlements • Subaccount Assignments**. Select **Entitlement • Entity Assignment** when the directory feature is available.

2. Select your subaccount from the dropdown list, and choose **Go**.

3. Choose **Configure Entitlements** and then **Add Service Plan**.

4. From the service catalog, select **Transport Management,** as shown in Figure 10.14. Under **Available Service Plans**, select the **Standard** checkbox. Choose **Add 1 Service Plan**.

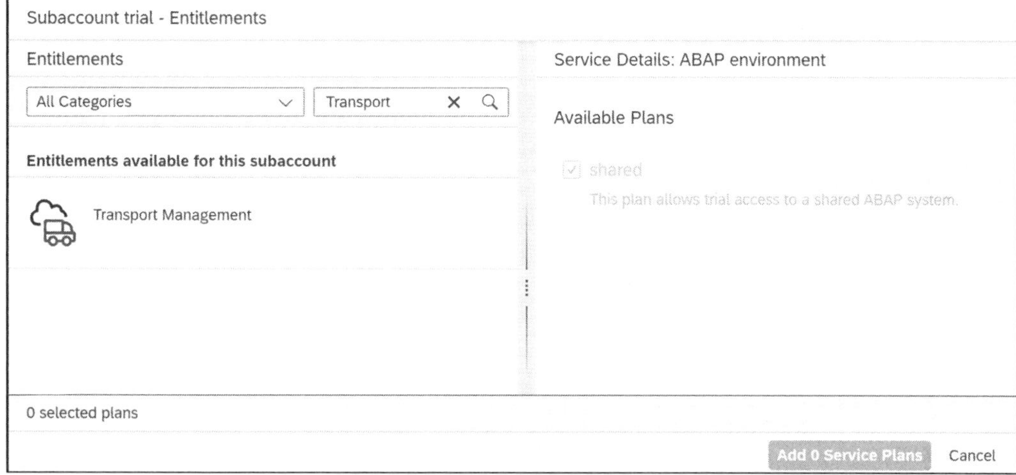

Figure 10.14 Configure Transport Management Entitlements

5. **Save** the entitlement selection.

6. Navigate inside the subaccount in which the service has been added.

7. In the subaccount, choose **Subscription**. The entitled subscriptions are shown.

8. Select the **Transport Management** tile and choose **Subscribe**. The subscription is activated.

9. Now, assign users to required roles for SAP Cloud Platform Transport Management. If you already have a role collection created, the next step is to assign the role collection to predefined roles in SAP Cloud Platform Transport Management.

10. Choose **Manage Roles** in SAP Cloud Platform Transport Management. All roles available for SAP Cloud Platform Transport Management are displayed.

11. Select the role that you want to add to the role collection created before, and choose **+ (Add to a Role Collection)**.

12. Select the role collection you created before and then **Add**, as shown in Figure 10.15.

13. Repeat the steps for other roles you want to create.

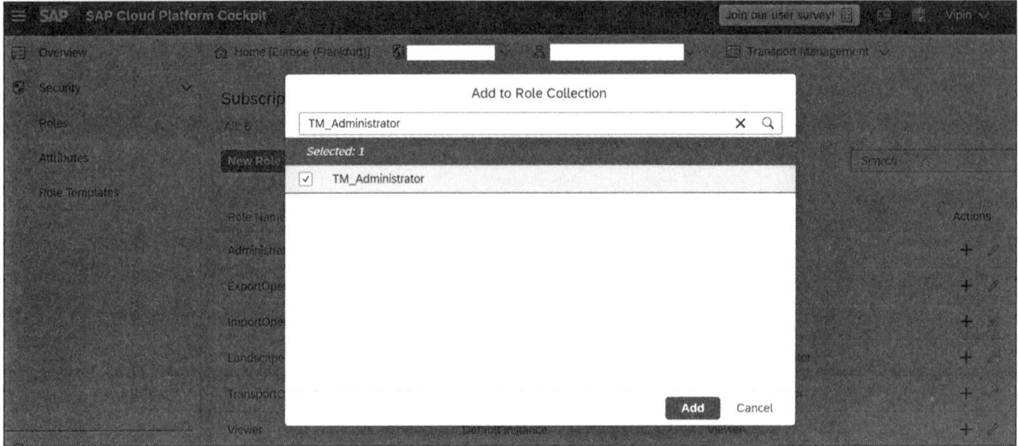

Figure 10.15 Add Role Collections in Transport Management Service

14. Make sure the role collection is added to your user ID from **Security • Trust Configurations**. Select an identity provider and add the role collection to the user.

15. Select the **Transport Management** tile and then **Go to Application**. Enter your login details. This will open the landing page, as shown in Figure 10.16.

Figure 10.16 SAP Cloud Platform Transport Management Service Home

16. In the subaccount, go to **Spaces** and select the space you created before to open it.

17. In the space, choose **Services • Service Marketplace** and choose the **Transport Management** tile.

18. Choose **Instances • New Instance**. In the **New Instance Creation** wizard, follow through the steps and give a **Name** to the service instance.

19. Click **Finish**. You now have a service instance created.

20. Navigate inside the service instance and choose **Service Keys • Create Service Key**.

21. Enter a name for the service key and choose **Save**. This will create the service key. The connection details in the service key will be used later to configure destinations for programmatic access to SAP Cloud Platform Transport Management.

Configure the Service

In this section, we'll talk about how to configure SAP Cloud Platform Transport Management to transport cloud applications between different subaccounts.

Before we go ahead with the configuration of SAP Cloud Platform Transport Management, let's go over common terminology used in this service:

- **Transport destination**
  Transport destinations are used to address the target endpoints of the deployment process. A destination contains the address of the target endpoint, as well as user credentials.

> **Note**
>
> You create transport destinations in an SAP Cloud Platform Cloud Foundry subaccount, not within the SAP Cloud Platform Transport Management application. For information on how to create a transport destination, refer to the SAP help document at *http://s-prs.co/v515726*.

- **Transport nodes**
  Transport takes place between transport nodes. Thus, transport nodes represent source and target endpoints of a deployment process—for example, subaccounts. A prerequisite for creating transport nodes is that you have defined a transport destination. Transport nodes are created within SAP Cloud Platform Transport Management.

- **Transport routes**
  Transport routes are used to connect transport nodes. Transport routes are created within SAP Cloud Platform Transport Management.

Inside SAP Cloud Platform Transport Management, there are two ways to create transport nodes and transport routes:

1. **Using the Transport Landscape Wizard**

   Here you use a wizard to create a simple landscape using templates. In Figure 10.17, you'll find a simple three-tier landscape created using the **Transport Landscape Wizard**. The created landscape can be edited using **Landscape Visualization**.

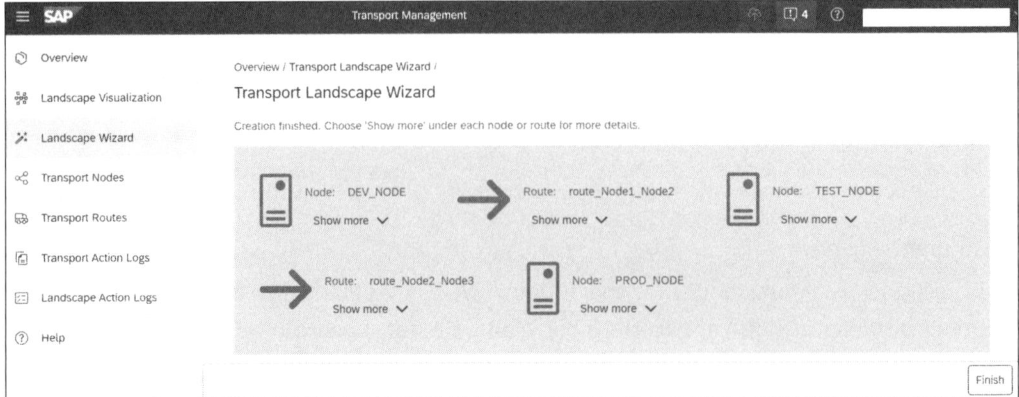

Figure 10.17 Landscape Wizard Used to Create Three-Tier Landscape

2. **Using Landscape Visualization**

   You can use **Landscape Visualization** to configure transport nodes and transport routes to your landscape, as well as edit the existing landscape. Figure 10.18 shows a landscape that was created using **Landscape Visualization** and the landscape that was created using the **Transport Landscape Wizard**.

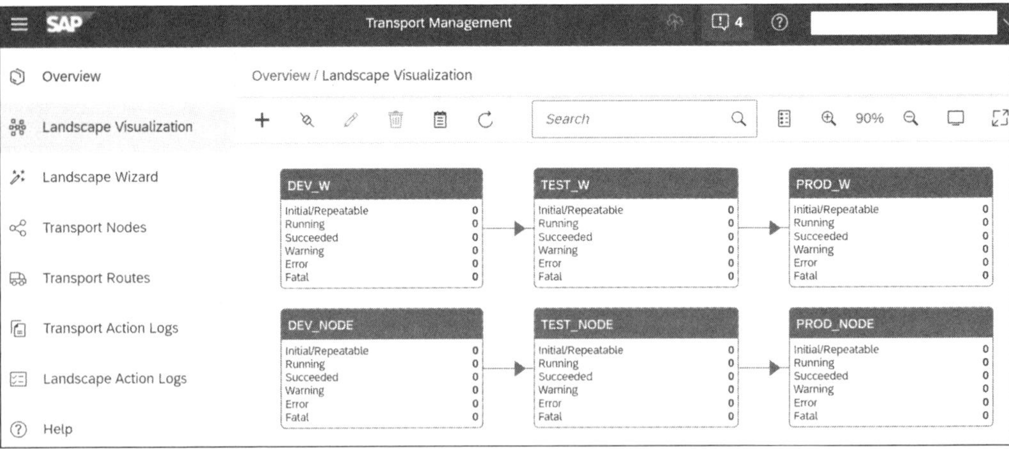

Figure 10.18 Landscape Visualization Used to Create and Edit Nodes

More complex landscapes can be configured using transport nodes and transport route selections within SAP Cloud Platform Transport Management.

> **Note**
>
> The help documentation for the **Transport Landscape Wizard** is at *http://s-prs.co/v515727*. The help documentation for **Landscape Visualization** is at *http://s-prs.co/v515728*.

SAP Cloud Platform Transport Management supports the following two scenarios, and configurations of the service are dependent on which scenario you choose to use:

1. **Transport of content archives directly into another application**
   In this scenario, another application with content archives you want to transport is directly integrated with SAP Cloud Platform Transport Management. You can transport content archives directly from within the application's source subaccount to the target subaccount of the application. You use SAP Cloud Platform Transport Management to import the content archives into the target subaccount.

> **Note**
>
> The direct upload from the SAP Cloud Platform cockpit is now available only for solutions built for the Neo environment with some limitations. This process would require you to create a destination service from the source subaccount in Neo in which your application is running to call the access to SAP Cloud Platform Transport Management using the service keys created previously.
>
> The deployed contents can be exported to SAP Cloud Platform Transport Management from **Solutions** in the SAP Cloud Platform cockpit. For understanding the limitations, refer to the SAP help document at *http://s-prs.co/v515729*.
>
> For applications based on the Cloud Foundry runtime, we'll cover how the content can be directly uploaded and transported in this section and the next.

2. **Transport of content archives available in your local file system**
   In this scenario, the content archives that you want to transport between subaccounts are downloaded into a local file system. SAP Cloud Platform Transport Management is used to upload the content archives to an import queue of the target subaccount and then import them into the target subaccount.

   The contents from the file systems are then uploaded into the import queue of the transport node defined in SAP Cloud Platform Transport Management, as shown in Figure 10.19.

> **Note**
>
> Information about the process of importing the contents and information about the required roles to do this in SAP Cloud Platform Transport Management can be found in the help document at *http://s-prs.co/v515730*.

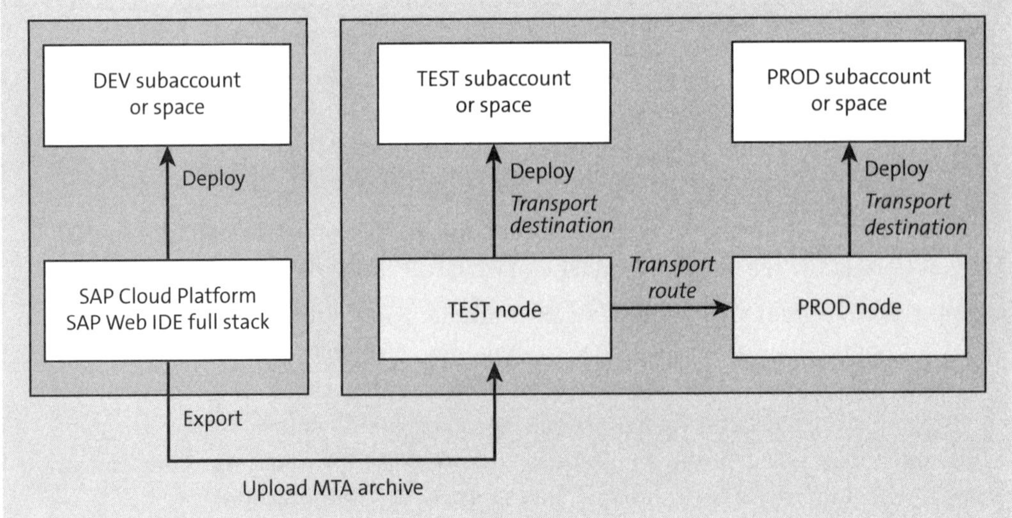

Figure 10.19 Transport MTA Archive from Local File System

10.4   Project "Piper"

In this section, we'll discuss templates for project "Piper," which will substantially ease setting up continuous integration and delivery in your project using SAP technologies. These tools are best suited for developers and operators already aware of continuous integration and delivery technologies, including Jenkins, an open source automation server.

In the following sections, we'll first provide an introduction to project "Piper" and then show you how to configure the SAP Cloud SDK pipeline and integrate with SAP Cloud Platform Transport Management.

10.4.1   Overview

You get the following artifacts with project "Piper":

- A set of ready-made continuous delivery pipelines for direct usage in your project, as follows:
  - A general-purpose pipeline that supports various technologies and programming languages and is suited for building standalone SAP Cloud Platform applications.
  - An SAP Cloud SDK pipeline. If you're building an application with SAP Cloud SDK or SAP Cloud Application Programming Model, this pipeline helps you quickly build and deliver a high-quality app. The SAP Cloud SDK helps you efficiently build, test, and deliver extensions of SAP solutions such as SAP S/4HANA. We'll discuss this in detail in the following sections.

- A shared library containing steps and utilities that are required by the pipeline. This enables you to customize preconfigured pipelines or to build your own customized ones.

- A set of Docker images to set up a CI/CD environment in minutes using sophisticated lifecycle management.

> **Note**
>
> For more information about the topics discussed in this section, consult the following resources:
>
> - SAP General Purpose Guideline Reference: *https://sap.github.io/jenkins-library/stages/introduction/*
> - SAP Cloud SDK Pipeline: *https://sap.github.io/jenkins-library/pipelines/cloud-sdk/introduction/*
> - SAP Cloud SDK Reference: *https://community.sap.com/topics/cloud-sdk*
> - Shared library: *https://github.com/SAP/jenkins-library*
> - DevOps Docker Images: *https://github.com/SAP/devops-docker-images*
> - More about Jenkins and its documentation: *https://jenkins.io*
> - More about Docker and its documentation: *https://www.docker.com*

Before we get into details of configuring project "Piper" to enable continuous integration and delivery for SAP Cloud Platform projects, we'll take a look at two important files, shown in the project structure created using SAP Web IDE in Figure 10.9. The details of this structure are shown in Figure 10.20:

- **Jenkinsfile**
  - Jenkinsfile is a text file that contains the definition of the pipeline and is checked into source control (e.g., Git) along with the application code.
  - The default Jenkinsfile consumes the master branch of the pipeline. For productive usage, we recommend using a released version and updating it as new versions are released. This can be done by changing `pipelineVersion` to the respective Git tag.

- **pipeline_config.yaml**
  - This is the configuration file available within your project to maintain the configuration required to run the pipeline.
  - The file has different common configuration parameters to define various configurations—for example, for the deployment environment.
  - You need to change this configuration manually to reflect a project structure: *.pipeline/config.yml*.

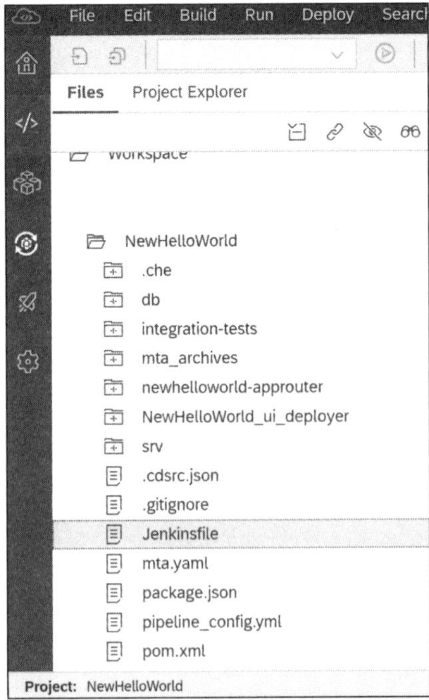

Figure 10.20 "Piper" Configuration Files

> **Note**
>
> You can read more about Jenkinsfile here: *https://jenkins.io/doc/book/pipeline/jenkinsfile/*.
>
> Different configuration parameters and library steps for pipeline_config.yml can be found here: *https://sap.github.io/jenkins-library/configuration/*.

10.4.2   Configuring SAP Cloud SDK Pipeline

The SAP Cloud SDK pipeline is based on project "Piper" and offers unique features for assuring that your SAP Cloud SDK-based application fulfills the highest quality standards. SAP Cloud SDK has the following components to achieve this:

- Project archetypes that help in creating project templates
- Preconfigured Jenkins CI and CD server, available via DockerHub
- SAP Cloud SDK-specific CI and CD pipelines and libraries for building, testing, quality checking, and deploying applications

Ahead, we'll cover two options to install the Jenkins tool to run "Piper" for the SAP Cloud SDK. This will let you run SAP Cloud Platform Continuous Integration and Delivery applications based on the SAP Cloud Application Programming model.

Preconfigured Jenkins in SAP Cloud SDK

As previously mentioned, SAP Cloud SDK provides a preconfigured Jenkins server as part of the package. You need to meet the following hardware requirements to install this Jenkins server:

- Linux system with at least 4 GB of memory
- Have the latest version of Docker installed
- Access in the system to *github.com*

The SAP Cloud SDK comes with a lifecycle management tool called Cx Server to bootstrap the preconfigured Jenkins instance within minutes, with all required plug-ins and libraries automatically included. This server is based on Docker images provided as part of the project "Piper" package. To get started, initialize the Cx Server by using the docker run command as in Listing 10.1.

```
docker run -it --rm -u $(id -u):$(id -g) -v "${PWD}":/cx-server/mount/ ppiper/
cx-server-companion:latest init-cx-server
```

Listing 10.1 Initialize Cx Server

This will install two new files, *cx-server* and *server.cfg*, in the current working directory. These files help manage the lifecycle of a Jenkins server. The Jenkins server can be started via the command in Listing 10.2.

```
chmod +x ./cx-server
./cx-server start
```

Listing 10.2 Jenkins Cx Server Start

> **Note**
>
> For more information on configuring the Cx server and customizing your Jenkins instance, consult the help document at *http://s-prs.co/v515731*.

Bring Your Own Jenkins Instance

If you have your own Jenkins instance running in your landscape, you can use the same instance and configure it to use "Piper." In this section, we'll cover the basic configuration for your own Jenkins instance. You first need to meet the following requirements:

- Java runtime environment 8
- Installation of Jenkins version 2.60.3 or higher
- Jenkins plugin for "Piper," as described at *https://sap.github.io/jenkins-library/customjenkins/*

To set up the "Piper" library to be used in your Jenkins installation, you need to perform the following steps after logging in with your administration user credentials:

1. In Jenkins, navigate to **Manage Jenkins • Configure System**.

2. Navigate to the **Global Pipeline Libraries** section and add a new library by clicking the
    Add button.

3. Set the library **Name** to "piper-library-os", as shown in Figure 10.21.

4. Set the **Default version** to the branch or tag you want to consume.

5. Set the **Retrieval Method** to **Modern SCM**.

6. Set **Source Code Management** to **Git**.

7. Set the **Project Repository** to "https://github.com/SAP/jenkins-library".

8. Save the changes.

Figure 10.21 "Piper" Settings in Jenkins Configuration

Now you can create a new pipeline job in Jenkins that retrieves a pipeline from SCM.
The library configured in the previous step can now be used in the Jenkinsfile by adding
code as in Listing 10.3.

```
@Library('piper-library-os') _
```

Listing 10.3 Using Piper Library

Jenkins will download the library and compile it during the execution of the Jenkinsfile. A project structure with all the required configuration and files will look like Listing 10.4.

```
├── Jenkinsfile
├── .pipeline
│    └── config.yml
├── web-app  // web application, not required
├── database   // only if database module exists
├── integration-tests // integration tests for module
│    ├── pom.xml
│    └── src
│        └── test
├── mta.yaml
├── package.json
├── pom.xml
└── srv
     ├── pom.xml
     └── src
         ├── main
         └── test  // Unit-Tests for this service
```

Listing 10.4 Project Structure with Jenkinsfile

10.4.3   Integrate SAP Cloud Platform Transport Management with Pipeline

In this section, we'll cover how to integrate SAP Cloud Platform Transport Management with your pipeline. The prerequisites for this process are as follows:

- You have SAP Cloud Platform Transport Management set up and a service key generated as explained in Section 10.3.

- You have "Piper" set up in Jenkins as the CI server and set up to build an MTA archive.

- You have an MTA project set up with a folder structure corresponding to the standard MTA structure.

The scenario and set up of this process chain for a sample three-tier landscape is outlined in Figure 10.22.

To implement this scenario, the configuration to upload the MTA archive needs to be defined in the configuration file (*.pipeline/config.yml*). The configuration file needs to include the tmsUpload library step. Table 10.1 outlines the parameters included in this library step.

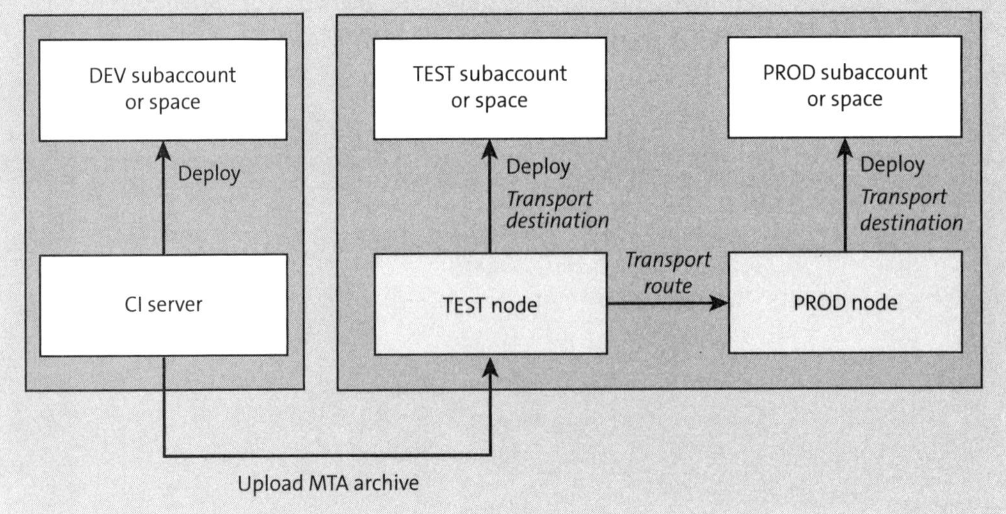

Figure 10.22 Integrate SAP Cloud Platform Transport Management with "Piper"

Name	Mandatory	Description
credentialsId	Yes	These are credentials to be used for the file and node uploads to SAP Cloud Platform Transport Management. This is the service key JSON file generated as mentioned in Section 10.3.2. The service key (JSON) must be stored as a secret text within the Jenkins secure store.
customDescription	No	This is a description of a transport request. The default is Git Commit-ID, which will be overwritten.
mtaPath	Yes	This is the path to *.mtar for upload to SAP Cloud Platform Transport Management.
nodeName	Yes	This is the name of the SAP Cloud Platform Transport Management node to which *.mtar is to be uploaded.
proxy	No	This is the proxy used for communication with the SAP Cloud Platform Transport Management backend.
script	Yes	This is the common script environment of the running Jenkinsfile.
stashContent	No	If specific stashes should be considered, their names are passed via the stashContent parameter. Has a default of [buildResult].

Table 10.1 Parameter Definitions for Pipeline Configuration File

Name	Mandatory	Description
verbose	No	This option determines whether you will print more detailed information to the log. Possible values include true, false.

Table 10.1 Parameter Definitions for Pipeline Configuration File (Cont.)

With the mandatory parameters, a basic configuration example of *.pipeline/config.yml* would look like Listing 10.5, which is referenced from the Jenkinsfile.

```
steps:
  tmsUpload:
    credentialsId: 'TMS_SERVICE_KEY'
    nodeName: 'TMS_NODE_NAME'
    mtaPath: <path to *.mtar file>
    customDescription: <description>
```

Listing 10.5 Configuration Example in pipeline_config.yml

10.5   SAP Cloud Platform Continuous Integration and Delivery

In this section, we'll cover SAP Cloud Platform Continuous Integration and Delivery, which lets you create and run CI/CD pipelines that automatically test, build, and deploy code changes to speed your development and delivery cycles.

SAP Cloud Platform Continuous Integration and Delivery supports the SAP Cloud SDK pipeline with predefined continuous integration and delivery pipelines, which we covered in earlier sections. These are used for developing extensions using the SAP Cloud Application Programming Model framework.

> **Note**
>
> This service is available in the Cloud Foundry runtime, so a prerequisite to activate this service is that you need a subaccount created with Cloud Foundry enabled in a Cloud Foundry region.

In the following sections, we'll look into how you can configure this service, further access the service, connect with your enterprise GitHub repository, and access the pipeline from your development environment, the SAP Web IDE full-stack service.

10.5.1   Service Configuration

To configure SAP Cloud Platform Continuous Integration and Delivery, you need to follow these steps:

1. Navigate to SAP Cloud Platform's Cloud Foundry subaccount via **Subaccount • Subscriptions** and enable **Continuous Integration & Delivery**, shown in Figure 10.23.

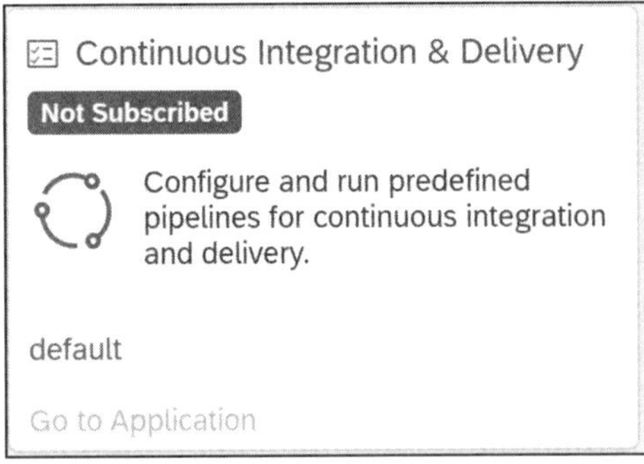

Figure 10.23 SAP Cloud Platform Continuous Integration and Delivery Service

2. Now manage your roles to give access to the service. SAP Cloud Platform Continuous Integration and Delivery has two predefined roles: administrator and developer.

3. In your subaccount, navigate to **Security • Role Collections**. Then create a new role collection by clicking on +, if a role collection doesn't yet exist.

4. Navigate inside **Role Collection**, click **Edit**, and click **+ (Add Role)** to add a predefined role.

5. Now add users to the role collection. For this, within your subaccount, navigate to **Security • Trust Configuration** and choose the configured trust.

6. For the SAP ID service, enter the email ID of the person you want to provide access to and then choose **Assign Role Collection** (if the button is disabled, choose **Show Assignments** first).

7. Repeat these steps as needed to add other role collections and users.

8. To enable programmatic usage of the service, inside your subaccount, choose the space where you want to enable this usage and navigate to **Services • Service Market Place**.

9. Choose **Continuous Integration and Delivery** and create a **Service Instance**.

10. Navigate inside the service instance you created, choose **Service Key**, and create a new service key.

10.5.2   Access the Service

To access SAP Cloud Platform Continuous Integration and Delivery after enabling it as described in the previous section, follow these steps:

1. In **Subaccount • Subscriptions**, access the **Continuous Integration and Delivery** tile enabled in the previous step and choose **Go to Application**. Log in to the service using the login email address. You'll see the home page as shown in Figure 10.24.

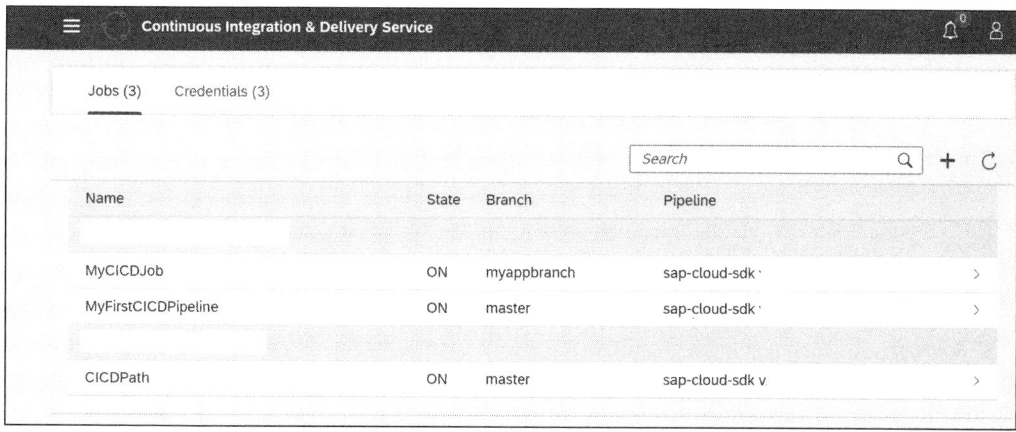

Figure 10.24 SAP Cloud Platform Continuous Integration and Delivery Service Home

2. You can create a new job from the **Job** tab and click the **+** sign. Enter all required details as shown in Figure 10.25.

3. GitHub must be connected, and you need credentials. This is set up from the **Credentials** tab, where you need to click the **+** sign. You can maintain all the necessary credentials here.

4. Choose **Add**.

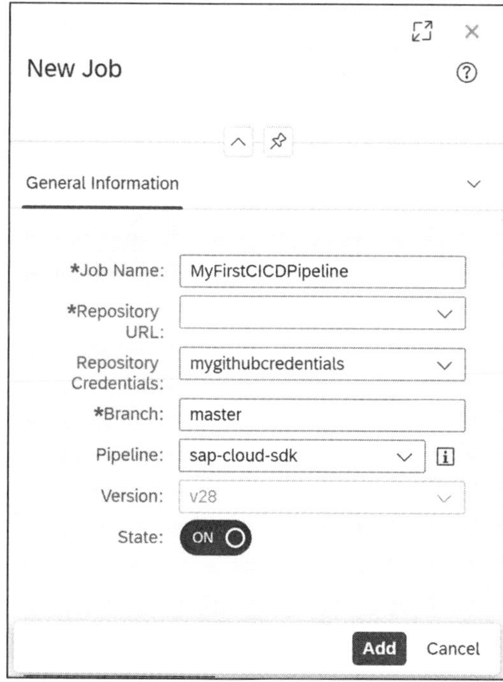

Figure 10.25 New Job in SAP Cloud Platform Continuous Integration and Delivery

10.5.3   Configure GitHub Hook

The webhooks in GitHub send push events to SAP Cloud Platform Continuous Integration and Delivery, which then triggers jobs. To set this up, follow these steps:

1. In the created job, you can find the **Webhook Data** for GitHub, as shown in Figure 10.26.

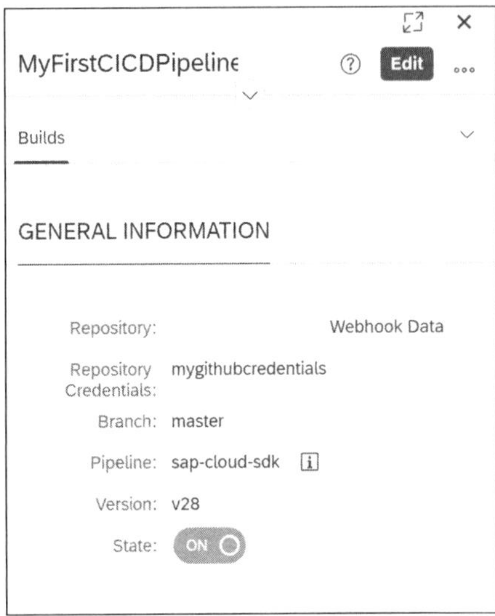

Figure 10.26 Webhook Data inside SAP Cloud Platform Continuous Integration and Delivery Job

2. Inside your GitHub, open the **Settings** tab and choose **Hooks**, as shown in Figure 10.27. Enter the **Payload URL**, **Content type**, and **Secret**.

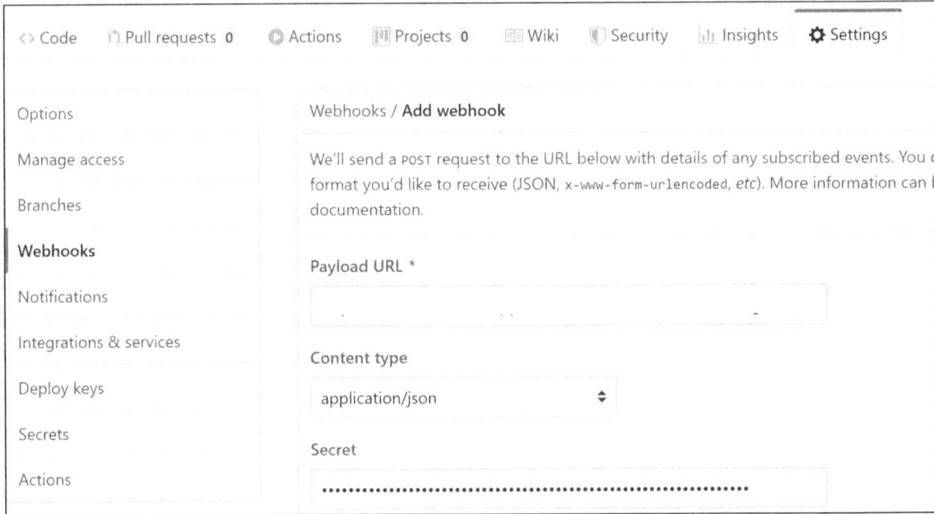

Figure 10.27 Webhooks Configuration from GitHub

10.5.4   Access SAP Cloud Platform Continuous Integration and Delivery from SAP Web IDE

You can access SAP Cloud Platform Continuous Integration and Delivery from SAP Web IDE, and in this section, we'll cover the configuration required to do so. First, however, you must meet the following prerequisites:

- You have a subaccount in which SAP Web IDE is activated and in use.

- You have created a service key for SAP Cloud Platform Continuous Integration and Delivery.

The necessary configuration for this access is as follows:

1. Inside the subaccount in which SAP Web IDE is activated, create a new destination service from the **Connectivity • Destinations** section.

2. Name the destination "cicd" (exact naming required).

3. Select **OAuth2ClientCredentials** for the authentication

4. From the service key created for SAP Cloud Platform Continuous Integration and Delivery, enter the following details for the destination:
   - **URL**
   - **Client ID**
   - **Client Secret**
   - **Token Service URL** with `/oauth/token?grant_type=client_credentials` appended

5. Inside the SAP Web IDE service, activate the plug-in as shown in Figure 10.28 by navigating to **Preferences • Extensions**.

Figure 10.28 SAP Web IDE Plug-in for SAP Cloud Platform Continuous Integration and Delivery

6. Once activated, you'll see a new icon for **SAP Cloud Platform Continuous Integration and Delivery** in the right-hand navigation bar within SAP Web IDE.

10.6 Summary

In this chapter, we covered an overview of continuous integration and delivery concepts and the need for a strategy in this arena to accelerate your innovation cycles without disrupting the solution's end users.

We also covered different strategies to manage your landscape within SAP Cloud Platform, giving guidance on different ways to manage your landscape. We then discussed developing SAP Cloud Platform business applications using MTA and generating a project structure to consider for continuous integration and delivery.

Finally, we covered various services available in SAP Cloud Platform to enable and empower your continuous integration and delivery based on governance practices that exist in your organization.

Chapter 11
Continuous Monitoring

In the previous chapter, we discusses how to enable continuous integration and delivery practices in your organization for quick, agile, and flexible business innovation lifecycles and to adapt to future needs. In this chapter, we'll look into various continuous monitoring practices for your applications and extensions using different services on SAP Cloud Platform. This will help you run, monitor, and log an application in the cloud for faster recovery.

Continuous monitoring is about dealing with operational aspects of applications after delivery/go-live for end users. Continuous monitoring capabilities are essential for operations and the future development of applications in the cloud to bring innovations to end users. Continuous monitoring along with continuous integration and delivery contributes to building a DevOps plan in organizations.

In this chapter, we'll talk about various continuous monitoring tools and capabilities available in SAP Cloud Platform, which includes monitoring and alerting capabilities for applications in the cloud. Specifically, we'll discuss logging, alerts, and remediation.

11.1　Logging

Logging is an important requirement when developing applications in the cloud. Logging provides capabilities in the platform to understand how an application is performing and to analyze the application's health, status, and any issues. In this section, we'll look at the various logging capabilities provided in SAP Cloud Platform and how to use them. We'll cover both the Cloud Foundry and Neo environments and then discuss application performance monitoring with Dynatrace.

11.1.1　Cloud Foundry

In the Cloud Foundry runtime environment for SAP Cloud Platform, the application logging service is used to create, access, and analyze application-related logs, container metrics, and custom metrics. The service uses Elastic Stack (open-source logging platform Elasticsearch, Logstash, Kibana [ELK]) to store, parse, and visualize the application log data.

You have both application logs that originate from the Cloud Foundry router by default and logs explicitly issued by the application itself. For your application to stream logs to this service, the application needs to be bound to a service instance of the service. Before you begin, you must meet the following prerequisites:

- You have administrator rights at the global account level.
- You have a subaccount and organization created in the Cloud Foundry runtime.
- You already have a space created inside your subaccount.

To activate this service, follow these steps:

1. At the global account level, navigate to **Entitlements • Subaccount Assignments**.
2. Enter the **Subaccount** name and click **Go**.
3. Click **Configure Entitlements** and then click **Add Service Plans**.
4. Select the **Application Logging** service and the available plans. Select **Add 1 Service Plans**.
5. Select **Save** on the **Assignments** screen to save the changes.
6. Navigate inside the subaccount and then to **Spaces**. Now navigate to **Service • Service Marketplace** and find the **Application Logging** tile, as shown in Figure 11.1.

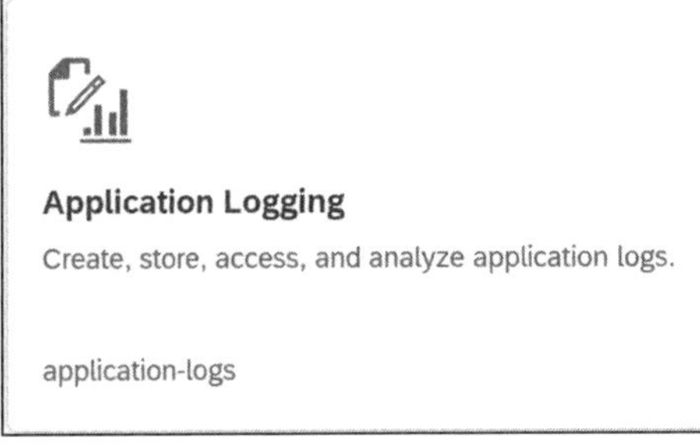

Figure 11.1 Application Logging Service Tile inside Spaces

Once you've activated the service for usage, the next step is to bind the service to an application for usage. You must already have an application running and deployed inside the space before you begin.

There are three ways to bind the service, as follows:

1. **Via SAP Cloud Platform cockpit:**
   - Navigate inside the space. From **Applications**, select the application to which the application logging needs to be bound.
   - In the application, navigate to **Service Bindings** and click **Bind Service**.
   - A pop-up will open. Choose **Service from Catalog** as the **Service Type**. Click **Next**.
   - In the **Choose Service** window, select **Application Logging** as shown in Figure 11.2. Click **Next**.

- Select the **Service Plan** to use and click **Next**. You can choose to create a new instance or reuse an existing instance (if you created one previously).
- Optionally, specify parameters. You can browse for the app's JSON file. Click **Next**.
- Enter an **Instance Name** and click **Finish**.

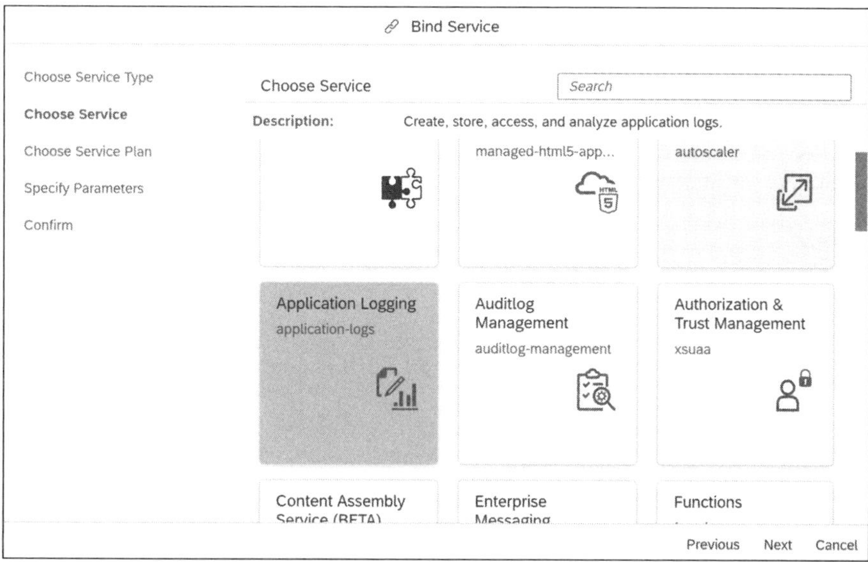

Figure 11.2 Application Logging Service

Note

Alternatively, you can do this inside the space from **Services • Service Marketplace**. Choose the **Application Logging** tile.

Inside the service overview, navigate to **Instances • New Instance**. Follow through the steps and select **Application** from the dropdown, as shown in Figure 11.3.

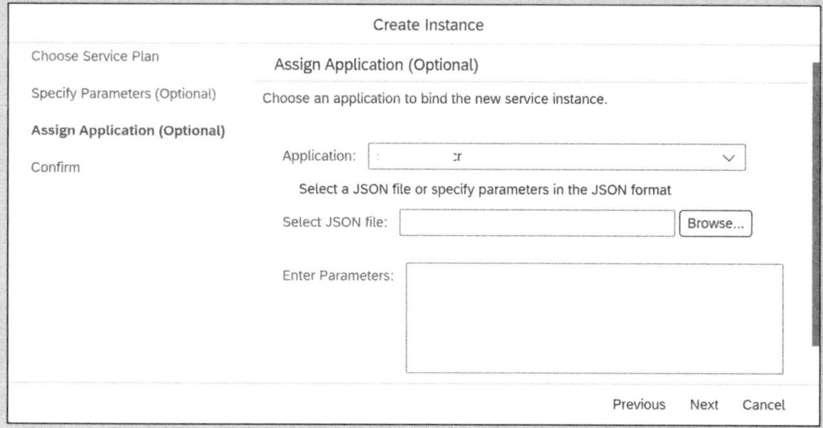

Figure 11.3 Select Application to Bind to Service Instance

2. **Via SAP Cloud Platform's Cloud Foundry command-line interface**
   Before you begin, you must already have installed the Cloud Foundry command-line interface. In the command prompt, enter the following commands in sequence:
   - First access the API endpoint of your region. The API endpoint for your subaccount is `cf api <api_end_point>`. So, for example, you might use `cf api https://api.cf.eu10.hana.ondemand.com`.
   - You'll be prompted to enter your user name and password.
   - Once logged in, you need to target the organization and space where your application is and where you want to create the application logging service instance by entering the following:

     ```
     cf target - o ORG_NAME -s SPACE_NAME.
     ```
   - Now create the service instance with the following command:

     ```
     cf create-service application-logs <SERVICE_PLAN> <SERVICE_INSTANCE_NAME>.
     ```

     So, for example, you might use the following:

     ```
     cf create-service application-logs lite logging_ucr.
     ```
   - Bind the app to the service instance created with following command:

     ```
     cf bind-servce <APP_NAME> <SERVICE_INSTANCE_NAME>.
     ```

     So, for example, you might use the following:

     ```
     cf bind-service NewHelloWorld logging_ucr.
     ```

3. **Via configuration in an MTA descriptor file of the application**
   You can add or automate the service instance creation as part of the application deployment lifecycle by defining the service creation in the MTA descriptor file (*mta.yaml*) of the application. You need to specify the elements in Listing 11.1 as an example in your *mta.yaml* file under `resources`. Make sure that your MTA file is correctly indented.

```
resources:
- name: <my_service_instance_name>
    type: org.cloudfoundry.managed-service
    parameters:
        service: application-logs
        service-plan: <standard or lite>
```

Listing 11.1 mta.yaml Descriptor to Add Application Logs

Now you're ready to access the logging dashboard for the application, by following these steps:

1. Navigate inside the space, go to **Applications** and select the application to which the logging service is bound.
2. Inside the application, navigate to **Logs** and click **Open Kibana Dashboard**.

3. This will open the logging dashboard, as shown in Figure 11.4.

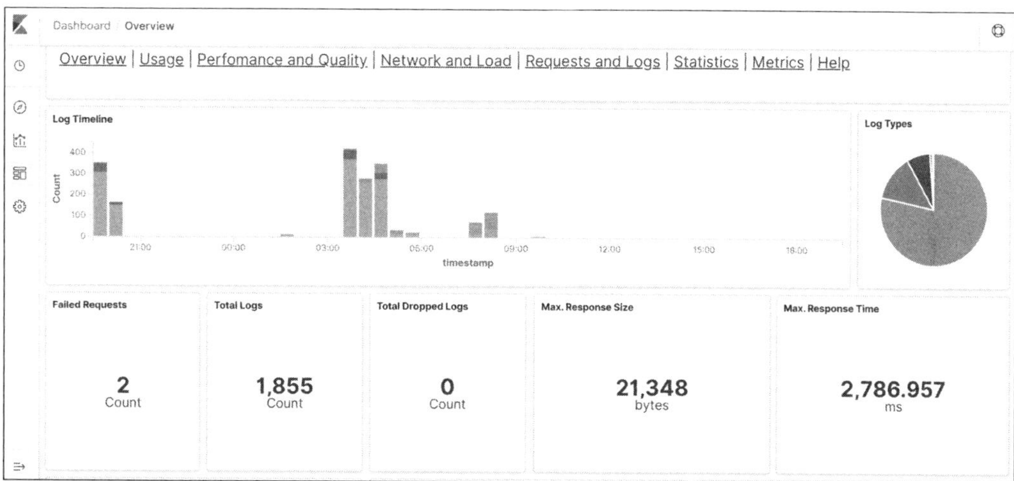

Figure 11.4 Kibana Dashboard Example

In the dashboard, you can see the following sections with information to analyze your application:

- **Overview**
  The total logs, failed requests, log sources, and KPIs
- **Usage**
  Total requests and timeline and request distributions across various components
- **Performance and Quality**
  Details about failed requests, error codes, and response times
- **Network and Load**
  Details about response payload time, size, and network traffic
- **Requests and Logs**
  Details for correlation of logs and requests using correlation IDs
- **Statistics**
  Details about log statistics, service plans, and dropped logs
- **Metrics**
  Details about the CPU and the memory percentage overview and distribution

11.1.2 Neo

In this section, we'll look at how to access and analyze the logs for applications in the Neo environment. Before you begin, you must have a subaccount in the Neo runtime and have a deployed application.

In the Neo environment, a monitoring service is enabled by default. The service provides current and past metrics, options to configure availability checks for your application, and JMX checks. Monitoring capabilities for Java, HTML5, and database systems are available in the Neo runtime.

Let's walk through how to access application monitoring details for a Java application deployed in the Neo subaccount:

1. Navigate inside your Neo runtime subaccount to **Applications • Java Applications**.

2. Select the application for which you want to see the monitoring details. You'll be taken to the screen shown in Figure 11.5.

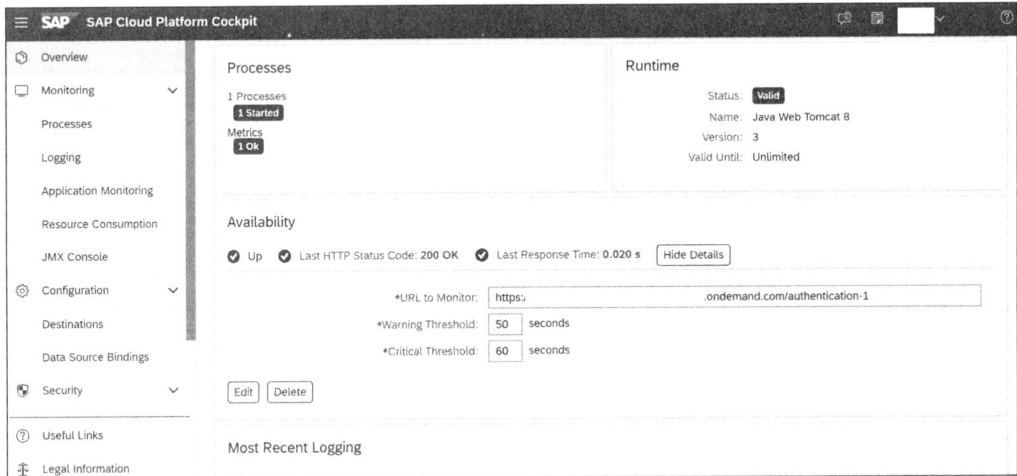

Figure 11.5 Java Application Monitoring in Neo

You can access monitoring information by navigating to the following on the left-hand sidebar:

- **Overview**
  Here you'll see a quick overview of maximum CPU consumption and requests in the last 24 hours and can configure an availability check for the application.

- **Monitoring • Processes**
  Here you'll see the metrics for the processes, like average response time, busy threads, CPU load, disk I/O, and so on.

- **Monitoring • Logging**
  Here you can view logs and change the log settings of the application. You can get the following types of logs: default traces, HTTP access logs, garbage collection logs, and Java Connector (JCo) logs.

- **Monitoring • Application Monitoring**
  Here you'll see an overview of various metrics at the application level, which includes all processes in the application. In the filter, you can choose a process to see metrics based on the individual process.

- **Monitoring • JMX Console**
  Here you can monitor and manage the performance of JVM and your Java applications.

> **Note**
>
> You can get more information about monitoring capabilities in Neo for Java, HTML5, and database systems at *http://s-prs.co/v515712*.
>
> You can find out more about logging capabilities in Neo at *http://s-prs.co/v515713*.
>
> The SAP JVM Profiler is another tool to help analyze resource-related problems for Java applications. Read more about it at *http://s-prs.co/v515714*.

11.1.3   Application Performance Monitoring with Dynatrace

SAP Cloud Platform provides the capability to connect with Dynatrace for application performance monitoring. This capability helps you with monitoring and management of application performance and availability. This section discusses how you can configure Dynatrace for application performance monitoring in the SAP Cloud Platform environment. We'll cover the steps for both the Neo environment and the Cloud Foundry environment. You must have a license for the Dynatrace SaaS monitoring environment in order to use this functionality. You must also generate an API token and a PaaS token in your Dynatrace environment.

Neo

The service capability allows you to connect Java applications in Neo to your Dynatrace SaaS environment for monitoring and end-to-end trace collection. To get started and activate Dynatrace for the Neo environment, follow these steps:

1. Inside the subaccount, navigate to **Service** and search for "Agent Activation for Dynatrace".
2. Select the tile and click **Enable**.
3. Select **Configuration** and enter the **Environment URL** and **PaaS Token** that you received from Dynatrace.
4. Choose **Test Connection**.
5. Select **Save**.

By configuring this step, you connect all applications that are running in this subaccount to the Dynatrace SaaS monitoring environment. You need to restart the Java application after configuration, at which point the agent collects the application data and sends it to Dynatrace.

Cloud Foundry

To activate Dynatrace-based application performance monitoring in Cloud Foundry, follow these steps:

1. Navigate inside the subaccount to spaces and **Services • User-Provided Service**.

2. Select **New Instance**.

3. Provide an **Instance Name** with dynatrace as the substring, and in **Credentials**, provide a JSON object with the parameters shown in Listing 11.2.

```
{
    "environmentid": "<ENVIRONMENT_ID>",
    "apitoken": "<API_TOKEN>",
    "tag:SAP CP": "",
    "tag:Region": "<Region you want to tag in Logs>"
}
```

Listing 11.2 Credentials for Dynatrace Connectivity with Cloud Foundry User-Provided Services Instance Creation

4. Select **Save**.

5. The service instance can then be bound to an application as explained in Section 11.1.2. Select **User-Provided Service**.

6. Application logs produced in Dynatrace after binding the service instance look like the example shown in Figure 11.6.

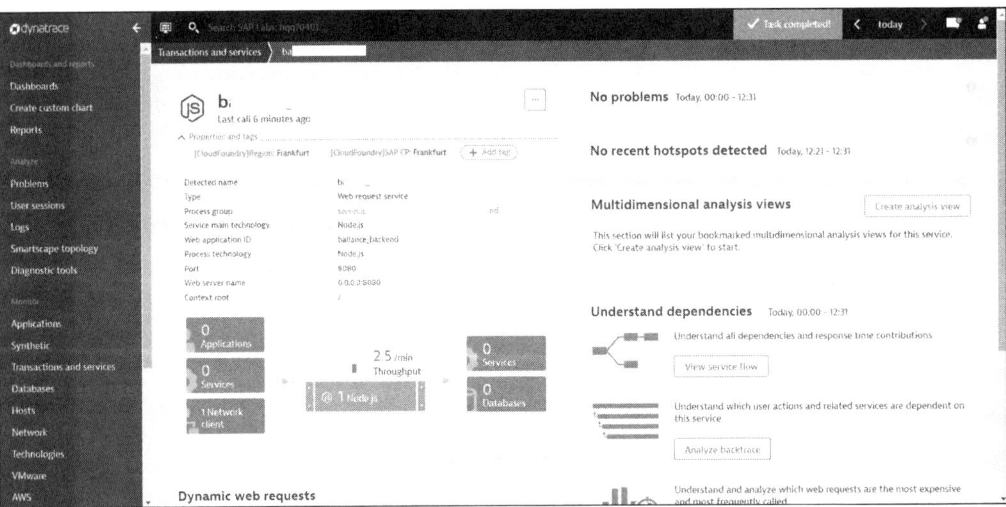

Figure 11.6 Dynatrace Application Log Screen

> **Note**
>
> You can get more information about getting started, finding applications in Dynatrace, and configuring application parameters at *http://s-prs.co/v515715*.

11.2   Alerts and Remediation

Getting notified of important events is critical in the cloud to continuously monitor your applications' health and status. Further, cloud applications should ideally be provided with capabilities to take immediate automatic remediation actions to recover the applications automatically without much manual intervention. In this section, we'll look at two capabilities and services that SAP Cloud Platform offers to help with this task: SAP Cloud Platform Alert Notification and SAP Cloud Platform Automation Pilot.

11.2.1   SAP Cloud Platform Alert Notification

The SAP Cloud Platform Alert Notification service is used for receiving and creating custom alerts and notifications for your resources and then further consuming/alerting them in a system of your choice for prompt action.

In this section, we'll look at how to enable and configure SAP Cloud Platform Alert Notification for your landscape and applications. This service is available in both Cloud Foundry and Neo runtimes. Before we explain the setup of SAP Cloud Platform Alert Notification, let's first go through the basic terminology used in this service:

- **Subscriptions**
  To work with this service, you create a subscription, which identifies a condition for which an alert needs to be issued and what action it needs to trigger.

- **Conditions**
  Conditions define the set of relevant events for different applications that SAP Cloud Platform Alert Notification listens for. You can specify multiple conditions per subscription.

- **Actions**
  Actions are the responses that need to be triggered when a condition is met. The action could, for example, trigger an email, trigger an incident in ServiceNow, or trigger an update in Slack or Microsoft Teams. Actions also provide webhooks to trigger an external URL address. You can specify multiple actions per subscription.

> **Note**
>
> When multiple conditions are specified, the following is true:
>
> - Conditions with different properties follow AND logic.
> - Conditions for the same event property follow OR logic.
>
> The catalog of available events in the Cloud Foundry environment is found at *http://s-prs.co/v515716*.
>
> The catalog of available events in the Neo environment is found at *http://s-prs.co/v515717*.

Let's now look at the initial setup you must do in each environment.

Neo

For the Neo environment, the following are the steps you must take before setting up SAP Cloud Platform Alert Notification:

1. To activate the service in Neo, navigate to the subaccount using the Neo environment.
2. Go to **Services** and search for the **Alert Notification** tile.
3. Click the tile and select **Enable**.

Cloud Foundry

For the Cloud Foundry environment, the following are the steps you must take before setting up SAP Cloud Platform Alert Notification:

1. Navigate inside **Space** to **Services** • **Service Marketplace** and search for "Alert Notification".
2. Click the tile and navigate to **Instances** • **New Instance**. In the pop-up to create a new instance, enter the necessary details and click **Finish**.
3. Select the created service instance name, navigate to **Service Keys** • **Create Service Keys**, and enter a name and authentication method (**BASIC** or **OAUTH**) to create a service key to enable programmatic access to this service.

The following are the steps you must take to set up SAP Cloud Platform Alert Notification:

1. For the initial setup, navigate inside the service instance created before for Cloud Foundry.
2. Inside the service instance, navigate to **Subscription** • **Create**. Enter a name for the subscription in the pop-up and click **Create**.
3. In the next step, select conditions. To create new conditions, select **+ Create Condition**.
4. Enter a name and condition. Click **Create**.
5. You can create multiple conditions from here and have them selected for the subscription. Alternatively, if you already have conditions created, you can select them.
6. After you have created the conditions, select **Assign**.
7. In the next step, select actions. To create a new action, select **+ Create Action**.
8. Select an action you want to perform and click **Next**.
9. Enter a name for the action and complete the required fields depending on the action selected in the previous step. Click **Create**.

10. You can create multiple actions here, as mentioned previously. Choose the required actions and select **Assign**.

11. A new **Subscription Created** screen is shown, which mentions the conditions and actions you have created/selected. You can close this window.

12. A tile with the created subscription name is shown (see Figure 11.7). Click the tile to see the details of this alert with the selected condition(s) and action(s).

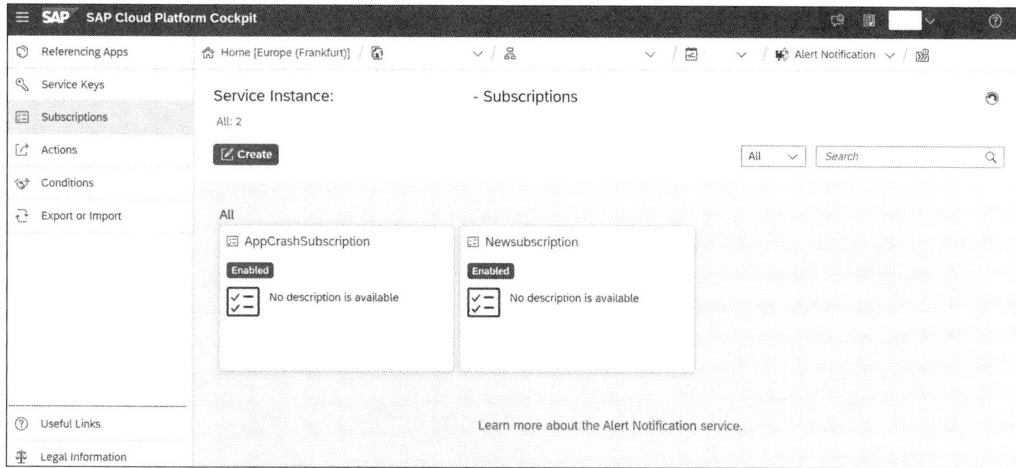

Figure 11.7 SAP Cloud Platform Alert Notification Subscription

You can create alert notifications to get proactive notifications about your applications. Some of the use cases and examples are as follow:

- Get notifications about SAP Cloud Platform integration flow failure using the SAP Cloud Platform Integration Suite's Cloud Integration capability.

- Get notifications about the API target service endpoint that is unavailable, managed using the SAP Cloud Platform Integration Suite API Management capability.

- Get notifications about import actions of a transport using SAP Cloud Platform Transport Management Service.

- Get notifications about application- and database-specific lifecycle events—for example, application recovery, restart, and restage, and database restart.

11.2.2   SAP Cloud Platform Automation Pilot

A continuous monitoring cycle is completed and closed by having an effective capability provided within the cycle to react to alerts, logs, and updates, taking necessary corrective actions like applying patches, restarting/restaging applications, and many more.

SAP Cloud Platform Automation Pilot service provides the capability to automate these complex manual efforts. Thus, the service helps to reduce the manual operational effort and effective continuous monitoring and update of applications. It provides access to a wide variety of provided procedures—for example, HTTP requests, bash scripting, monitoring, application and database lifecycle management, and Kubernetes cluster management.

Before getting into the details of the initial setup and configuration of the service, let's go over important terminology used in this service and referred to in the configuration:

- **Tenant**
  A tenant in SAP Cloud Platform Automation Pilot is an onboarded subaccount in which the service is subscribed/activated. Each tenant is isolated from the others and has a unique ID starting with T*.

- **Catalog**
  Catalogs consist of commands and inputs. SAP Cloud Platform Automation Pilot comes with predelivered catalogs, called *provided catalogs*, which offer commonly used commands and inputs. These cannot be edited in a tenant, but you can create your own catalogs specifically for your tenant.

  SAP Cloud Platform Automation Pilot also provides the following two catalogs (these are available out of the box and are owned by your tenant):
  - The Welcome catalog provides examples of basic SAP Cloud Platform Automation Pilot mechanisms.
  - The Recommended Actions catalog provides example commands for recovering databases and applications.

- **Input**
  An input is a collection of key-value pairs. Commands use its values as input data. Each inputs belong to exactly one catalog and has a unique name and version combination in that catalog. Each key input must have a unique name and data type.

- **Command**
  A command represents a series of steps that are executable/executed. Each command belongs to exactly one catalog. Every command has a set of input keys for which it receives input data when the command is started, and a set of output keys to produce the output when the command is completed. An output key must have a unique name and a data type. Commands can have multiple steps between the input and output steps, which are executed in sequence. A command is triggered to start an execution.

> **Note**
>
> This service is available only in multicloud data centers. Note the availability of the service in data centers and create a subaccount in a specific data center in which the service is available.

To begin this setup, follow these steps:

1. Navigate to your Cloud Foundry subaccount and to **Subscriptions**.

2. Search for "Automation Pilot" and click **Subscribe**. Wait for the service to be subscribed.

3. Now provide users with permissions and roles to access the service. SAP Cloud Platform Automation Pilot offers the predefined role templates described in Table 11.1.

Role Template	Description
Executor	Provides access to read all resources and execute them
Maintainer	Provides access to read and modify catalogs, commands, and inputs
Observer	Provides read-only access to all resources
Security maintainer	Provides access to maintain security configurations

Table 11.1 Role Templates in SAP Cloud Platform Automation Pilot

4. If you have an existing role collection, add the required role to it. If you do not have a role collection created, you can create one by navigating to **Security • Role Collections**.

5. After adding the roles to the role collection, provide user access to the role collection. For this, navigate to **Security • Trust Configuration** and select the SAP ID service.

6. Provide your email address and click **Assign Role Collection**. If this isn't active, click **Show Assignments** first.

7. Navigate back to **Subscriptions**, choose the **Automation Pilot** tile, and click **Go to Application**.

8. Log in using your credentials, and this will launch the home screen for SAP Cloud Platform Automation Pilot, as shown in Figure 11.8.

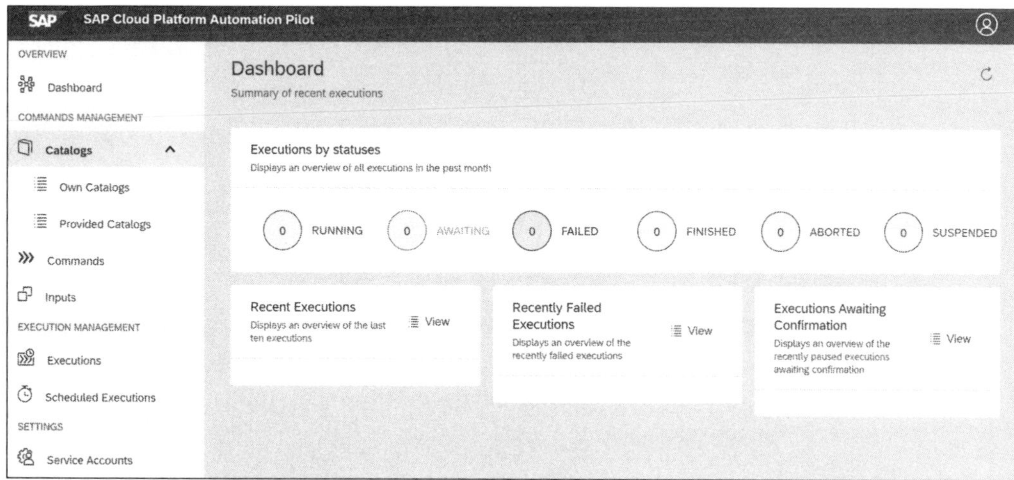

Figure 11.8 SAP Cloud Platform Automation Pilot Home Screen

As shown in Figure 11.9, SAP Cloud Platform Automation Pilot comes with several catalogs provided out of the box, which you can find under **Provided Catalogs**. The catalogs are as follows:

- **SAP CP Application LM (applm-sapcp)** provides commands for lifecycle operations for applications in SAP Cloud Platform, for both Cloud Foundry-based (restart, stop, restage, or recover process instances) and Neo-based (restart, Java application process stage) Java applications.

- **Automation Pilot (autopi-sapcp)** provides commands for managing SAP Cloud Platform Automation Pilot entities like create input, check for input existence, update draft input, and update input.

- **Cloud Foundry (cf-sapcp)** provides commands for Cloud Foundry-specific operations like managing service instances, get organization and space details, and app service bindings.

- **SAP CP Database LM (dblm-sapcp)** provides commands for lifecycle management of database systems in SAP Cloud Platform.

- **Dynatrace (dynatrace-sapcp)** provides commands to gather information from Dynatrace, like getting dynatrace events and problem feeds and statuses.

- **HTTP Operations (http-sapcp)** provides commands used to execute HTTP requests and get a token for the OAuth 2.0 flow.

- **Kubernetes (kubernetes-sapcp)** provides commands for Kubernetes-specific operations like rolling restart of Kubernetes deployments, pod management, list Kubernetes resources, and so on.

- **SAP CP Metadata (metadata-sapcp)** provides inputs to hold various metadata for SAP Cloud Platform, like Cloud Foundry and Neo region-based data.

- **SAP CP Monitoring (monitoring-sapcp)** provides commands to get various metrics from the Neo monitoring service and Cloud Foundry app service events.

- **Script Operations (script-sapcp)** provides commands to execute scripts.

- **Utility Operations (utils-sapcp)** provides different utility commands (like `ForEach` or `JsonPath`). These commands come in handy when creating your specific scenario commands.

> **Note**
>
> SAP Cloud Platform Automation Pilot can be integrated with SAP Cloud Platform Alert Notification to provide end-to-end monitoring and continuous operations capabilities for your applications running in SAP Cloud Platform.
>
> Every time a specific event is created by SAP Cloud Platform Alert Notification, you can configure SAP Cloud Platform Automation Pilot to execute a specific set of commands to react to events and start remediation and correction measures for your application to recover it. You can read about the setup instructions in the help document at *http://s-prs.co/v515718*.

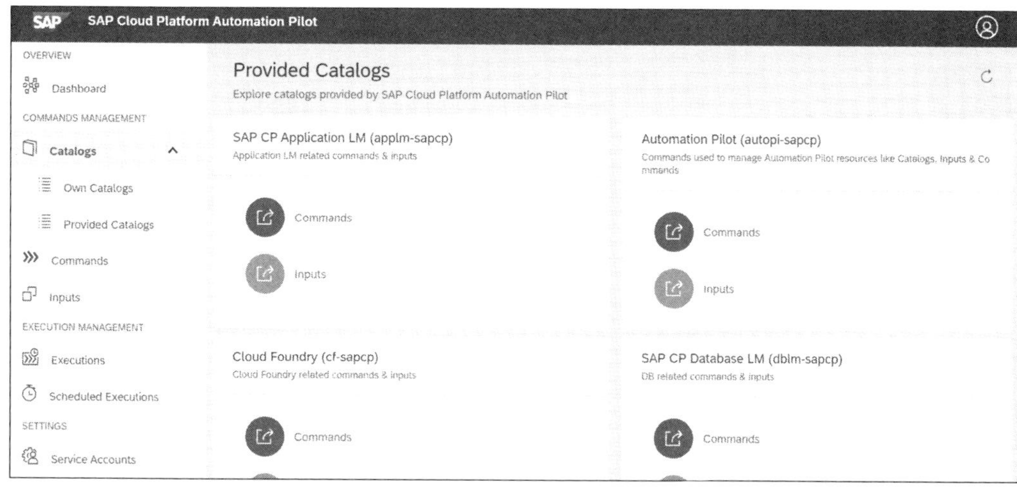

Figure 11.9 SAP Cloud Platform Automation Pilot Provided Catalogs

11.3   Summary

In this chapter, we covered various monitoring and alerting capabilities in SAP Cloud Platform. We suggest enabling these functionalities in your application development lifecycle to enable continuous monitoring capabilities in your landscape.

We covered application logging functionality to log, analyze, and visualize the logs produced by your application. We also covered various alerts produced by your application. For these alerts, you can promptly notify relevant stakeholders, like the 24/7 support team, using SAP Cloud Platform Alert Notification. We also described how you can leverage SAP Cloud Platform Automation Pilot to take remediation actions for your application in case of alerts. You can use this service to further schedule and automate regular updates in your organization.

Finally, we covered how to enable and do initial setup for these services in your landscape, including a few example use cases. You can use these various capabilities in SAP Cloud Platform to automate and run your application in the cloud and prepare for various uncertain events in the cloud well in advance.

Chapter 12
Developing Cloud-Native Applications

It's very important for your applications in the cloud to run in a high-availability mode for the best experience for your business and end users. In this chapter, we'll look into various best practices and resiliency patterns you need to follow while building your extensions and applications on SAP Cloud Platform.

In previous chapters, we've discussed how you can build enterprise-grade, secure, cloud-native applications and extensions using SAP Cloud Platform; apply continuous integration and delivery cycles for quick innovation; and have continuous monitoring of applications to plan for quick alerts and remediation. SAP Cloud Platform provides features and capabilities that would help you build your applications, extensions, and integrations with speed and agility to meet your business demands. It's extremely critical for applications, extensions, and integrations in the cloud to bring value to end users and to be available all the time. If the software isn't available, it can create severe loss for the business and have a negative impact on end user satisfaction.

Software designed for the cloud is core to the success of your organization and business. Designing software for the cloud requires certain cloud-native guidelines. For example, such software should be able to work with the least latency for users, no matter where across the globe they access the application; should be able to meet the varying user load and requests from your business; and should be able to self-recover from failures. Thus, building applications, extensions, and integrations in the cloud with cloud-native guidelines is not a feature requirement, but a mandatory requirement.

Cloud application developers need to consider developing applications that can respond to failures gracefully, run in a healthy state 24/7, deliver new features without any downtime, and recover from major nontransient wide-scale incidents either in the infrastructure or in a data center location. This is quite the contrast from developing applications for on-premise systems, in which expensive server infrastructure could be built into an on-premise data center to handle some of these aspects.

In this chapter, we'll investigate best practices that can be followed to make your applications, extensions, and integrations resilient in the cloud and fail-safe. You can build these capabilities into the application to make it resilient, scalable, and performant in the cloud by applying capabilities provided by SAP Cloud Platform and following cloud-native architectural principles. We'll start the chapter by discussing resiliency and then move onto high availability and multi-data center failover.

12.1   Resiliency Principles

Resilience is the capacity to recover quickly from difficulties. It's toughness, or the ability of a substance or object to spring back into shape. In software design, this is all about the ability of the application to respond to failures. You want to design applications that can be available all the time and spring back from any failures. It's important to understand that resilience is not about avoiding failures, which may not be possible, and failures may happen often with hardware. It's about designing applications that take into consideration these failures that can occur and nullify or at least reduce the impact of these failures. Thus, it's about designing applications so as to increase their availability in cloud.

In the following sections, we'll talk about service-level agreements (SLAs), how availability is calculated, and various resiliency patterns to increase application availability.

12.1.1   Service-Level Agreements

Before going into the need to develop applications with high availability, let's discuss how service-level agreements for different services can have an impact on application design and architecture. This helps explain how applying various resiliency principles and patterns discussed in this section can improve your application's availability.

An SLA defines a system availability percentage during each month for productive systems. This percentage is calculated as follows:

System Availability Percentage = ((Total Minutes in Month – Downtime) ÷ (Total Minutes in Month)) × 100

Each SLA percentage will mean a different maximum allowed downtime for your service or application, as explained in Table 12.1 for some common SLA values.

SLA	Downtime per Week	Downtime per Month	Downtime per Year
99%	1.68 hours	7.2 hours	3.65 days
99.5%	54.6 minutes	3.64 hours	1.82 days
99.9%	10.1 minutes	43.2 minutes	8.76 hours
99.95%	5 minutes	21.6 minutes	4.38 hours

Table 12.1 SLA Values and Downtime Metrics

You need to identify and differentiate between two types of SLAs while developing and designing your application:

1. **Service-level SLA**

   This is a service-level agreement between your organization and your cloud provider—in this case, SAP. When purchasing cloud solutions from SAP, there is a published service-level agreement signed as part of your contract in which application and service SLAs are documented and provided. This is not something your organization can influence; these terms are set and defined by cloud providers.

2. **Application SLA**

   This is a service-level agreement with your application end users. As part of developing and designing your application, you want to consider your business demand and requirements in an SLA for your business and end users. This is an SLA you can set to improve based on different resiliency patterns, which we'll discuss in this chapter. You must consider the operating constraints of the SLA between your organization and your cloud provider.

Let's discuss how the second SLA type, the SLA defined for your business, is calculated when you develop an application using different services on SAP Cloud Platform. As you have already seen in previous chapters, many organizations use multiple services on SAP Cloud Platform for business applications, extensions, and integrations. To understand the application SLA, let's consider it in the context of an application architecture/solution as defined in Figure 12.1.

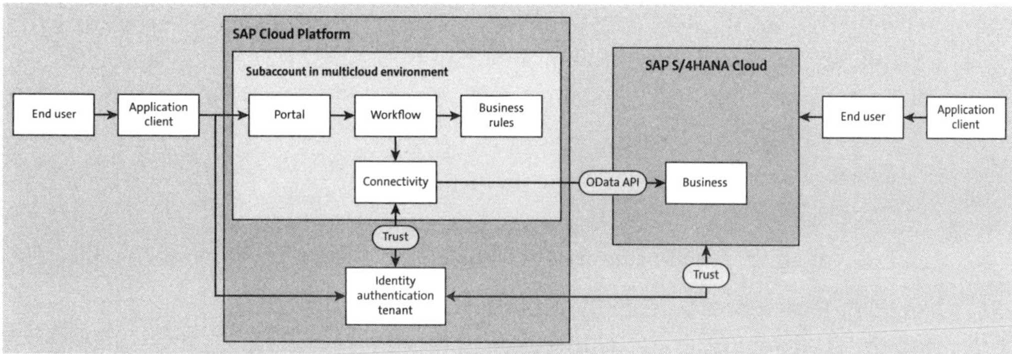

Figure 12.1 Sample Application Architecture

> **Note**
>
> The sample application architecture is part of a use case for automating business partner creation using SAP Cloud Platform, which you can access via *https://www.sap.com/products/cloud-platform/use-cases/workflow-business-rules.html*.

For example, let's say an organization uses five different SAP Cloud Platform services. Let's assume that each of the services has a service-level SLA of 99.9%.

> **Note**
>
> For the actual SAP Cloud Platform service-level SLA, see the service-level agreement referenced in your contract and published at *https://www.sap.com/about/trust-center/agreements.html*.

Now, your application SLA would be calculated as follows:

Application SLA = Product of the individual service-level SLAs

In this case, it would be calculated as follows:

Application SLA = 99.9 × 99.9 × 99.9 × 99.9 × 99.9 = 99.5%

Thus, your application SLA is lower than the individual service-level SLA. This is because if any service fails independently, the whole application might fail, bringing down the application SLA to a lower level than individual service SLAs. In this example, taking note of the potential downtime for the application SLA of 99.5 % from Table 12.1, you commit to a much lower level of availability of your application to end users. Hence, you need to consider this impact to your application-level SLA while developing applications in the cloud to set the right expectations for your business. Utilizing different services also exposes the application to more potential failure points.

All this means you should explore possibilities to improve your application SLA and provide higher availability for your end users to meet business expectations. The way to do that is to consider building resiliency into the application, to fallback in case of an individual potential failure path to increase availability. To achieve higher availability for your application, you need your application to have self-healing capabilities. Of course, the costs involved in building such an application can be a decision point to decide on the right application SLA. We'll further investigate these various resilience patterns in the next section.

12.1.2   Availability and Resiliency Patterns

Now, let's look at how availability is calculated. Availability can be defined as follows, with the details explained in Table 12.2:

Availability = MTTF ÷ (MTTF + MTTR)

Terminology	Description
Mean time to failure (MTTF)	This is the average time between the start of system operations and the occurrence of a failure.
Mean time to recover (MTTR)	This is the average time between the occurrence of a failure in the system and the point at which the system is back in operation.

Table 12.2 Availability Definition Parameters

Thus, in this equation for availability, the numerator (MTTF) defines the time during which a system works properly and the denominator (MTTF + MTTR) is the total time. Hence, the value of availability per this equation varies between 0 (not available at all) and 1 (always available). Thus, there are following approaches to building an "always available" application:

- **Maximize the value of MTTF**

  In this approach, you increase the value of MTTF to such an extent that MTTR becomes insignificant. This approach follows the principle to identify failure sources in advance and take adequate measures to reduce the probability of failure.

  Such an approach can be implemented at the infrastructure level by setting up redundant hardware that can failover in a seamless way, ideally within the same network connection and using highly available system clusters.

- **Minimize value of MTTR**

  The second approach to build an "always available" application is to minimize MTTR. This would mean that an application can recover as quickly as possible after a failure, even to such an extent that end users don't notice the failure.

  This approach considers at every step that failure can happen in an application at any point, and the only value to make the system available all time is by the ability to handle those failures gracefully. This is the core principle behind resilient software design.

You can develop and design a highly available application by following both mentioned approaches. However, also note that for developing applications in the cloud, MTTF is a factor that the application developer doesn't have much influence over, so the prime consideration is to minimize MTTR. There are fundamental principles you can apply while designing and developing your applications, extensions, and integrations to target for high availability. Let's take a look at these various patterns you can apply in your application architecture to build resiliency, as follows:

- **Retry (transient fault handling)**

  Temporary service interruptions are quite common while developing applications in the cloud. There could be multiple reasons for this, from network connections and connection timeouts to database cluster node failover operations in progress or efficient load balancing on the server side due to high loads. These frequent transient errors need to be handled during application design and development in a self-healing way to avoid any bad experiences for your end users.

  Hence, instead of throwing back an error to an end user, applications can implement an automatic retry mechanism wherever there is possibility of such transient errors. Thus, the application recovers from the error on the second or third try without the end users even noticing the failures, as shown in Figure 12.2.

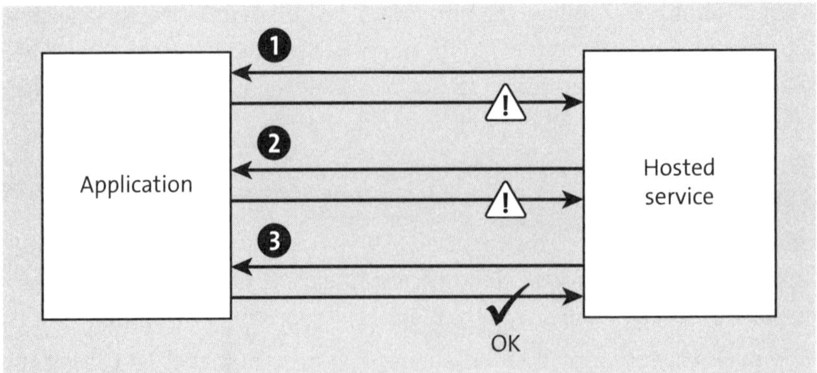

Figure 12.2 Retry (Transient Fault Handling) Principle

- **Circuit breaker**

  You just learned how retries can help an application handle transient errors in the cloud. However, too many retries can be counterproductive and bring down the application. Too many users attempting continuous retries can potentially degrade other users' experiences, making them wait longer for the application to respond. It's also essential to understand whether the error in the application that resulted in a retry was because of an actual service error or was just transient in nature.

  The solution to this problem is to keep track of the frequency, amount, and time between retries, and then fail requests gracefully if errors are persistent rather than transient.

  The circuit breaker principle focuses on implementing retry thresholds in your app, after which an application stops retrying and adopts another pattern or principle to fail the request gracefully for end users.

- **Bulkhead**

  A single cloud application or extension can have multiple different services. It's a best practice to develop applications in the cloud using microservice architecture. These microservices interact and communicate with each other to form the complete application. Failure in one microservice could have a cascading effect on other services, bringing down the whole application, as illustrated in Figure 12.3.

  To prevent cascading failure, the failure unit in the application needs to be isolated. Thus, while developing applications, you can divide your application into different autonomous units based on technical or business needs like expected technical load, availability requirements, and ability to operate multiple instances of a single service. This can be achieved by striving for replaceability of services in an application, instead of reuse, which often leads to bad design.

  Features of the application served by a failed service may not be available, whereas other services and functionalities will continue to work.

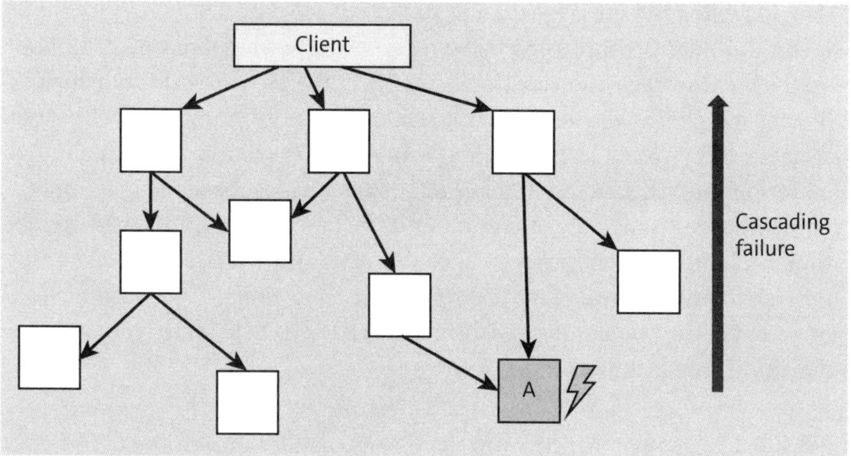

Figure 12.3 Cascading Failure Impact

- **Throttling**
  Huge load in a system could result in system-wide failure of the application, impacting all consumers. These loads on the application could be highly dependent on business scenarios in your organization, like peak loads during certain business events. Your applications should be able to handle such peaks without crashing an entire application. Thus, you need to design and architect your application to handle such high loads in the system.

  There are multiple ways to implement such a design pattern, as follows:

  - Applications should be able to spin up more instances automatically based on loads to handle such sudden loads.
  - You can rate-limit the number of requests coming to your application, thus rejecting additional requests coming from certain consumers after the rate limit.
  - You can queue your application requests to handle them asynchronously.

  Thus, you can handle and scale according to loads in your application and not allow the application and system to crash.

There are multiple other patterns you can apply in your application architecture. We won't cover them in detail here; they've already been covered in some of the earlier chapters. However, a few important patterns worth mentioning are as follows:

- **Publisher/subscriber patterns**
  In a cloud context, multiple applications need to interact with each other to complete a business process by exchanging information or events. The publisher/subscriber pattern helps applications communicate with each other asynchronously by decoupling the sender application from different consumers and thus avoid blocking of the sender for responses. Such a pattern can be implemented using SAP Cloud Platform Enterprise Messaging, which we covered in Chapter 5, Section 5.3.5.

- **Load balancing using the message queues pattern**
  We covered this under the throttling pattern earlier. However, here we'll talk about how introducing a message queue between applications can help scale your application and make it highly available. This could be ideal for communication with a service that needs to be called from different senders. With such a varied high load, your application can increase the number of queues besides the consumer application itself being able to increase the number of instances. In addition, the sender application can continue sending messages to queues even when the consuming application isn't available, which is ideal for scenarios in which asynchronous processing is acceptable for business needs. Thus, this pattern help scales your application and maximize availability.

- **Redundancy**
  Redundancy is a resilience principle that you can consider during the design and architecture phase of your application. With this pattern, you're ideally looking to maintain a complete redundant application instance to which you can switch over if the primary application instance goes down completely. This is a powerful pattern that can be implemented to achieve application resiliency, but it requires careful consideration and significant work upstream during the application design phase.

- **Caching**
  In addition to having a data store for your application, like the SAP HANA database, you can implement a data caching mechanism. The obvious advantage of this pattern is to improve application performance by holding data accessed quite frequently in the cache. The cache needs to be updated back and forth with your data store, which can be triggered on demand. The caching strategy also provides a fallback strategy in your application to read and write to the cache in case of failure in the data store.

> **Note**
>
> We've covered various resiliency patterns that you can consider during your application development. Note also that during application design, one or more of these patterns can be applied together to achieve the required resiliency in the application and thus increase its availability and application SLA.

In the following sections, we'll cover best practices for using SAP Cloud Platform to help you define different architectural principles and provide higher availability while building your applications and extensions.

12.2   High-Availability Best Practices with SAP Cloud Platform

In this section, we'll look into setting up and using various services in SAP Cloud Platform. We'll examine various best practices and architectural services that can be applied across different SAP Cloud Platform services to build a highly available cloud-native application using SAP Cloud Platform.

12.2.1   Cloud Connector High Availability

In Chapter 5, we discussed the SAP Cloud Platform Connectivity's cloud connector as an essential service to build secure connectivity with your on-premise system. While installing the cloud connector in your DMZ in a specifically sized virtual machine, you have the option to install and set up the cloud connector in a high-availability setup.

In the cloud connector, this high-availability setup is called a *master shadow instance*. While installing the cloud connector, you can set it up as either a master or shadow instance, as shown in Figure 12.4.

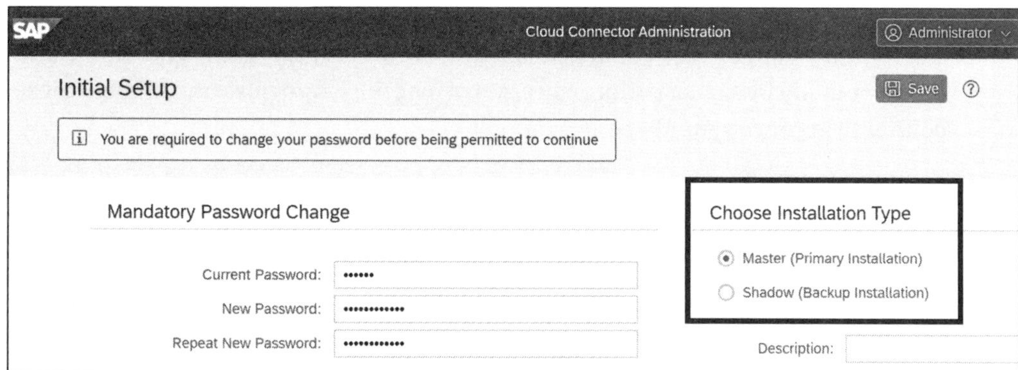

Figure 12.4 High-Availability Configuration for Cloud Connector

The master and shadow instance are set up in different virtual or physical machines so that failover can happen in case of infrastructure failure. To configure this setup, you need to enable high availability in the master instance and enable the shadow instance to connect to the master instance. Once high availability is set up, all the connections mentioned in the master instance are mirrored to the shadow instance. This enables your shadow to take over any connectivity requests if the master goes down. Installing the cloud connector in this master shadow mode offers the following advantages for your application:

- Establish high availability in case of failure at the infrastructure level
- Upgrade the cloud connector to new versions without impacting your application connectivity

We recommend that you set up your cloud connector in this high-availability mode, at least for your applications and extensions in production mode to increase their availability.

12.2.2   Caching Setup

We mentioned in Section 12.1.2 that caching is an important pattern. It provides the advantage of higher performance in an application when data is accessed from the cache, and when used along with a data store, it provides an additional capability to plan against any failure of the data store. There are different levels at which you can set up caching based on your application setup.

In the following sections, we'll discuss different levels and services for which caching can be applied as best practice architecture while building your applications, extensions, and integrations in SAP Cloud Platform.

SAP Cloud Platform Integration Suite: API Management Caching Policy

Chapter 5 discussed utilizing the API Management capability of SAP Cloud Platform Integration Suite to secure, govern, and manage APIs in your organization.

API Management provides caching policy templates. This helps cache your API calls to enable your API for higher performance. API Management provides two different cache policies to improve your API performance:

1. **Response cache**

   This policy caches data from backend resources. You use this policy to return a cached response instead of forwarding the request every time to the backend server. The response cache must be attached to both the request and response flow of the API proxy.

2. **Metadata cache for OData API**

   The metadata cache for OData API policy template is of the *response cache* type, which is published in SAP API Business Hub for reuse purposes. It can be reused across API proxies to improve the performance of OData APIs by caching the $metadata response for the API. This also helps to reduce the load on the target system by returning the cached metadata response.

SAP Cloud SDK Caching

In Chapter 9, we briefly covered using SAP Cloud SDK to develop applications and extensions. When using SAP Cloud SDK, you can introduce caching to your application. You can cache repetitive calls and parameterized calls to the OData service.

SAP Cloud SDK makes it easy to cache your requests because it handles the complexity under the hood. SAP Cloud SDK includes tools for maintaining and monitoring your tenant- and user-specific caches for your SAP S/4HANA connections. You can specify in your cache configuration the duration for which to store the cache and can store the cache in relation to a specific parameter/ID so that calls associated with a specific ID can be retrieved.

SAP Cloud SDK uses JCache (JSR-107) as its underlying caching technology. So the application developer can use the caching library of their choice, so long as it supports

JCache (JSR-107). To use the library, you just need to add it to your application dependency file. The data thus cached is stored and is available in the scope of the application container.

> **Note**
>
> To learn about using SAP Cloud SDK caching in your application, refer to the SAP Developer tutorial at *https://developers.sap.com/topics/cloud-sdk.html#services.*

> **Note**
>
> In the previously mentioned scenarios for caching calls, the most important design consideration while using a cache is to define its size and lifetime. Both should be carefully designed to avoid the application accessing outdated data or requiring higher memory consumption.
>
> Either caching mechanism provides various policies, settings, and configurations to help invalidate the cache, define the cache lifetime, and skip the caching.

Redis on SAP Cloud Platform

Redis on SAP Cloud Platform is a service that can be used for caching important application calls. A Redis cache can be used to cache your most important data in memory across the application container scope and also can be used for lightweight message broker capabilities.

It's recommended to use the Redis service in conjunction with a datastore in your application to cache the most frequently accessed data. This not only improves the application performance and latency, but also provides an added cushion to account for potential data store failure.

It supports data structures such as strings, hashes, lists, sets, and so on and stores them in key/value format. The cache also helps to remove unnecessary load from the database by directing the read to the cache instead. You can implement a reactive or proactive caching mechanism using Redis, which is dependent on your application business requirements. A proactive mechanism is employed when, for example, the cache is updated at same time as the database; a reactive mechanism is followed when data in the cache is updated as and when data is requested for a read. The latter approach is also referred to as a *cache-aside* or *lazy loading pattern* and is explained in Figure 12.5. The flow is explained as follows:

❶ The application instance first tries to read data from the cache. If the data is available in the cache, the data is returned.

❷ If the data is not available in the cache, the data is directly read from the database.

❸ The read data is then updated in the cache.

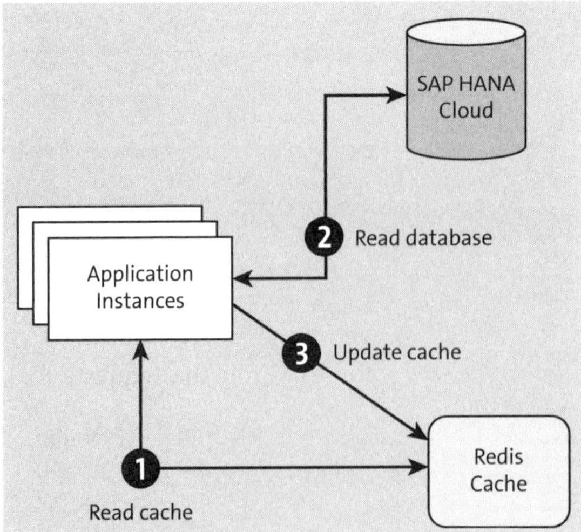

Figure 12.5 Cache-Aside/Lazy Loading of Cache

Tip

SAP Cloud SDK provides capabilities to call the Redis cache with a dedicated cache configuration, with default resiliency settings maintained in SAP Cloud SDK.

12.2.3   Rate Limiting

Your applications could have sudden spikes in traffic, and you want to protect and guard your applications against such increased load. A common resilience pattern used to protect your application is to limit the calls to the application. This principle is called *rate limiting* and is the process of controlling the traffic received or sent to a particular endpoint.

API Management provides various capabilities and policies to rate-limit your APIs. The following policies are available from API Management to address various rate-limiting requirements.

- **Quota policy**
  This policy enables rate-limiting based on the number of request messages over a given time period. The policy is applied in context of each request message, and the time period can be defined based on application requirements, like daily, monthly, or even hourly.

  Thus, this policy will help you restrict the number of calls made to an API. This could mean specifying a limit such as 100 calls to an API per day. API Management maintains a counter that keeps track of the number of calls made to the API. Once the counter is hit, all calls made after this point are rejected. The counter is reset after the defined time range ends.

- **Spike arrest policy**
  This policy enables rate-limiting for sudden spikes in calls from the point in the processing flow at which the policy is attached as a processing step. The policy limits the API calls based on the last matching message that was processed within a specific time interval; that is, it spaces and process the API requests evenly during the time interval. If two requests come close together within the time interval, one request will be rejected. Every successful request through the spike arrest policy updates the spike arrest's last processed count.

  You can attach the policy at the proxy endpoint, where the policy will limit the inbound requests, or at the target endpoint, where it will limit the requests forwarded to the backend service.

- **Concurrent rate limit**
  This policy enables rate-limiting on a number of concurrent connections to the target endpoint or to your backend services from API proxies. This policy manages to restrict traffic to backend services to keep them up and stable by not flooding them with too many requests.

  Additional requests after the concurrent rate limit is reached are returned with HTTP response code 503: Service Unavailable.

12.2.4   Application Autoscaler

Another way to handle increases in load to an application is to scale the application. You can have applications scale vertically or horizontally.

Vertical scaling means a change in the quota in which you can change the resources that are applied to all app instances. In a Cloud Foundry-based environment, you do this by assigning additional application runtime to your application.

Horizontal scaling is about increasing or decreasing the number of application instances. In a Cloud Foundry-based environment, this is done by increasing or decreasing the application instance.

SAP Cloud Platform provides the Application Autoscaler service, which you can bind to your application to automatically scale up or scale down your bound application instances. This automatic scaling option triggers based on increased load on your application. It increases the number of application instances and thus guards your application against failing. As the application load reduces, a dynamic scale down is triggered, which ensures that your application utilizes optimal resources.

The Application Autoscaler service supports both dynamic and schedule-based scaling based on the following application metrics:

- The *memory consumed* metric monitors memory consumption in megabytes.

- The *response time* metric monitors response time in milliseconds.

- The *throughput* metric monitors requests per second.

It's mandatory to have at least one policy of any specific type for scaling, which you pass as a parameter during service instance creation of the Application Autoscaler service in JSON format. You also specify in the policy the maximum number of instances to be created.

To begin your setup, follow these steps:

1. Navigate to the space inside your Cloud Foundry environment subaccount.
2. Inside the space, navigate to **Services • Service Marketplace** and click the **Application Autoscaler** tile.
3. Choose **Instances • Create New Instance**.
4. In the **New Instance Creation** wizard, choose a plan and click **Next**. To specify parameters, provide the parameters to define the scaling policy in JSON format. An example is provided in Listing 12.1 to scale up/down based on memory consumption. Click **Next**.

```
{
    "instance_min_count": 1,
    "instance_max_count": 5,
    "scaling_rules": [
        {
            "metric_type": "memoryused",
            "threshold": 90,
            "operator": ">=",
            "adjustment": "+1"
        },
    {

            "metric_type": "memoryused",
            "threshold": 30,
            "operator": "<",
            "adjustment": "-1"
        }


    ]
}
```

Listing 12.1 Application Autoscaler Service JSON Parameters

5. Choose the application that you want to bind the service to and choose **Next**.
6. Provide an **Instance Name** and click **Finish**.

Figure 12.6 shows how instances are created automatically based on memory consumption.

Instances

State	Since	CPU Usage	Memory Usage		Disk Usage	
RUNNING	17 Jul 2020, 10:19:17 (GMT+05:30)	98%	299.6 MB		164.2 MB	
STARTING	17 Jul 2020, 10:18:31 (GMT+05:30)	97.8%	339 MB		164.2 MB	
RUNNING	17 Jul 2020, 10:19:15 (GMT+05:30)	97.2%	281.9 MB		164.2 MB	
STARTING	17 Jul 2020, 10:18:31 (GMT+05:30)	97.3%	303.6 MB		164.2 MB	

Figure 12.6 Application Autoscaler Service Starting New Instances Based on Memory Consumption

12.2.5   Application Resiliency

You can embed various resiliency principles within your application coding. Such principles are best practices to follow while developing your applications and extensions. These resiliency principles allow you to fail your application in a safe way rather than it just crashing, which results in a bad experience for your end users.

SAP Cloud SDK builds upon the Resilience4j library to provide resilience to your cloud applications. Resilience4j is a lightweight fault-tolerance library similar to Netflix Hystrix. It has many modules to protect your application interfaces, lambda expressions, or method references from failures like circuit breakers, rate limiters, retries, or bulkheads. You can use more than one of the following modules for interfaces, lambda expressions, or method references:

- **CircuitBreaker**
  If the remote service call fails too many times, Resilience4j will open/trip the breaker, which means any further calls made to same remote service are automatically stopped. Resilience4j will periodically check if the service is available again and, if available, will close the open breaker accordingly.

- **Bulkhead**
  This allows for configuring the maximum amount of parallel executions allowed for the remote service. If the requests are exceeded, the bulkhead enters a saturated phase and any further requests attempting to enter the saturated bulkhead are stopped for a defined wait duration.

- **TimeLimiter**
  This allows you to set time limits for every remote service call. If the response time exceeds the set time limit duration, the call is considered a failure and you can decide to whether cancel the running call.

The resilience library for SAP Cloud SDK is under the `com.sap.cloud.sdk.cloudplatform.resilience` package. Two important classes under this package are `ResilienceConfiguration`, in which you perform various resilience configurations, and `ResilienceDecorator`, which provides methods to simply wrap the provided function and return a new function.

SAP Cloud SDK also enables you to provide fallback functions. This helps to handle the calls that fail. For example, when a bulkhead is saturated or circuit breaker opens, the call sequence will be directed to the fallback method and call it automatically. This way, you can handle the failures gracefully, such as by reading the data from the cache when a backend service isn't available or providing an error message to users saying "Please try again later" when trying to write data into the database.

> **Note**
>
> SAP Cloud SDK documentation is provided in the help document found at *http://s-prs.co/ v515701*.

12.2.6   SAP Cloud Platform Application Deployment

An important aspect of application resiliency is the ability to roll out new application features with no or little downtime for existing applications.

The philosophy of zero downtime deployment always asks for applications to be available. This concept of zero downtime deployment is called *blue-green deployment*, and SAP Cloud Platform supports it through the route service. The blue-green deployment concept is illustrated in Figure 12.7. The steps are as follows:

❶ You have both application versions, the old version (called blue) and new version (called green), both mapped to the router. The users of the application might be randomly directed to either version.

   At this time, based on load, reduce the number of instances of the blue version.

❷ As the number of calls to the blue version reduces, unmap the blue version from the router service.

❸ Delete the blue version.

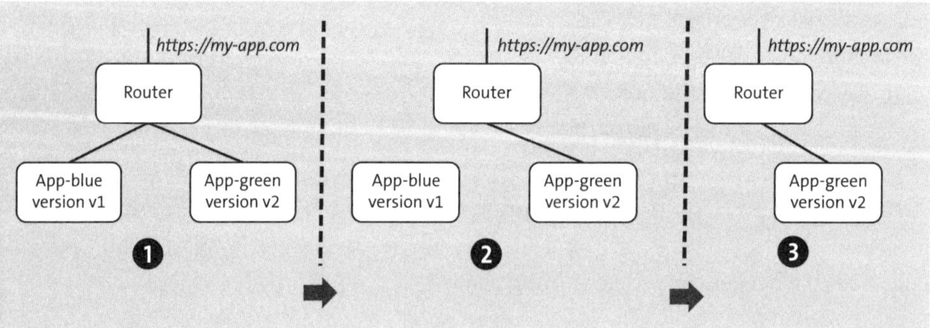

Figure 12.7 Blue-Green Deployment in SAP Cloud Platform

In addition to the blue-green deployment capability, there is also the SAP Cloud Platform Feature Flags service. This service allows for enabling or disabling new features without redeploying or restarting the application. The feature flags service helps to decouple deployment and release of application functionality.

The service provides the ability to centrally synchronize releasing new features for your cloud application. It provides a centralized dashboard in which every feature flags service instance is displayed separately and enables managing individual flags. To use feature flags service, the following prerequisites must be met:

- Your application should be bound to a feature flags service instance. This binding updates the environment variable for your application, which holds the credential data and URI for the service instance.

- The application code for the new functionality to be introduced has to be wrapped inside a toggle point (like an if-else statement or switch-case statement) to execute based on the active/inactive state of a feature flag.

To begin your setup for the feature flags service, follow these steps:

1. Navigate inside the space of your Cloud Foundry subaccount.

2. Navigate to **Services • Service Marketplace** and choose the **Feature Flags** tile.

3. Navigate to **Instances • Create New Instance**.

4. In the New Instance Creation wizard, choose a plan and select **Next**.

5. Skip the next step and click **Next**. On the **Assign Application** screen, select the application to which you want to bind the service. Click **Next**.

6. In the next step, select an instance name and click **Finish**.

7. Next to the created service instance, under **Actions**, click **Open Dashboard**. This opens the **Feature Flags Dashboard**, as shown in Figure 12.8.

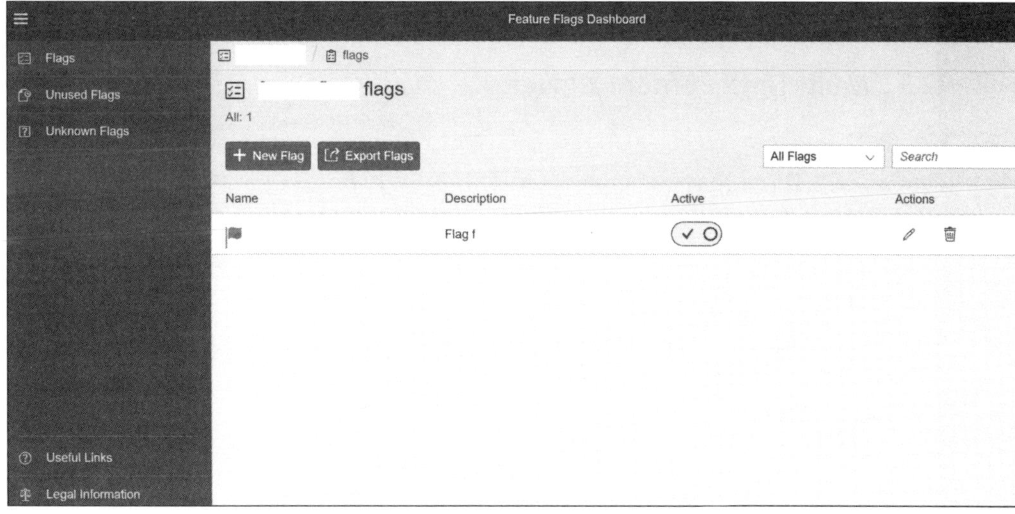

Figure 12.8 SAP Cloud Platform Feature Flags Dashboard

8. You can create and manage new flags from this central place and activate/deactivate such flags.

9. Once you click a feature flag, you can click **Code Snippets** to view sample code snippets provided to integrate feature flags into your application.

10. From **Monitoring** inside the feature flags service, you can monitor the usage of feature flags.

Tips

Here are a few tips for using the feature flags service:

- Minimize the number of toggles.
- Avoid intertoggle dependencies.
- Test with all toggles off and all toggles on.
- Remove feature flags when they're no longer needed; that is, create feature flags with short lifetimes.

12.2.7   Logging

Finally, logging is quite important while building cloud-native resilient applications. We covered various logging capabilities related to continuous monitoring offered by SAP Cloud Platform in Chapter 11.

Use these logging capabilities extensively and make them mandatory in your application to understand errors, warnings, and various information from the application. This helps you monitor the application behavior, HTTP requests, and responses on a continuous basis to provide visibility and adjust the application as required.

12.3   Multi-Data Center Failover

Another important high-availability pattern to ensure application availability is multi-data center failover for your application. This helps to protect your application in case of a failure or outage in one data center.

Warning

Implementing such a scenario involves high cost and complexity. Thus, you want to consider such a scenario implementation quite early while designing and architecting your application.

Note

To implement such a scenario, the application should be based on stateless architecture.

Multi-data center failover is a process in which you switch from one primary data center or system to another redundant secondary data center automatically in case of planned or unplanned downtime in your primary data center, making your application available all the time for your business. Such a scenario operates via an active/passive setup, which means the traffic is always directed to your primary data center until an outage, at which time the traffic is routed to a secondary data center.

Proceed as follows to implement multi-data center failover for your application:

1. **Decide on two or more data centers where your application needs to be deployed.**
   You first need to decide which data center you need your application needs to failover to. You can make this decision based on the following decision criteria:
   - Region where your end users will access the application
   - Planned maintenance windows for regions

2. **Deploy your applications in the regions selected in Step 1.**
   Once you have selected the regions where your application needs to be, deploy your application to both regions.

   We recommend that you use SAP Cloud Platform Continuous Integration and Delivery, SAP Cloud Platform Transport Management, or delivery pipelines, as discussed in Chapter 11, to achieve this automatically, letting you deploy new versions of the application and keep the application versions in sync.

3. **Implement the failover.**
   To implement the failover, you have to perform the following functions:
   - Decide how and when to define if the application is down in the primary data center.
   - If the application is down in the primary data center, route the call automatically to the secondary application instance.

   You can achieve this through usage of rule-based performance solutions such as Akamai GTM.

4. **Plan a failback to the primary data center.**
   The last step is to plan the failback of your application to the primary data center when the primary data center is back up and running.

12.4   Summary

In this chapter, we discussed why building applications and extensions with required resiliency is important for your business. We also discussed how your application SLA is impacted by service-level SLAs and how this is helpful in setting up the right expectations for businesses.

Next, we discussed various important resilience patterns and the issues they address while building applications in the cloud. Then we discussed best practices for the usage of various SAP Cloud Platform services to plan for high availability and resiliency. We also discussed various capabilities and functionalities provided by SAP Cloud Platform that can help you build your cloud-native applications and extensions.

Final Steps

Chapter 13
Trailblazing Enterprise Innovations

This chapter describes how a future-ready enterprise can use innovation to address whitespaces and gaps in business, using industry-standard guidelines and specifications. It also describes the necessity of using workshops and discovery sessions to review your architecture and landscapes taking into consideration both internal and external viewpoints as well as future market needs.

In this chapter, we'll cover innovation as one of the critical stages of the future-ready enterprise framework. This stage is about remaining competitive in your industry and continuously innovating and transforming your organization. It builds on the previous stages.

To achieve continuous innovation and to keep your application and system landscape agile and flexible to adapt to changing business requirements, you need to build your organization's application and technology architecture based on the platform capabilities and methodologies covered in previous chapters.

This chapter describes industry trends and challenges. We shows examples of reimagined business process that blend digital transformation with SAP Cloud Platform solutions, highlighting specific industry-specific solutions. Also, this chapter covers how to address whitespaces and gaps over discovery workshops and architecture-led engagements that SAP offers, which eventually helps to build resilient architecture and a proven IT landscape. The right scoping and determining prioritization are explained in the following sections.

13.1 Industry Trends and Technology Trends

Alongside the emergence of new technologies and continuing digitization, the IT strategy of companies is fundamentally changing. Figure 13.1 describes the journey of IT revolutions, starting from industrial automation, followed by business and digital transformation. With technology trends maturing each year, the current focus is building an intelligent enterprise to that uses both experience and operational data along with SAP Cloud Platform innovation services such as SAP Intelligent SAP Robotic Process Automation, machine learning, etc.

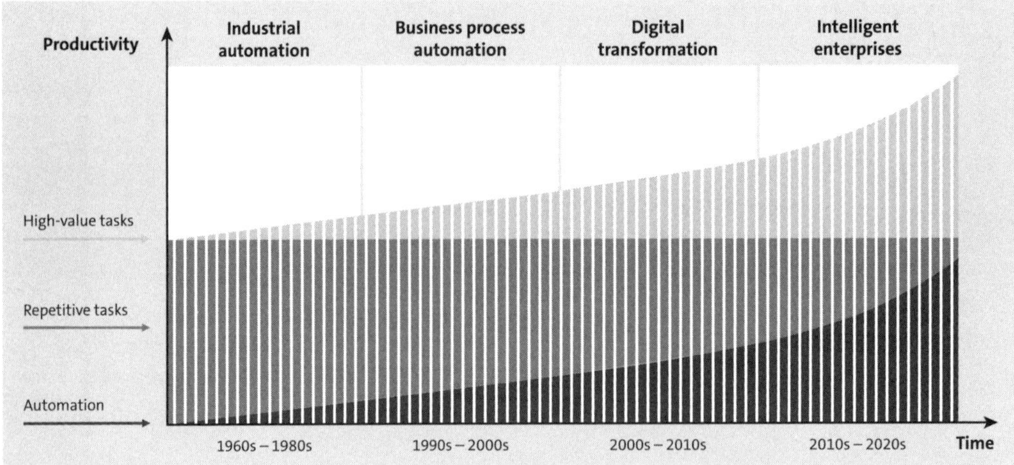

Figure 13.1 Intelligent Enterprises Elevate Organizations to Focus on Higher-Value Tasks

By adopting intelligent technologies like the Internet of Things (IoT), artificial intelligence (AI), machine learning, and advanced analytics, leading companies transform into event-driven businesses, go to market faster, and are more competitive in the marketplace. Event-driven businesses automate repetitive tasks, enable employees to focus on higher-value tasks, and allow companies to invent new business models and revenue streams by monetizing data-driven capabilities and applying core competencies in new ways.

Hyperautomation is leading the way with task automations and reduction of manual repeated tasks. With digital services available, automation can be built using artificial intelligence and machine learning approaches.

With multiple mobile gadgets and devices available, consumers expect to be able to access and consume services anywhere, anytime. Operational data derived from machine sensors and equipment comes from different sources and in multiple formats. Users are looking for the ability to work with operational data and experience data at the same time, all with multitouch technology. Similarly, in the development arena, where traditional development is still done in silos, developers' expect to be able to fast track development and reuse components. This can be achieved by, for example utilizing an AI model or easy-to-configure development tools specifically designed to integrate with AI capabilities.

Lastly, data transparency and the ability to track information quickly and accurately will only become more and more important. This will allow for improved trust and establishing more ethical relationships with business partners and your own employees or contractors. The pillars to achieve this are work ethics, maintaining integrity, openness with clear commitments, accountability, and ensuring completeness and consistency.

> **Note**
>
> Gartner's top 10 strategic technology trends highlight trends that will drive significant disruption and opportunity over the next five to 10 years. For more information, visit *https://www.gartner.com/en/doc/432920-top-10-strategic-technology-trends-for-2020*.

13.2   Industry Challenges and Reimagined Business Processes

Intelligent enterprises have the visibility, focus, and agility to overcome industry challenges and deliver outcomes of changes in productivity, redefined end-to-end customer experiences, and transformed workforce engagement. SAP's strategy using SAP Cloud Platform helps every company to become an intelligent enterprise using an intelligent suite, a digital platform, and intelligent technology.

Reimagining businesses as intelligent enterprises is demonstrated in the following sections with reference to a few selected industries. The suggested reimagined business processes help companies overcome challenges found in traditional business process using SAP Cloud Platform services.

13.2.1   High-Tech Industry

With digital trends shifted across different segments of customers, high-tech industries especially play an important role in offering the right solution at the right time. It's a continuous process to meet the challenges of disruptions, regulation changes, increased consumption, and more. Reimaging business processes to support a capital expenditures or operating expenses model or move to a subscription-based model has different positive outcomes in business revenue models. Dynamic changes and monetizing early is supported by using SAP Cloud Platform for building faster and sustainable business models, which is a game changer. Table 13.1 shows the trends and challenges in this industry, along with an example of a reimagined business process that addresses those challenges.

Trends	Challenges	Reimagined Business
Consumption preference: customers prefer hybrid consumption options	Selling measurable results	Subscribing to outcomes
Agility: scaling and safeguarding without sacrificing efficiency	Limited demand visibility, counterfeit products	Realizing an intelligent supply network

Table 13.1 High Tech Industry Poised for Change

Trends	Challenges	Reimagined Business
Software is king: software being embedded in every manufactured product	Turning products into smart solutions	Providing digital smart products
Changing regulations: privacy laws, trade policies, and accounting rules	Disruption to operational efficiency	Achieving customer intimacy

Table 13.1 High Tech Industry Poised for Change (Cont.)

13.2.2 Consumer Products Industry

With the advent of e-commerce using digital platforms, consumer product segment companies have grown and expanded their operations many times over. They now reach many consumers of different statures with the ease of direct delivery and end-to-end integration with business-to-business (B2B) access, suppliers, sellers, shipping organizations, and others. Reimagining the business to adopt focused campaigns based on customer experience data ("X" data) and further using machine learning and AI, supported in SAP Cloud platform, is a key differentiator for faster time to market and retaining a large consumer base. Table 13.2 shows the trends and challenges in this industry, along with an example of a reimagined business process that addresses those challenges.

Trends	Challenges	Reimagined Business
The "Amazon effect": change in shopping patterns and customer expectations	New competitive landscape created by digital marketplace	Direct shipment from supplier to consumer
"Born digital" innovators: delivering personalized outcomes	Start-ups provide better personalization and flexibility	Flexible, consumer-centric approach
"Right now, right here, just for me" consumers: competing as an ecosystem	Complex silos and information gaps	Using consumer insights to create offers with partners

Table 13.2 Consumer Products Industry Poised for Change

13.2.3 Industrial Machinery and Components Industry

The industrial machinery and components industry business model is building industry components that are consumed by different industries and business units. Now with industry trends changing, it's important that industrial components have more digital smart capabilities and can transfer machine data for future processing or analytical consumption. SAP Cloud Platform supports SAP Internet of Things and SAP Edge Services, for example, which can integrate smart data with applications or analytics layers. Table 13.3 shows the trends and challenges in this industry, along with an example of a reimagined business process that addresses those challenges.

Trends	Challenges	Reimagined Business
Mass customization: tailor-made products in a "pay-as-you-go" and usage-based pricing model	Customers demand products that fit their exact needs at competitive prices	Serving the "segment of one"
Differentiation on value: digital capabilities and value-added services	Customers, partners, and suppliers are entering into competition	Digital smart products
Globalization and right-shoring: make best use of regulations and local advantages	Manage supply fluctuations or changes in the configuration of a customer order	Digital supply chain and smart factory
Monetizing asset data: shifting from selling products to provide complete solutions	Organizational setup and systems, engineering capabilities, and right talent	Value-added service components and new business models

Table 13.3 Industrial Machinery and Components Industry Poised for Change

13.2.4   Travel and Transportation

Travel and transportation is a powerful industry segment that allows the supply chain to move goods efficiently from source to destination. Goods storage in warehouses and distribution centers is mapped based on demands and further linked with production generation for supply. A reimagined business model allowing digital exchanges of data for new partnerships and adopting new services is essential. SAP Cloud Platform builds extension capabilities that allow ecosystem partners to use an integrated digital platform for consumption of information for greater agility of their business. Table 13.4 shows the trends and challenges in this industry, along with an example of a reimagined business process that addresses those challenges.

Trends	Challenges	Reimagined Business
Digitize customer experience: by offering new digitized ways of engaging with the customer	Avoid being commoditized by new digital entrants	■ Customer centricity ■ Extended partnerships and new services

Table 13.4 Travel and Transportation Industry Poised for Change

Trends	Challenges	Reimagined Business
Generate new business models: offer premium services and go beyond the current offering with the help of new digital tools	Provide new services and increase the portfolio offered to customers	■ Next-generation analytics ■ Intelligent portfolio planning
Digitize operations: growing use of digital technologies to better forecast demand and using integrated technology platforms to manage operations holistically	Offer an integrated technology platform that allows for innovative, effective end-to-end processes to optimize operations	■ Integrated technology platform ■ Intelligent asset management

Table 13.4 Travel and Transportation Industry Poised for Change (Cont.)

13.2.5   Banking Industry

The banking industry has undergone a paradigm shift to adopt digital transformation. Being an early adopter of information systems, compared to other industry segments, there is continuous process improvement in banking to strive for better customer experiences. Reimagined businesses focus on acquiring more data insights about customers and using machine learning to provide a more focused digital experience to customers. SAP Cloud Platform's machine learning services like SAP Data Intelligence allows for this kind of transformation journey, benefitting banking customers. Table 13.5 shows the trends and challenges in this industry, along with an example of a reimagined business process that addresses those challenges.

Trends	Challenges	Reimagined Business
Customer expectations: bank customers want and expect more from their banks	Attracting, cultivating, and retaining customers by enabling an integrated, multichannel environment	Seamless connectivity
Data as the new currency: create a single view of each customer	Fragmented and piecemeal data analysis	Data-driven intelligence
"Platformification" of banking: multiple participants connect to the bank; interact, create, and exchange value	Fragmented and manual banking operations with no embedded compliance and risk management	Reimagined operational effectiveness

Table 13.5 Banking Industry Poised for Change

Trends	Challenges	Reimagined Business
Banks as technology companies: banks reviewing and changing organizational structure, technologies, and cultures to run more like technology companies	Complex regulatory environment	Agile and responsive to requirements for financial insight and control

Table 13.5 Banking Industry Poised for Change (Cont.)

13.2.6   Telco Industry

Telco is one of the most competitive industry segments, in which disruptions are frequent and large. This industry segment is facing huge challenges in competition, changing customer needs, the expectations of youth communities or middle-aged customers, and so on. Better revenue stream diversification needs experience data from social media, along with survey data, which is important for building the right products and solutions. SAP Cloud Platform and Qualtrics work with big data, which can be processed using SAP Data Intelligence to generate more insights to address a customer-first strategy. Table 13.6 shows the trends and challenges in this industry, along with an example of a reimagined business process that addresses those challenges.

Trends	Challenges	Reimagined Business
Disruption and competition: nontraditional market entrants driving margin pressure and speeding up commoditization of core communication services	Focus on delivering a highly personalized customer experience while excelling in operational efficiency	Customer-first
Business model innovation: market saturation accelerating diversification of revenue streams	Quickly meet customer demand, deliver new services, and make new business models work with an ecosystem-focused approach	Revenue stream diversification
Next-generation network: enable other industries to reinvent their business processes	Connect and monetize a sensor-based world with the combination of 5G, IoT, edge computing, AI, and machine learning	Intelligent connectivity

Table 13.6 Telecommunications Industry Poised for Change

Trends	Challenges	Reimagined Business
Slow growth: average revenue growth has been slowing down continuously in the past decade	Increasing operational expenses and expected huge capital expenditure	Intelligent connectivity

Table 13.6 Telecommunications Industry Poised for Change (Cont.)

13.2.7   Utilities Industry

With large customer bases, utilities organizations have effectively used digital transformation to support the needs of their customers. Using mobile assets, data can be seamlessly collected to a central IT processing unit; a good example of this is image processing for meter readings and extracting the units consumed by the consumer. Table 13.7 shows the trends and challenges in this industry, along with an example of a reimagined business process that addresses those challenges.

Trends	Challenges	Reimagined Business
Resource democratization	Partnering with customers to constantly optimize energy streams, decrease costs, and offer new products	Distributed energy resource operation
Field automation	Massive investments required in infrastructure to support decentralized energy generation	Smart and efficient distribution
Diversification of products and services	Collection and analysis of granular consumption and production data	Demand and supply balancing services
Customer centricity	The emergence of prosumers with increasing environmental and energy awareness	Omnichannel retail to digitalized consumer

Table 13.7 Utilities Industry Poised for Change

13.2.8   Engineering, Construction, and Operations Industry

Engineering, construction, and operations (EPC) companies are facing challenges in global markets. ECP companies can use technologies like Digital Twins to plan for future assessments of construction projects. SAP Cloud Platform is a good PaaS solution to integrate with event-based data to consume for various needs, such as marking completion of a task and initiating a financial transaction or triggering dependent activities for the next sequence of activities. Table 13.8 shows the trends and challenges in this industry, along with an example of a reimagined business process that addresses those challenges.

Trends	Challenges	Reimagined Business
Lean: modern "construction-manufacturing site" with prefabrication	Labor shortage with flat productivity	Connected construction
Agile: scale with automation of repetitive tasks	Increasing project complexity	Building information modeling-based collaboration
Innovate: new business models	Complex silos and information gaps	Event-driven business
Digital: advanced technology as strategy enabler	High IT TCO with low adoption	Digital twins

Table 13.8 Engineering, Construction, and Operations Industry Poised for Change

13.2.9   Automotive Industry

The automotive industry is adopting smart technologies to gather machine data and analyze it for customers; this information is shared with customers as a service. This trending need is leading to reimagine business to find opportunities to implant smart products and allow data to be collected using smart mobility functions. SAP Internet of Things is a perfect cloud service to integrate machine data, which is a big data scenario for manufacturers. Big data can be processed using SAP Data Intelligence, which contains capabilities for implementing machine learning, deriving analytical views, and so on. Table 13.9 shows the trends and challenges in this industry, along with an example of a reimagined business process that addresses those challenges.

Trends	Challenges	Reimagined Business
Connected: new business models like shared mobility, transportation as a service, and so on	Use a car as a platform to deliver innovative services	Smart mobility
Autonomous: fully autonomous vehicles to "give time back" to the consumer	Innovative vehicle systems with embedded sensors and software	Smart products
Shared mobility: flexible car-sharing services and other on-demand services	Manage customer's experience throughout the entire lifecycle of the relationship	Customer centricity

Table 13.9 Automotive Industry Poised for Change

Trends	Challenges	Reimagined Business
Digitization: synchronization of end-to-end processes across the ecosystem	Respond quickly to new market expectations	Digital supply chain and smart factory

Table 13.9 Automotive Industry Poised for Change (Cont.)

13.2.10   Retail

A few of challenges traditional retail organizations are facing include competition from newcomers who are more technologically equipped and the need to provide smart retail stores. Reimagined businesses help retailers to detect, predict, and anticipate changes with the right forecast. Using smart devices and sensors can provide real-time data that's useful for connecting with stores and customers. SAP Cloud Platform Integration Suite allows seamless integration capabilities in B2B and B2C scenarios, with tools that are able to support faster adoption of a reimagined business process. Table 13.10 shows the trends and challenges in this industry, along with an example of a reimagined business process that addresses those challenges.

Trends	Challenges	Reimagined Business
Demographic changes: customer centricity	Challenge of traditional channels	Retailers detect, predict, and anticipate unspoken needs
Personalization: Segment of one	One size fits all, static assortments, campaigns and offers	Contextual marketing delivered in real time
Changing role of the store: smart retail stores	Traditional brick-and-mortar experience	Connected store and consumers, endless aisles with instant delivery
New competitors: changed retail landscape with new players	Beginning disruption in traditional market set-up	Leverage data and predictive analytics, allowing new revenue-generating offers

Table 13.10 Retail Industry Poised for Change

13.3   Addressing Whitespaces and Gaps

Identifying whitespaces allows you the opportunity to take a look at the full scope of your landscape, up and down the value chain, under a new lens. It can help uncover opportunities that are not obvious; it can identify new openings yet untouched, or it can be considered part of what was traditionally deemed a remote, different industry or outside the boundaries of the customer organization.

Discovery of whitespace is also an important outcome of customer inquiries or discovery via an architecture-driven approach and other discovery processes. It leads to new profit growth opportunities by defining potential gaps in existing markets. The process can be used to identify both your as-is landscape and your to-be landscape, or it can be used to map incremental innovation in products or services. SAP Cloud Platform expert teams are available to execute workshops with customers to identify whitespaces and provide architectural views and roadmaps using SAP Cloud Platform solutions for extension and extended scenarios.

Whitespace mapping for building enterprise application architecture can be achieved with the following objectives:

- An externally focused perspective caters to an ecosystem that includes suppliers, vendors, and other business partners, with information made available or exchanged via business transactions. In today's world with high globalization, information sharing is critical to execute operations and at the same time have information transparencies across business stakeholders. Any whitespace or gaps will reduce the agility in business operations, so it's extremely important to take into account external factors and needs while building an enterprise system.

- An internally focused perspective can also identify whitespaces in the current system landscape, which is very important. Unless internal systems are built to hold adequate data, they cannot be scaled to cater to external entities. Internal systems carrying all functions should be built to ensure they are harmonized and redundant data and information are not stored.

- The future-focused perspective offers agility to adopt changes for market needs, thus reducing the time to market. The ability to scale and integrate is of the essence, so regular exercise of architecture reviews across global views focusing on industry trends and challenges to overcome is important to spot and identify whitespaces and factor them into the enterprise architecture.

Whitespace finding and mapping is usually executed over a number of workshops for revaluating processes, one that places a focus on finding new contexts, beyond traditional assumptions.

Chapter 14 describes the farming framework and approach, which has six pillars to demonstrate how companies can benefit from building a success story via building a resilient systems architecture.

An architecture-driven approach and establishing design thinking workshops, as explained in Chapter 2 and Chapter 14, respectively, also helps to identify whitespaces or gaps in the enterprise landscape systematically. These workshops are typically driven by business process discovery sessions, followed by data collections for getting analytical data insights. Based on the outcomes and information gathered, recommendations are made to overcome the whitespace that exists in the landscape. Based on the nature of the scenarios, recommendations for solutions can be addressed using SAP Cloud Platform services for extension and integration scenarios, tying back to the on-premise, cloud, or legacy systems.

13.4   Frameworks and Prioritization

SAP provides expert methodologies and proven frameworks that can help you systematically determine your priorities when it comes to innovation needs. In this section, we'll look at the innovation engagement framework and some of the tools used when executing a process using the framework.

13.4.1   Innovation Engagement Framework

An innovation framework is designed to support global needs, visualize creating a scalable platform, and harness the creative talents of a team. This in in line with a company's vision and mission statements in alignment with corporate strategies. It's a strategic approach to innovation from the inside out that provides a way to act on new ideas.

Planning your innovations using the framework allows you to deliver tangible outcomes; otherwise, what you'll have is just a collection of ideas, which may not provide concrete results. Figure 13.2 shows the innovation framework, which is built on the following principles:

- Big ideas and quick wins, from customer workshops understanding business challenges and use cases

- Business sponsorships, with executive sponsorship important for the next steps and executing workshops

- Create plan and roadmaps, with the business and IT team collaborating to identify and prioritize use cases

- Validations, in which key information gathered needs technical validations with architects, data scientists, implementations teams, and important key team members

- Final approach, which outlines value drivers that are tangible and nontangible along with business case recommendations

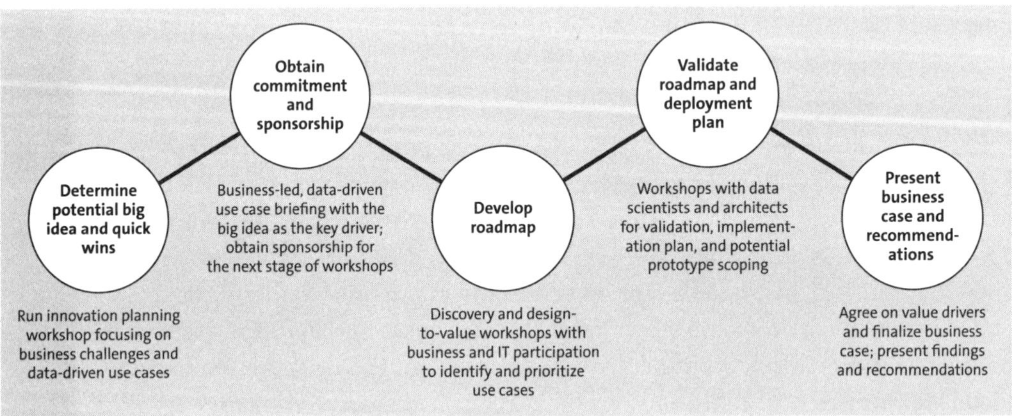

Figure 13.2 Innovation Framework

Let's now look at the execution steps of this framework. The execution is broadly covered in seven steps, which are performed in multiple workshops across each stage (see Table 13.11), as follows:

1. Innovation planning workshop

2. Executive briefing workshop

3. Discovery workshop

4. Design-to-value workshop

5. Digital prototyping workshop

6. Architecture and implementation planning workshop

7. Executive presentation and proposal review

The execution framework describes the following for each stage, each of which will be represented in Table 13.11:

- **Activities**
  Details of activities planned with key goals and expected outcomes

- **Outcomes**
  Described clear expectations of the outcomes to ensure the right participants are available and introduced for discussions

- **Customer participant details**
  The right teams are mapped and available to make decisions and set the right expectations

- **Deliverables**
  Maps and defined deliverables are the final outcomes of the discussion for the identified tracks

The extension framework includes information about suggested timelines, which can be used for planning purposes. This ensures identified resources are correctly allocated for the desired period and can accomplish activities successfully.

Innovation Engagement Framework	
Step 1: Workshop—Innovation Planning Workshop	
Activities	<ul><li>Understand customer's business model, challenges, and opportunities</li><li>Identify relevant data-driven use cases and customer stories</li><li>Plan for executive sponsored innovation-led collaborative engagement</li></ul>
Outcomes	<ul><li>Three to five big ideas/themes and use cases to form basis of innovation-led agenda</li><li>First draft of innovation-focused use case storyboard</li></ul>
Time commitment	<ul><li>One-hour optional customer coach/sponsor contribution, guidance, and feedback</li></ul>

Table 13.11 Innovation Engagement Framework in Detail

Innovation Engagement Framework	
Deliverables	■ Initial big idea, use cases, and customer stories ■ Draft use case storyboard
Step 2: Executive Briefing Workshop	
Activities	■ Secure executive sponsorship on engagement outcomes, scope, time-lines, and resource allocation ■ Agree on engagement governance model ■ Schedule interviews, follow-up meetings/workshops, and resources
Outcomes	■ Clear expectations on business impact and engagement process ■ Alignment on resources ■ Agreed executive sponsorship and progress review cadence
Time commitment	■ Two hours for executive meeting/workshop
Deliverables	■ Input to engagement plan ■ Input to governance model
Step 3: Workshop—Discovery Workshop	
Activities	■ Execute workshops spanning one to two LOBs using design thinking approach ■ Analysis of key capabilities required to address customer challenges and opportunities
Outcomes	■ Five-plus high-value use cases per LOB ■ High-value use cases with defined requirements to turn use cases into reality
Time commitment	■ Three to four hours for business and IT workshops
Deliverables	■ Prioritized use cases ■ Refined storyboard ■ Initial value and deployment roadmap
Step 4: Workshop—Design to Value	
Activities	■ Validate and quantify business value by use cases/business processes ■ Discuss technical architecture ■ Agree on target use case and prototype project ■ Present findings to executive sponsor
Outcomes	■ Defined business value, use case architecture, and roadmap for success-ful deployment of target use cases ■ Confirmed prototype use case project

Table 13.11 Innovation Engagement Framework in Detail (Cont.)

Innovation Engagement Framework	
Time commitment	■ Three to four hours for findings validation ■ One to two days for technical workshop
Deliverables	■ Prototype scope ■ Prototype resources, timings, and project plan ■ Prototype success criteria
Step 5: Workshop—Digital Prototyping Workshop	
Activities	■ Initiate prototype project, ideally in a sandbox or trial account ■ Present results of prototype project to executive sponsor and decision makers ■ Agree on next steps to secure buy-in and execution of roadmap
Outcomes	■ Landscape established for prototype project ■ SAP/partner/company undertakes prototype project
Time commitment	■ Three to six weeks (elapsed time, not dedicated customer time)
Deliverables	■ Prototype findings and recommendations ■ Refined use case deployment roadmap and implementation plan
Step 6: Workshop—Architecture and Implementation Planning Workshop	
Activities	■ Develop enterprise architecture roadmap and implementation plan ■ Develop broad enterprise architecture and implementation roadmap
Outcomes	■ Architecture roadmap ■ Data science/advanced analytics implementation plan ■ Point of view document with business value and business case
Time commitment	■ Two weeks (elapsed time) for workshops for enterprise architecture roadmap and implementation plan
Deliverables	■ Architecture roadmap and implementation plan ■ Input to business case and point of view document
Step 7: Workshop—Executive Presentation and Proposal Review	
Activities	■ Validate business value by use cases/processes and confirm impact to the business ■ Agree on next set of use cases and projects
Outcomes	■ Validation of the business value, use case architecture roadmap, and implementation plan ■ Financial proposal
Time commitment	■ Two hours for validation, discussion of findings, and next steps

Table 13.11 Innovation Engagement Framework in Detail (Cont.)

Innovation Engagement Framework	
Deliverables	■ Input to business value and financial proposal

Table 13.11 Innovation Engagement Framework in Detail (Cont.)

13.4.2   Tools

Table 13.12 outlines different tools and artifacts used to execute the innovation framework in each stage and how to capture output data in identified artifacts. The collection of data used as input for tools and the derived outcomes are important to refer to in the next stage, processing and analysis.

Experts within an enterprise benefit by using represented data derived from tools, which is further used for setting recommendations, and can analyze and share their inputs before proceeding to the next stage. Tools and templates are designed by SAP experts using industry vertical experiences, ensuring the criteria and dimensions set are accurate as needed to drive the right integration or extension scenarios using SAP Cloud Platform.

Innovation Planning Workshop	Executive: Use Case Storyboard
Executive briefing workshop	■ Case studies from pace layering perspective ■ Use case prioritization ■ Use case roadmap implementation timing ■ Use case roadmap from pace layering perspective
Discovery workshop	■ Use case ideation ■ Use prioritization ■ Topic summary
Design to value workshop	■ Use case roadmap ■ Estimate financial benefits ■ Ideas to priorities to next steps ■ Technical architecture
Digital prototyping workshop	■ Prototyping execution
Architecture and implementation planning workshop	■ Use case architecture
Executive presentation and proposal review	■ Business priorities ■ High-level architecture ■ Roadmap planning

Table 13.12 Innovation Framework Workshop Data

A simplified example of the innovation framework in action is shown for the agricultural industry in Figure 13.3.

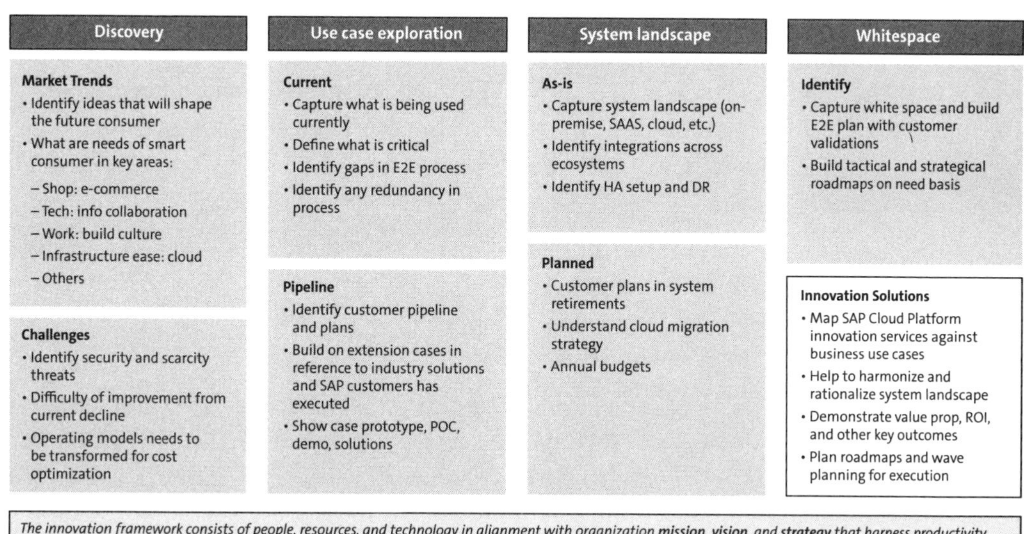

Figure 13.3 Simplified Innovation Framework

Continuing the example, the information from the innovation framework can then be mapped to SAP Cloud Platform services as shown in Figure 13.4.

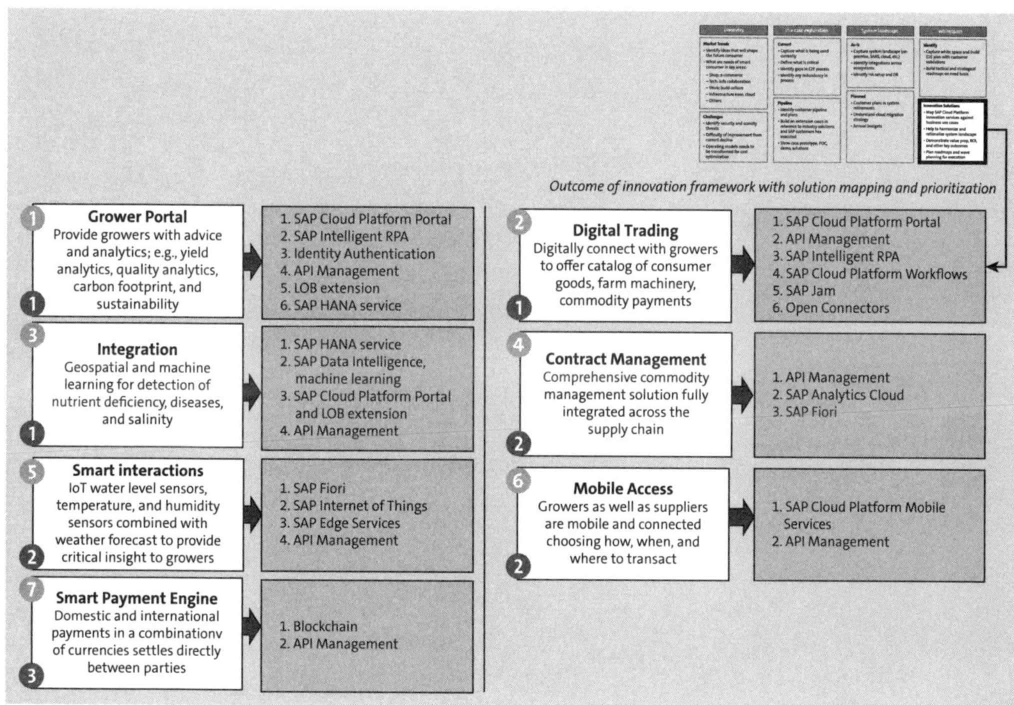

Figure 13.4 Example of Industry Extension Framework Leveraging SAP Cloud Platform Services

13.5   Summary

With the rapid changes in market trends and digitization, organizations are becoming more conscious of adopting digital transformations for their enterprises. With the agility required to move at a faster rate, cloud services are a good option for building smart digital assets.

Due to large on-premise bases, companies are still looking for hybrid solutions or sidecar strategies using the cloud. With cloud-first strategies being adopted as a future option, companies have started planning to adopt cloud solutions and phase out on-premise solutions over a period of time.

SAP Cloud Platform solutions and tools help customers identify their shortcomings, highlighting whitespaces and then taking necessary actions for remediation. You can conduct discovery workshops to help build component-based architecture and appropriately map SAP Cloud Platform solutions to overcome process gaps.

Chapter 14
Measuring Success

This chapter describes various approaches to measure success and offers insight into the design of a future-ready enterprise program for enterprises planning a digital transformation journey. It covers how to implement a transformation strategy with effective measurements using the farming framework.

As businesses evolve, the IT landscape becomes increasingly complex. Today, the IT landscape and all enterprise applications need to provide greater agility and keep pace with emerging technology trends and demands. SAP-based applications are no exception. With many larger corporations running their core ERP systems on SAP, the associated applications must be more intelligent, flexible, faster, and *future proof*. SAP Cloud Platform contains many services and provides support for different application development scenarios. Some of these are as follows:

- Building innovative applications using the latest technologies like AI, blockchain, IoT, and so on

- Building mobile and portal applications that can connect seamlessly with on-premise and cloud applications

- Extending cloud or on-premise applications

- Integrating cloud and on-premise applications

Evaluating how future-proof your landscape is can be an important measure of success, and each section in this chapter will discuss ways to think about future-proofing. Two examples of future-proofing are as follows:

1. **SAP Cloud Platform Integration Suite**
   With older solutions like SAP Process Integration and SAP Process Orchestration nearing the end of maintenance in a few years, it's advisable to move on-premise process integration and process orchestration tools to SAP Cloud Platform Integration Suite, which supports integration in a holistic manner with prebuilt packages, which can be leveraged for accelerated deployments and fast-track implementations.

2. **Multicloud architecture**
   Enterprises are advised to move toward multicloud platforms. SAP architecture supports open-source frameworks and "bring your own language" options: Java, Node.js,

"

and SAP HANA extended application services, advanced model (SAP HANA XSA). This allows companies to build a system landscape that's future proof to support multiple program languages and various runtime capabilities.

This chapter explores various approaches to measure success and offer value during a future-ready enterprise design project. You'll learn about the farming framework, dive into opportunities for harmonization and optimization, and discuss the value of having an innovation framework. The chapter will conclude with a brief look at value realization engagement.

14.1   Farming Framework

The farming framework is a set of methodologies that helps enterprises execute step-by-step systemic processes to discover and determine what the enterprise needs. They then can come up with appropriate solutions to fit business needs and priorities. An enterprise can utilize the six pillars of the farming framework to achieve success. For SAP Cloud Platform solutions, an enterprise can choose the right methodologies, tools, and approaches under the guideline of the farming framework, which is measurable based on project outcomes.

The future-ready enterprise paradigm is supported by the farming framework. This section will describe how the farming framework can help you to design new architecture during an enterprise design project.

Businesses are evolving quite fast to meet consumer demands. The six pillars of the farming framework are as follows:

1. Line of business-driven engagement
2. Architecture-driven engagement
3. Harmonization and optimization
4. Reimagined business processes
5. Pilot-driven engagement
6. Value realization engagement

These are the key foundations to help you determine the right architecture for your company and allow you to understand how to build an SAP enterprise landscape to meet future requirements while becoming an intelligent enterprise.

The farming framework's essential pillars—LOB-driven engagement and architecture-driven engagement—have been described in earlier chapters. LOB extensions such as SAP S/4HANA extensions are quick wins, using SAP Cloud Platform for building company-specific processes and customizations thus closing whitespaces in the enterprise landscape. Suggested SAP frameworks and tools help to identify current architecture and processes running across a heterogeneous system and evaluate pros and cons in terms of expandability, time to adopt change, resilient architecture, maintenance overhead, and so on. The framework has accelerators that help to determine the architecture-to-be. The farming framework helps you achieve the following measures of success:

- Agility to adopt changes faster and have an easier time during go-to-market activities

- Keeping the core clean using LOB solutions, with seamless extensions for a unified UX strategy

- Building solutions using resilient architecture

- Lower costs for operation and maintenance

14.2   Harmonization and Optimization

Harmonization offers great value during a future-ready enterprise design project. For every enterprise that wants to drive transformation, there are typically a few hurdles. A major one could be legacy and archaic technology landscapes, including systems and data in the enterprise that would create a bottleneck in the journey toward becoming future ready. Therefore, the enterprise must revisit the existing landscape and look at their options to replatform, to get from the current disparate, siloed set of systems to setting a stage for a future-ready platform. Many enterprises have complex operational landscapes that support fragmented processes over disparate and aged (in some cases, expired) systems.

SAP Cloud Platform helps you to harmonize end-to-end processes and to streamline processes for application to application, business to business, business to government, and IoT for SAP and non-SAP processes on premise or in the cloud. Also, using tailored interactions for customers, employees, and partners allows you to deliver omnichannel experiences across devices and processes with full context. Integration across diverse application landscapes and ecosystems allows harmonization across enterprise landscapes via a single-point cockpit for multiple deployment environments and more than 1,300 integrations, 900 APIs, and 150 connectors.

Some of the tools and technologies SAP Cloud Platform offers for harmonization and optimization are as follows:

- Prebuilt integration packs and business events to unify access to SAP and partner APIs through the SAP API Business Hub, leverage ready-to-use data model transformations and adapters, and sense and respond to business events

- Ready-to-use business services, AI-driven data orchestration, and pipelining to reuse existing business processes and data to compose applications faster and leverage prepackaged content

- API-first approach to assemble, publish, and monetize reusable APIs and integrations for reduced app development time and ease of maintenance and development

Let's now look some different aspects of harmonization and optimization, from harmonizing your system for DevOps to replatforming your system landscape.

14.2.1   System Harmonization

System harmonization for data symmetries across an enterprise is very important. Using the SAP Cloud Platform master data process initiated in SAP Fiori or SAPUI5, applications can be processed for validation and process verifications and finally pushed to SAP S/4HANA. To support building harmonized landscapes, the following tools are used:

- SAP Cloud Platform Integration Suite for A2A, B2B, B2C, and B2G integrations

- SAP Cloud Platform Integration Suite: Open Connectors to access data from non-SAP applications

- SAP Cloud Platform Integration Suite: API Management for all the security best practices and governance policies to be applied for harmonious APIs

When successful, companies will have their data harmonized across the landscape and have a single source of truth. Loss of data harmonization leads to various versions of master data or missing information in the target system. This can create or erroneous critical transactional information, resulting in poor business execution and more lead time to reach to consumers or business partners.

> **Note**
>
> To learn more about business partners and harmonization, you can check out the following resources:
>
> - Creating business partners: *http://s-prs.co/v515750*
> - Reading business partners: *http://s-prs.co/v515751*
> - Replicating business partners from nonleading consuming applications: *http://s-prs.co/v515752*
> - Replicating business partners from leading systems: *http://s-prs.co/v515753*
> - Updating business partners: *http://s-prs.co/v515754*

14.2.2   DevOps for Continuous Integration and Continuous Delivery

The DevOps approach described in Chapter 11 is good for measuring success in the areas of optimizing tasks, procedures, and the effort of handling the complete lifecycle of applications (see Figure 14.1). Key measurable successes using the DevOps approach are as follows:

- Provide insights into the landscape relevant for a specific app, enriched with analytical data that can point your company to the right areas that might need improvement

- Faster collaboration options for the DevOps team, allowing contributions to an optimization that is aligned and can come together in one place

- The integrated management of relevant DevOps procedures, allowing you to take immediate action

- Allowing faster handling of large deployments and release cycles, which with helps automatic detection of dependencies and allows for a risk-free deployment package

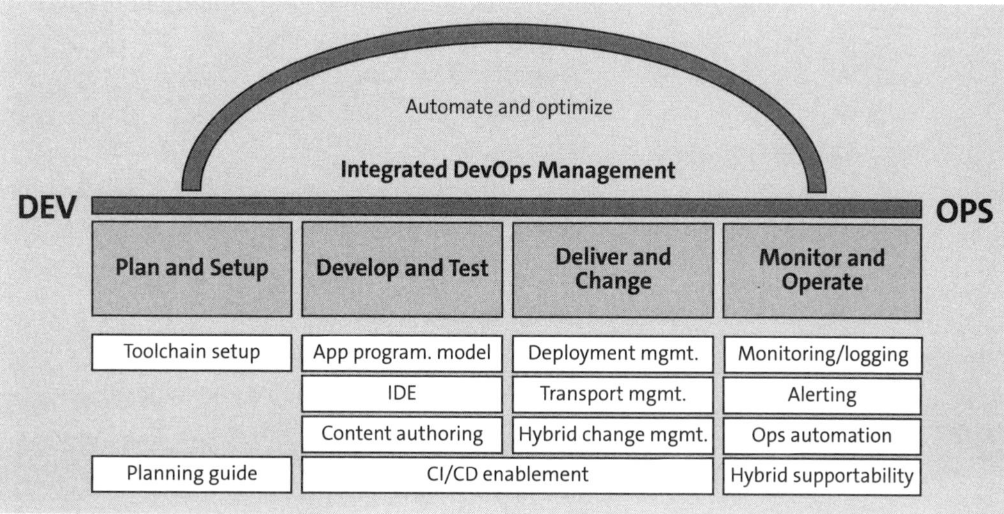

Figure 14.1 DevOps Framework

The DevOps model, shown in Figure 14.2, is able to support distributed development as well as decentralized development environments. The DevOps models also supports the *one project, one system* approach in a distributed development environment. Continuous integration allows for automatic testing of each change and the developer receives direct feedback. It also supports branch-based development, which supports different releases, quick maintenance, and situations to separate development of new functions. To ensure SAP Cloud Platform developers can focus more on business requirements than on technical aspects, several offerings and best practices are in place, such as the following:

- The multitarget application (MTA) approach eases handling of interdependencies of the software modules and artifacts that comprise applications and which are the basis for an automated deployment on SAP Cloud Platform.

- It is a web-based tool with an intuitive UI. It is well-integrated with SAP Web IDE which is the main development tool for SAP Cloud Platform.

- The SAP Cloud Application Programming Model offers a consistent end-to-end programming model to guide developers' best practices and tool recommendations.

- Offerings around project "Piper" and SAP Cloud SDK ease the adoption of continuous integration via downloadable pipeline templates. These templates are extendable via provided step libraries or via opinionated ones that allow a very easy and quick adoption for specific use cases.

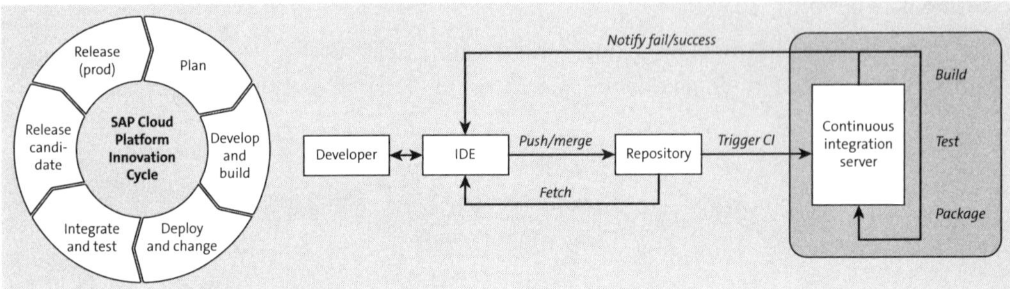

Figure 14.2 SAP Cloud Platform Continuous Integration and Delivery Pipeline Extended with Delivery Management Planning and Lifecycle Management

14.2.3   Replatforming the System Landscape

By using SAP Cloud Platform services to replatform, an organization can enjoy the common characteristics of the cloud, including increased stability, reliability, easier user and role management, and maintaining operational control, enabling DevOps to have common management and monitoring tools.

Many of the applications that have yet to move to the cloud are client-server based, with the vast majority built upon one server for many clients, utilizing the end user's workstation and network for communication. The drive toward operational efficiency—delivering more with less—and increased mobility, combined with tighter compliance regulations, means that is no longer a practical delivery method.

In a scenario in which on-premise or legacy solutions are planned to be out of support in the future, replatforming to the cloud, adopting a cloud-first strategy, is an effective solution.

Figure 14.3 shows replatforming of portal, integration, and mobility solutions to SAP Cloud Platform equivalent services.

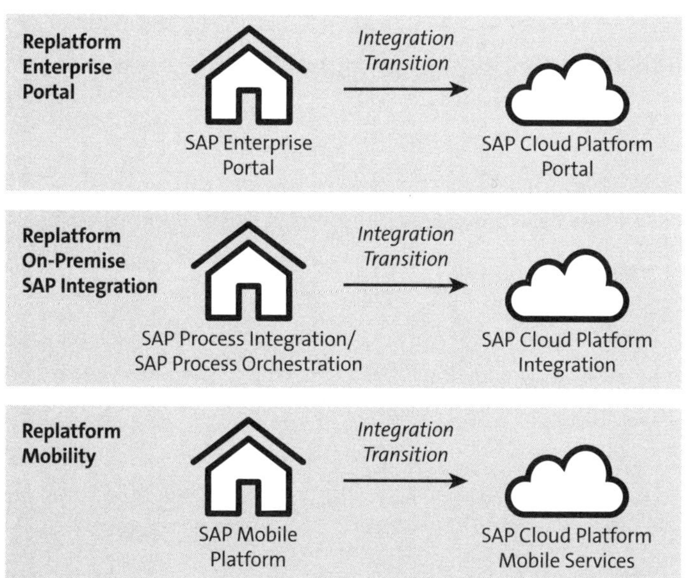

Figure 14.3 Common Scenarios for Replatforming from On-Premise Solutions to SAP Cloud Solutions

14.3   Innovation Framework

Innovation is an important framework for companies to adopt identifying emerging trends and solutions, as explained in Chapter 13. It's important for organizations to remain competitive in their market spaces. With a cloud-first strategy, companies choosing SAP Cloud Platform are able to quickly develop solutions using one of the best architected PaaS offerings. The plethora of cloud services allows developers to enjoy a low code or no code approach and perform rapid deployments.

Coinnovation strategies are also available with SAP, allowing companies, partners, and business communities to work with SAP directly or form a consortium to address market needs using SAP Cloud Platform intelligent technologies such as SAP Intelligent Robotic Process Automation, machine learning, SAP Data Intelligence, etc. A good example of a business consortium is found in medical science, where blockchain technology is used for validating drugs being sold in medical retail markets.

Figure 14.4 illustrates the latest innovations for the automobile industries for connected cars using cloud solutions available via SAP Digital Vehicle Hub. Innovation with OEM partners is a classic example of how twin technology can use operational data or experience data for diagnosis or analytical purposes. Adopting innovation as part of your business strategy can provide a number of benefits, as follows:

- Accelerate business innovation using SAP Cloud Platform innovation services and build smart digital products

- Build key differentiators in the business model to keep ahead of the competition
- Build business disruptions to win new customers across different segments

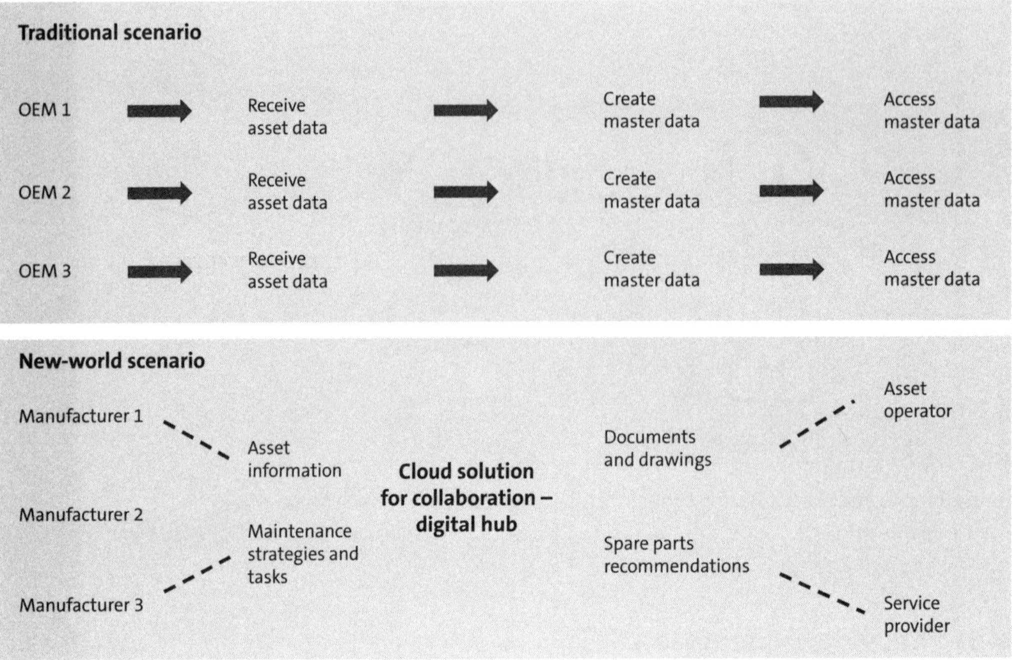

Figure 14.4 Innovations in SAP Digital Vehicle Hub for Automobile Industries

14.4   Value Realization Engagement

It is important that to calculate the total cost of ownership (TCO) and return on investment (ROI) to determine the benefits of using the farming framework.

The following sections describe how the success of a project can be measured. These measurements are done in a modular approach specific to key segments like business/social, IT, and employees/external business associates. Also, using the farming framework, companies can identify their business roadmaps and architecture solutions supporting the business goals. This helps companies to come up with a phase-based project plan to achieve goals and targets.

14.4.1   Business Benefits

A future-ready enterprise project is expected to provide business benefits to an organization in such a way that investments made have a direct ROI (with tangible and intangible returns). Using the following as a reference, you can come up with measurable outcomes and realize the benefits of using SAP Cloud Platform:

- **Business/social**
  - ROI within three years
  - Decrease operational inefficiencies
  - Reduce operational costs with extended visibility to partners in ecosystem
  - Analytical view of historical traceability data
- **Information technology**
  - Flawless integration with B2B and A2A scenarios
  - Integration brings high speed of data from multiple sources
  - More automated tasks to reduce human processing and improve accuracy
  - Built architecture that connects to other systems, such as SAP Streaming Analytics, SAP Business Warehouse (SAP BW), SAP BusinessObjects Business Intelligence, and equipment data for reporting and analyzing big data
- **Human empowerment**
  - Company benefits from real-time transparency to determine value and non-value-added operation times more accurately
  - Reduced workforce and cost with automation
  - Indicators and error codes received from operational data like plants and machineries ensure the maintenance team can predict the condition of the equipment and respond to failures immediately

14.4.2   Planning Tool

Planning tools are important factors to leverage and perform planning. Typically, companies undergo workshops to conduct business process identification, requirements, business priorities, and compliance data to plan opportunities based on business requirements.

The main output of the workshop is a matrix that charts opportunities based on their value to the organization versus the difficulty to implement the solution based on technical capabilities and the business' ability to change. SAP's value-added accelerators act as a catalyst to fast track building a prioritization matrix like the one in Figure 14.5. The matrix then helps to drive a discussion with the between business leads and IT representatives to validate (and often change) the relative prioritization of opportunities.

After plotting the opportunity prioritization matrix for every line of business application, the company and its consultants need to plot the road ahead and plan for implementing SAP Cloud solutions in a phased manner.

The solutions in the quick wins segment in the opportunity prioritization matrix could be implemented in Wave 1, followed by strategic tasks and alternatives in Wave 2 and low-priority tasks in Wave 3.

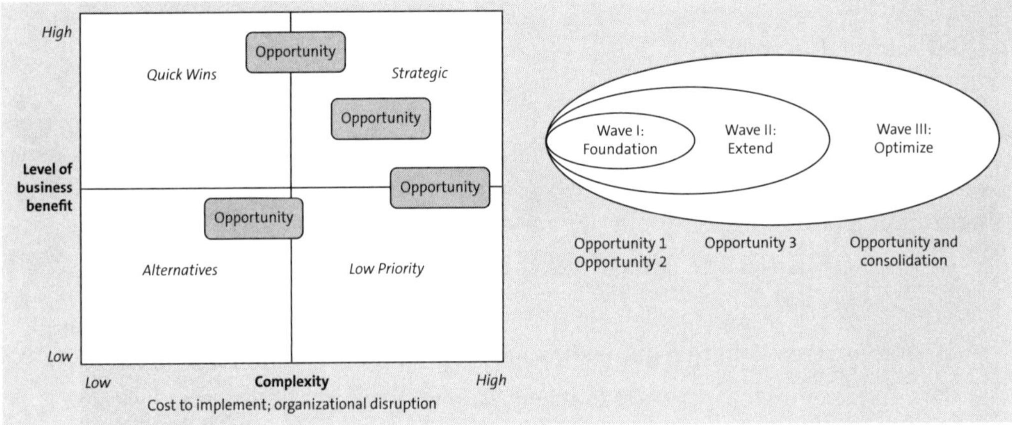

Figure 14.5 Opportunity Prioritization Matrix and Roadmap Planning

14.5   Summary

Success only gets rewarded if the outcome of project is measurable and quantifiable against the business KPI set. Achieving targets and goals is sometimes a Herculean task, so it's important that your organization chooses the right approach to identifying business processes and architecture using the farming framework.

We suggest using the farming framework to allow your organization to validate its system landscape and ensure deployed SAP Cloud Platform solutions are future proofed in alignment with your IT strategies and roadmaps. Suggested success measuring criteria explained in this chapter be should be defined and validated during design, solutioning, and project-planning stages. Setting the right prioritization of projects and clearly outlining the sequencing for execution, considering interdependencies, is important for measuring success.

The Authors

Paresh Mishra is the vice president of the SAP Cloud Platform customer success team for the Asia Pacific and Greater China (APAC) regions. He is responsible for driving customer adoption and supporting customer success during all phases of SAP Cloud Platform implementation (adoption, extension, and optimization). His team helps companies leverage SAP Cloud Platform for digital transformation by building new applications and extending standard applications to better meet business needs.

Previously, Paresh was head of the Cloud Innovation Center for the Asia Pacific region and was responsible for driving the adoption and implementation of cloud transformation services. He provided architectural guidance to companies and partners, helped them adopt the right SAP cloud solutions, and provided them with a portfolio of planning, implementation, and optimization services. He was a significant contributor to the regional SAP cloud practice and helped launch programs like the SAP Cloud SWAT team.

Paresh is a frequent speaker on a variety of cloud topics at events such as SAP All Cloud Connect, SAPPHIRE, and SAP TechEd. He is based in Melbourne, Australia.

Vipin Varappurath is the director of product success for SAP Cloud Platform in the SAP P&T Customer Success organization, helping to drive product success for SAP customers on SAP Cloud Platform, focused on strategic engagements. In his previous role, he was the head of the Digital Customer Success Team for EMEA and MEE and responsible for driving customer adoption for large customer segments in different phases of project implementation. In this role, Vipin helped companies derive value from their investments in SAP Cloud Platform during their transformation journey. In this role, Vipin also worked with large SAP ecosystems to help scale SAP Cloud Platform to a larger SAP customer base.

In an earlier role, Vipin lead the Customer Success and Engagement team in North America, helping large companies in North America succeed in their digital transformations using SAP Cloud Platform. In this role, Vipin has extensively worked with and helped companies derive value from SAP Cloud Platform for both their SAP and non-SAP applications and drive value transformation discussions. His team helped companies derive maximum value from SAP Cloud Platform for their digital transformation journeys, build new agile applications, and personalize/extend applications for their needs.

Prior to that, Vipin was with the SAP Consulting team and worked on multiple customer projects, helping and advising them as they architected, developed, and implemented cloud-native architecture and resilient application development, integrations, and extensions. He helped companies understand and evaluate the overall landscape and business priorities, thus helping develop cloud transformation plans for the short, medium, and long term. He has also been part of the development organization in SAP, in architect and developer roles for SAP Cloud applications.

Vipin is based in Germany.

N

O

P

Q